U.S. Marine Corps Physical Readiness Training for Combat

Plus

Marine Rifle Marksmanship

Department of Defense

Pentagon
Publishing

UNITED STATES MARINE CORPS
Marine Corps Combat Development Command
Quantico, Virginia 22134-5001

29 January 1988

FLEET MARINE FORCE REFERENCE PUBLICATION............0-1B

MARINE PHYSICAL READINESS TRAINING FOR COMBAT

1. PURPOSE

Fleet Marine Force Reference Publication (FMFRP) 0-1Marine Physical Readiness Training for Combat is intended for use by all Marines. It provides the information and references necessary to establish and conduct physical conditioning programs to prepare Marines for the physical demands of combat.

2. APPLICABILITY

Although not directive, this handbook is intended for use as a reference by all Marines in developing unit physical readiness training for combat. This FMFRP does not establish policy for the conduct of physical fitness training. Refer to Marine Corps Order 6300.3_ for applicable Marine Corps policy on physical fitness. It may also be used as a basis for instruction conducted by unit leaders or as a reference for instruction or training.

3. RECOMMENDATIONS

To improve this manual, the MAGTF Warfighting Center seeks comments and recommendations from the user. Officers and noncommissioned officers with constructive comments are encouraged to respond. Lack of response will be considered as concurrence. The User Suggestion Form which follows can be detached and forwarded to - -

Commanding General
Marine Corps Combat Development Command (WF 06)
Quantico, Virginia 22134-5001

4. DISTRIBUTION

Initial distribution of this FMFRP 0-1B is shown below. Addressees on this list should ensure that appropriate subordinate organizations are provided copies.

JOHN I. HOPKINS
Brigadier General, U.S. Marine Corps
Assistant Chief of Staff, Warfighting
Marine Corps Combat Development Command
Quantico, Virginia

DISTRIBUTION: TDJ

From:

To: Commanding General, Marine Corps Combat Development Command (WF 06), Quantico, Virginia 22134-5001

Subj: FMFRP 0-1B, MARINE PHYSICAL READINESS TRAINING FOR COMBAT, RECOMMENDATIONS CONCERNING

1. In accordance with the Promulgation Page to FMFRP 0-1B, which invites individuals to submit suggestions concerning this FMFRP directly to the above addressee, the following unclassified recommendation is forwarded:

Page________Article/Para No.________Line No.________Figure/Table No.________

Nature of Change:_____Add_____Delete_____Change_____Correct

2. Proposed New Verbatim Text: (Verbatim, double spaced; continue on additional pages as necessary.)

3. Justification/Source: (Need not be double spaced.)

NOTE: Only one recommendation per page.

PREFACE

MARINE PHYSICAL READINESS TRAINING FOR COMBAT

Chapter 1
PHYSICAL READINESS LEADERSHIP
Section I. Role of Physical Fitness in Combat

1101. PURPOSE OF PHYSICAL FITNESS TRAINING

Physical fitness training in the Marine Corps has one purpose: to prepare Marines to physically withstand the rigors of combat. All other goals of physical fitness training are subordinate to and must support attainment of this goal. The idea that only infantry or reconnaissance units and their attachments normally face physically demanding combat is wrong. This error must not influence the priority commanders of combat support, combat service support, aviation, and headquarters units give to physical fitness for combat. Physical fitness for combat has a high priority for all Marines. A sound, effective unit program, requiring limited time and material, offers a greater payoff in combat than many more expensive and time-consuming training programs. This handbook describes the concept and provides the ingredients of a program for commanders to use in carrying out one of their most serious responsibilities: ensuring that their Marines are physically ready for combat.

1102. SCOPE OF THE MANUAL

This handbook provides guidance for all leaders, trainers, and planners of physical training programs. It describes unit physical fitness for combat training in the following chapters:

a. Physical Readiness Leadership. Chapter 1 provides guidance to leaders in the conduct of physical combat readiness training.

b. Physical Readiness Training Programs. Chapter 2 discusses how to structure programs to meet physical readiness goals in a variety of combat training situations.

c. Physical Conditioning Activities. Chapter 3 discusses the primary physical conditioning activities which commanders may use to attain readiness for combat goals. Sections in this chapter address foot marches under load, strength-building activities, and activities which build confidence and the aggressive spirit needed in combat.

d. Combat Water Survival. As an amphibious force, Marines cannot ignore the importance of combat water survival training. Chapter 4 describes a program for commanders to conduct this training.

e. Competitive Conditioning Activities. Chapter 5 describes competitive activities which stress the relationship to combat actions, featuring team-building types of competition.

f. Evaluation of Performance During Training. Chapter 6 describes tests which measure individual and unit physical fitness for combat.

g. The Human Body. Chapter 7 covers the structure and functioning of the body.

1103. COMMANDER'S ROLE

Major General Commandant John A. Lejeune in the 1921 edition of Marine Corps Manual expressed our philosophy of leadership. Among other things, General Lejeune stated that "it will be necessary for officers ... to endeavor to enlist the interest of their men in building up and maintaining their bodies in the finest physical condition." It is significant that the Commandant prefaced these writings in 1921 with the statement that his thoughts were greatly influenced by the experience of World War I. We have never varied as a Corps from this belief that one of a commander's most serious responsibilities is to have Marines physically prepared for combat. Senior commanders and their staffs carry out this role primarily by providing command emphasis, including planning, support, and supervision. Commanders at the battalion/squadron and company/battery level execute a program of physical fitness for combat by close supervision and, most importantly, leadership by example.

1104. HISTORY OF MILITARY PHYSICAL READINESS

Every war has revealed our military physical deficiencies during the initial periods of mobilization. This realization followed the Civil War and has recurred regularly with each national emergency.

a. Training programs in each war were geared to the physical need of the era. Success was dependent upon the amount of time available during training to physically prepare Marines for battle conditions. Frequently, casualties in initial engagements were attributed to the inability of our Marines to physically withstand the rigors of combat over rugged terrain and under unfavorable climatic conditions. With adequate preparation, Marines have always handled the test of battle.

b. During World War II, the first physical conditioning doctrine that could be scientifically justified by testing was introduced. As the war progressed, this program was effective in the physical conditioning of millions of Marines for combat.

c. Postwar periods have traditionally been a time of consolidation. Unfortunately, some leaders considered the conditioning phase of training to be a wartime tool. With such a philosophy prevalent between wars, physical readiness was relegated to secondary importance resulting in a serious lowering of combat effectiveness. The initial commitment of Marines in Korea dramatically displayed this failure to recognize the extreme physical demands of warfare. Call-up of large numbers of reserve Marines and pressing them into the war within weeks after call-up mainly caused the loss in effectiveness.

d. Costly lessons learned from our military experiences over a period of years and the course of several wars led to an increasing interest in the physical conditioning of the individual Marine. No longer can we afford to emphasize physical fitness during wartime and de-emphasize it during peacetime. It is evident that, in spite of increased mechanization and modern weapons, physical readiness retains a vital place in the life of each individual Marine and in every unit within the Marine Corps regardless of mission.

e. Commanders are well aware of the need for rugged and well conditioned Marines. The daily demands of housekeeping, maintenance, support, training, operations, and other time-consuming tasks make it necessary for commanders to set aside time for frequent, regular, and vigorous exercise periods.

1105. PHYSICAL DEMANDS OF COMBAT

There are three primary elements of effective physical fitness for combat: lower body strength and stamina; upper body strength and stamina; and a competitive, combative spirit. Cardiovascular functioning is not treated as a separate element since it is a necessary foundation for the other three. Exercises and activities designed to improve Marines' condition in the three primary elements will foster good cardiovascular and respiratory condition.

a. Lower Body Strength and Stamina. Some common demands that combat places on lower body strength and endurance are marching long distances under load and functioning effectively at the destination; moving quickly and evasively under fire; and carrying wounded Marines to safety.

b. Upper Body Strength and Stamina. Some common demands that combat places on upper body strength and stamina are rapidly emplacing crew-served weapons; handling large-caliber ammunition for extended periods; climbing walls, cliffs, and other high obstacles; and performing field maintenance on air craft or heavy machinery.

c. Competitive, Combative Spirit. While part of this critical aspect of combat readiness is mental and emotional, robust physical condition and a training program which promotes physical aggressiveness greatly enhance a competitive and combative spirit. Activities which force Marines to overcome natural physical fear, which directly leads to fatigue, are particularly valuable.

1106. PHYSICAL EFFECTS OF COMBAT STRESS

It has been thoroughly documented that the added uncertainty and stress of combat have a major physical effect on Marines as well as the acknowledged psychological effect. In other words, in combat, fear equals fatigue. In training, we cannot easily reproduce this combat stress which reduces the effectiveness of individuals and units. However, we can produce fatigue and psychological doubt in training situations by developing a challenging physical training program which stresses the participants. This compels Marines to exercise their ability to continue to function under difficult and trying conditions. Such activities as long foot marches under load and difficult confidence courses train Marines to overcome their own fear and fatigue. Rugged competitive activities such as martial arts and pugil stick fighting prepare Marines psychologically to overcome an opponent as well as their own fear and fatigue. This sort of physically demanding training builds self-confidence and unit morale. The adage that "the unit that sweats more in peacetime will bleed less in battle" certainly applies to its physical training program. Athletic coaches have long followed this approach by such practices as having teams shoot foul shots after wind sprints in order to strengthen their mental conditioning and toughness. Although pushing Marines to their limit is necessary, commanders often set lower standards initially than ultimately desired. The program should then call for steadily increasing the difficulty. A good guide for a starting point is a standard that will be physically and psychologically demanding for no less than three-fourths of the unit. Expressed another way, no more than one-fourth of the unit should be working at less than their limit. The program should progress quickly, augmented, if needed, by remedial training for those unable to meet the unit standards, until the training is challenging for all but the most exceptional Marines in the command. Commanders must never forget that Marines expect to be challenged and thrive on being pushed to their limit.

1107. LIMITS OF PHYSICAL READINESS

A well-conditioned Marine unit has a significant advantage in combat, but this does not mean that physical conditioning can substitute for technical knowledge or good planning. For instance, units must, as discussed in chapter 3, pursue a continuing and challenging program of marches under loads. This does not disprove the axiom that fighting men, if required to carry over one-third of body weight, will generally become ineffective in battle. Leaders and planners must decide what is essential to be carried and must use an effective logistic distribution system, such as unit trains, to make remaining items available when they are needed. In a similar vein, the valid requirement to foster a competitive and combative spirit in Marines does not excuse leaders from the responsibility to practice tactics which will not needlessly risk lives. Physical readiness for combat is a vital goal, but it is only one part of a unit's overall readiness for the lest of combat.

Section II. Fundamentals of Physical Fitness

1201. FITNESS FOR MARINES

Total combat readiness includes technical proficiency and mental, emotional, and physical fitness. If any one of these attributes is lacking, combat effectiveness suffers proportionately. Without technical fitness, Marines lack the knowledge and skill to fight; without mental and emotional fitness, they lack the incentive and will to fight; and without physical fitness, they lack the physical ability and confidence to fight. Physical fitness in a Marine means a whole and healthy body, the capacity for skillful and sustained performance, the ability to recover from exertion rapidly, the desire to complete a designated task, and the confidence to face any eventuality.

1202. COMPONENTS OF PHYSICAL FITNESS

A sound body, free of disease and defect, does not in itself constitute physical fitness. Before an untrained Marine can be considered physically fit for combat, the following traits must be developed:

a. Strength. Every Marine needs enough strength to perform the heaviest task encountered in routine and emergency activities. The basic areas where strength is required are in the arm and shoulder girdle, abdomen, back, and legs. Muscles increase in size, strength, and firmness with regular and strenuous exercises. Without work, they grow flabby and weak.

b. Endurance. Each Marine needs sustaining power to maintain the maximum ability without undue fatigue. There are two types of endurance:

(1) Muscular Endurance. The Marine needs muscular endurance to fight the enemy under the most tiring combat conditions. Muscular endurance is the ability to perform continuous work over long periods of time. Endurance depends on the bloodstream's ability to deliver large amounts of oxygen and nutrition to the muscle masses and then carry away the waste products quickly.

(2) Cardiovascular-Respiratory Endurance. The development of cardiovascular-respiratory endurance ("wind") is necessary to maintain muscular endurance. Cardiovascular-respiratory endurance depends on the efficiency of the blood vessels, lungs, and heart. The maximum effort a Marine can exert over a period of time is limited by the capacity to absorb oxygen and expel carbon dioxide. The average Marine's cardiovascular-respiratory capacity can be greatly increased by exercise.

c. Agility. A Marine must be able to change direction quickly and as faultlessly as possible. The ability to react instantly and to maintain orientation during rapid changes of body position is important to survival. This important ability may be developed by conditioning exercises that require varied and rapid changes of body position on the ground and in the air, such as obstacle courses.

d. Coordination. Coordination is the ability to move all parts of the body in a smooth, efficient, and concerted effort (commonly called timing). A well-coordinated individual does not make useless movements. An individual who moves with precision and accuracy saves energy. Coordination is best developed by practicing diversified muscular activities and skills affecting all body parts.

1203. TYPES OF EXERCISE

Basically, there are two forms of exercise: isotonic and isometric. Both forms are contained in the activities found in this manual.

a. Isotonic. Isotonic exercises are those in which the expenditure of energy is regulated and released during concentrated efforts. The regulated expenditure of energy is controlled by both the mode of exercise and the individual's effort. This type of effort is common to most exercise and sports. To develop endurance, coordination, and flexibility, isotonic exercise must be used. Strength can also be developed or increased through this type of exercise.

b. Isometric. Isometric exercises are those in which maximum effort is applied and held until the engaged muscle energy is depleted during a single contraction effort. The individual exerts full force against an immovable object for a relatively short period of time (6 to 10 seconds) and then repeats the contraction several times with a short period of rest between each contraction. This type of exercise develops only strength; therefore, it has limited application.

1204. PRINCIPLES OF PHYSICAL CONDITIONING

Men and women vary in their physical makeup. Physiological (body) function and reaction also vary in proportion to heavy demands placed upon the body. To attain the maximum benefit without sacrificing Marines' welfare, fitness must grow with a careful program of conditioning. To allow for adjustments in body functioning as the conditioning program progresses and to ensure attainment of objectives, the following principles of physical conditioning must be applied:

a. Overload. As strength and endurance increase, the physical load must be increased until the desired level of fitness is reached.

b. Progression. In beginning stages, the load must be moderate. Gradual progression from this low state of fitness to a higher state is possible through application of a progressive program.

c. Balance. An effective program utilizes various types of activities and provides for the concurrent development of strength, endurance, and coordination as well as basic physical skills.

d. Variety. Some programs fail because the routine becomes boring. The most successful programs always include conditioning activities, competitive events, and military physical skill development.

e. Regularity. There is no easy or occasional way to develop physical fitness. Regularity of exercise is a must, with daily exercise preferred.

1205. THREE STAGES OF PHYSICAL CONDITIONING

Unconditioned or poorly conditioned Marines pass through the following stages in reaching the desired state of physical condition.

a. Toughening Stage. This stage is approximately 2 weeks long and is usually characterized by muscular stiffness and soreness followed by recovery.

b. Slow Improvement Stage. This stage is approximately 6 to 10 weeks long and is characterized by slow and steady improvement until the desired level, or a high level, of fitness is attained.

c. Sustaining Stage. This stage goes on indefinitely in order to maintain the level of conditioning achieved by passage through the previous stages.

1206. EFFECTS OF CLIMATIC CONDITIONS

Temperature, both atmospheric and body, affects the physical performance of Marines. Proper maintenance of body temperature through warm up exercise, proper dress in cold weather, and removal or adjustment of clothing in hot weather is necessary for effective performance and health. Climatic factors to be considered are the following:

a. Exercise In High Temperatures and High Humidity. Marines can endure strenuous physical activity in extremely hot temperatures if they are given an opportunity to become acclimated and if they take enough salt and water. It is essential to continue physical training programs in hot climates. Marines can better withstand high temperatures when they are well-conditioned. High humidity combined with high temperatures presents a serious danger. These conditions prevent the natural cooling of the body by the evaporation of perspiration. Training schedules should conform to the provisions of the base commander in respect to wet bulb conditions. Those who conduct training under these conditions should monitor weight loss and be careful to make fluid replacement adjustments.

b. Exercise at High Altitudes. Certain problems are encountered in conditioning Marines stationed in high altitudes because the heart undergoes greater exertion during exercise. It is particularly important that only light exercise be given initially at such altitudes. A Marine's body gradually adjusts to high altitudes within a few weeks. After this adjustment, progressively greater amounts of exercise are possible.

c. Exercise in Arctic Regions. Military duty in the arctic is so arduous that a high level of physical conditioning is essential. Because of the difficulties of carrying on physical conditioning exercises in extreme cold, Marines should be conditioned to the highest level possible before they arrive. A sustaining program will then maintain that level. When exercising in cold weather, Marines should be required to remove excess clothing to prevent them from becoming damp with perspiration.

1207. WARMING UP AND COOLING OFF

It is a fundamental physiological principle that Marines should warm up gradually before taking strenuous exercise. Such action speeds up the circulation to prepare the body to take an overload and helps to prevent injury to muscles and joints after exercising. Marines should be kept mildly active, walking, stretching, or performing some other mild muscular activity until breathing and temperature have returned to normal. Marines should never be allowed to cool off too rapidly; in cool or cold weather, they should put on additional clothing during the cooling-off period.

1208. PHYSICAL ACTIVITY AS AGE INCREASES

Combat makes severe physical demands on Marines. All Marines, regardless of age, must be physically ready to meet these demands. There is no physiological reason to cease exercise or exertion with age. Increased age usually brings increased responsibility which, in many instances, leads to a routine that can become almost devoid of physical activity. The key to fitness with increased age is to continue exercising at a reasonable level and to include exercise of a vigorous type in the daily routine. Older persons who have not regularly maintained a reasonable state of physical fitness require more time than younger persons to become fit. Such individuals usually require a longer period of time to recover from physical effort than younger Marines. If general health is good, evidence shows that older Marines can develop and maintain a degree of fitness which permits vigorous activity and proper performance of duties. It is both the individual's and the commander's responsibility that all Marines continue a daily sustaining exercise program.

Section III. Goals of Physical Readiness Training

1301. NECESSITY FOR PHYSICAL READINESS TRAINING

A very important objective of training is attainment and maintenance of operational readiness. Marines must be physically ready for operations at any time and under all conditions of climate and environment. A combination of training to develop proficiency in physical skills and conditioning to improve strength and endurance results in physical readiness for combat. The degree of physical fitness required of Marines can be acquired through physical exercises in a normal well-coordinated and closely monitored training routine. The performance of purely military duties, such as drills and marching, is not enough to build all the desired areas of fitness. Few recruits are physically fit for the arduous duties ahead of them. The softening influences of our mechanized civilization and the difficulties of conditioning Marines make physical fitness more important than ever before. If Marines are to be fully developed to and maintained at the desired standard of physical fitness, a well-conceived plan of physical readiness training must be a basic part of every training program. Marines cannot be adequately prepared in any other way for the hard work and arduous demands required on the battlefield.

1302. OBJECTIVE OF PHYSICAL READINESS TRAINING

The overall objective of the physical readiness training program is to develop individuals and units who are physically able and ready to perform their duty assignments or missions during training and in combat. To attain the objective of physical readiness, exercise activities must develop the following areas:

a. Strength and Endurance. Developing adequate strength to perform required duties and adequate endurance to sustain activity over a long period of time.

b. Muscle Tone. Developing muscle tone adequate to maintain proper posture and reasonable weight control.

c. Skills. Developing military physical skills which are essential to personal safety and effective combat performance. As skill is developed, agility and coordination are attained. The essential skills are-

(1) **Marching Under Load.** Marching with individual and unit weapons and equipment.

(2) **Running.** Distance and sprint running on roads and cross-country.

(3) **Jumping.** Broad jumping and vertical jumping downward from a height.

(4) **Dodging.** Changing body direction rapidly while running.

(5) **Climbing and Traversing.** Vertical climbing of ropes, poles, walls, and cargo nets. Traversing horizontal objects such as ropes, pipes, and ladders.

(6) **Crawling.** High crawling and low crawling for speed and stealth.

(7) **Throwing.** Propelling objects, such as grenades, for a distance with accuracy.

(8) **Vaulting.** Surmounting low objects, such as fences and barriers, by use of hand assists.

(9) **Carrying.** Carrying objects and employing man-carries.

(10) **Balancing.** Maintaining proper body balance on narrow walkways and at heights above normal.

(11) **Falling.** Contacting the ground from standing, running, and jumping postures.

(12) Surviving in Water (Or Other Specialized Situations) Using water survival techniques.

d. Character Traits. Jnstilling character traits which help accomplish military missions to include-

(1) Confidence. Confidence develops through achieving progressively more difficult tasks as physical ability grows.

(2) Aggressiveness. Participation in combative activities and contests develops desire and willingness to overcome an opponent.

(3) Reaction Under Pressure. Competitive contests and game situations train Marines to think and to act quickly while under pressure.

(4) Teamwork. Teamwork develops through competitive events in which a number of Marines must coordinate their efforts to accomplish a physical task.

1303. BENEFITS OF EXERCISE

The benefits of exercise are not always understood. Some of the more important results of exercise are listed below:

a. Improved Muscle Tone. Muscular tone improves and, at the same time, muscular strength and endurance are built up.

b. Cardiovascular-respiratory Endurance. Cardiovascular-respiratory endurance, or wind, improves through a process of opening up dormant lung capacity to absorb greater amounts of oxygen.

c. Circulation. Circulation of the blood speeds up and extends to a greater portion of the body as exercise forces the blood to service all parts of the body. The efficiency and effectiveness of the heart, lungs, and blood vessels improve.

d. Flexibility. A wider range of muscular movement is possible and rapidity in physical skills grows.

e. Elimination of Body Waste. Bending and twisting the body and the general speedup of body processes caused by exercise regulate and help eliminate body wastes.

f. Tension. Working off excess nervous energy and relief from daily worries and cares relieve tension. Participation in exercises leaves little time for worry.

g. Sleep. Sleep improves because muscles are healthfully tired after about of exercise. A by-product of sound sleep is relief of tension.

h. Obesity Control.. Control of obesity (fat) is made possible by using up excessive amounts of fat-producing food elements.

i. Injury Susceptibility. Susceptibility to injury is reduced through exercise. Muscles, tendons, and joints are strengthened. Injuries such as hernia, back strain, and joint sprains are less likely to occur if muscles are maintained in proper tone.

Section IV. Leadership Roles

1401. PSYCHOLOGICAL LEADERSHIP

The full development of a Marine's resources is not all physical. To be effective in developing physical readiness, leaders must realize that mind and attitude are also important to success. The more important psychological considerations are to—

a. Promote Understanding of the Value of Physical Readiness. A desire to be physically ready should be created in all Marines. Motivation is increased and Marines take greater interest in their individual physical fitness if they understand the value and benefits of vigorous exercise. When Marines realize their efforts are an investment in their own personal welfare, it should not be difficult to obtain their cooperation. Marines should understand the objectives, the benefits, and the value of each type of exercise activity in their program. They should also understand the relation of physical readiness to survival in combat.

b. Maintain a Positive Approach. Physical readiness training for combat is strenuous and demanding. It is a responsibility of leadership to create an atmosphere where all desire to participate fully. This attitude should be fostered. A negative approach must not be identified with physical readiness training, even with those having difficulty. Only in unusual cases should fear of punishment be the motivating factor behind good performance. For those few who cannot keep up or attempt to malinger, an effective remedial program is essential.

c. Seek Cooperation and Develop Morale. In a program placing maximum physical stress upon individuals, it is necessary to gain their cooperation. Favorable reaction is enhanced by proper planning and organization, challenging requirements, use of competition, and application of a progressive program resulting in physical fitness. As physical fitness grows, morale also grows.

1402. COMMAND AND SUPERVISORY FUNCTIONS

a. Command Functions. Commanders should take the following actions to support physical readiness training:

(1) Lead by personal example.

(2) Instill command interest and indicate to subordinate personnel the importance of this training to the welfare of the organization.

(3) Allot sufficient time for the achievement of objectives and monitor the use of such allotted time. The substitution of other training or routine duties for scheduled physical readiness training is unsound and unwise.

(4) Assign and properly utilize qualified personnel to supervise and conduct physical readiness training. If leaders are not competent, take action to ensure they become competent quickly. Reassign those who do not meet this standard to other duties.

(5) Make necessary facilities and funds available to support a program to develop physical readiness within all personnel.

(6) Measure the physical fitness of individuals and units in order to evaluate progress and to determine if the program is successful.

b. Supervisory Functions. Leaders responsible for planning, conducting, and supervising physical readiness training should take the following actions:

(1) Prepare physical readiness training schedules which apply the principles of physical conditioning and which aim for a particular type of program plan.

(2) Provide for wide participation of as many Marines as possible. All Marines, regardless of position or age, will benefit from regular exercise. In some instances, special efforts are necessary to overcome obstacles to regular and frequent training. Special effort is also necessary to ensure remedial conditioning. Such conditioning should occur for those who are physically substandard and after extended absence due to leave, sickness, injury, and travel.

(3) Prevent waste or unwise use of time allotted for physical readiness training. Time-wasters include unprepared instructors; assignment of one instructor to a group larger than a platoon; progression which does not keep pace with the physical development of the Marines; extreme formality; inadequate equipment or facilities which require waiting turns to exercise; and lengthy rest periods between exercises which interfere with the application of overload.

(4) Ensure that the program contains vigorous physical activity. Such activity places progressively greater demands upon the body during each exercise session and also over the duration of the training program. To be of benefit, exercise must tire the muscles and cause the heart to increase its rate of beat.

(5) Set an overall objective for each physical fitness program.Observe the training as necessary to ensure that the established objectives are being achieved.

(6) Observe physical readiness training to insure the use of a positive approach. To implement a positive attitude, small-unit leaders and instructors should personally set the example; have an understanding, fair, and sympathetic attitude; recognize individual differences; and motivate Marines toward their best effort.

(7) Guide and inform small-unit leaders and instructors concerning approved techniques, directives, and literature. As necessary, arrange for local training of instructors to include clinics, conferences, schools, and demonstrations.

(8) Determine the effectiveness of physical readiness training by personal participation in and observation of training, analysis of field inspection reports, and analysis of individual physical fitness test scores. Scores may be combined to reflect the fitness of the unit.

1403. SMALL-UNIT LEADERS AND INSTRUCTORS

a. Responsibility. The instruction and conduct of physical readiness activities are the function of company/battery commanders, platoonleaders and persons assigned as instructors. Experience has proved the effectiveness of physical fitness development when conducted in company-and platoon-size units under direct control of the leader with overall supervision by the parent-unit commander. For example, all the platoons of a company may exercise at the same time under the general supervision of the company commander, with each platoon conducting the assignment separately and under its own leadership.

b. Leader's Assignment. A small-unit leader or an instructor in a school or training activity is assigned to a combat unit or to a support unit. In this assignment, the leader is responsible for all training to include physical readiness training. In a different situation, the noncommissioned officer (NCO) or officer may beassigned as a full-time physical readiness instructor. This contrasts to a unit leader assignment where only part of the time is devoted to such training. In either case, the leader will hold an important and vital position for the physical fitness of Marines.

c. Leader's Training. Leaders may come to the assignment either fully or partially trained. It may be their first responsibility for the development of physical fitness. If they have had previous training through experience, make certain that their information is supplemented with study of this manual. If they have had professional training in physical education during civilian life, but no military experience, they should also use this manual to learn the methods used by the Marine Corps. A new leader should take advantage of various ways to learn including attendance at leader training courses, self-study, practice, and discussion with more experienced leaders.

d.Leader's Objective. As a physical readiness training instructor, the leader has two general objectives. The first is to motivate Marines to want to be physically fit. The second is to conduct a program that will develop a high degree of physical fitnss. Motivated Marines will react enthusiastically to such a program. It aids greatly in achieving local program objectives.

e. Leader's Personal Fitnes. A unit leader who must instruct and demonstrate physical activities must be in physical condition to do the job without undue physical stress.The leader should be able to do those things that must be demonstrated. The leader's strength, endurance, posture, and skill should set the example. This does not mean that the leader must excel, as other Marines do not expect championship performance. However, they do expect, and deserve, a creditable showing of fitness for the job.

f. Leader's Knowledge. The leader must have three types of knowledge to properly administer physical readiness training. Theyare- -

(1) Knowledge of Marines. The leader must understand Marines, know how to lead and motivate them, understand how they learn, and apply this knowledge wisely in the day-to-day training situation.

(2) Understanding of Body Functioning. A more intelligent exercise program results from understanding and applying the principles which govern physical conditioning of the body. The leader with such knowledge can better prescribe, adjust, and regulate exercise types, amounts, and progression to attain fitness.

(3) Understanding Exercise Activities. The leader needs to understand the contribution each type of physical activity makes to physical fitness, and how to use each activity to develop fitness. Skill to demonstrate and lead the various activities is a necessary part of technique and is invaluable to the instructor or small-unit leader.

Chapter 2
PHYSICAL READINESS TRAINING PROGRAMS
Section I. Development of a Program

2101.GUIDANCE FOR PLANNERS

This chapter instructs planners on physical readiness training procedures. It contains program planning guidance including factors to consider when developing programs; steps in assembling a program; definitions of activity packages and systems of exercise; selection of activity packages; and selection of systems used in implementing those packages.

2102. MARINE CORPS MISSION

The mission of the Marine Corps is to seize and defend advanced naval bases and to perform other missions as the President of the United States may direct. Traditionally, to perform this mission, Marines have been projected into the area of operations by Navy vessels and have attacked hostile beaches over the shore in landing craft, assault amphibian vehicles, and helicopters. Recently, performance of this task has been somewhat complicated by the introduction of the concepts of maritime prepositioning and the air-landed Marine expeditionary-brigade. Under these concepts, Marines, landed at airfields in foreign countries, will join their equipment at commercial ports. The distance between these airfields and ports is often 10 or more miles. In time of international crisis, it can be expected that transportation assets will not be adequate. Marines from all elements of a Marine Air-Ground Task Force may have to march from the airfield to the site of their equipment. All Marines must be physically conditioned to rigors of conducting foot marches with individual weapons and equipment.

2103. EVALUATION OF FITNESS

The Marine Corps mission is such that all Marines must constantly and consistently achieve a high level of fitness that prepares them for the demands of combat. While Marines have many different MOS's, all Marines must be prepared for the demands of marching under load and performing basic infantry tasks, such as rear security and patrolling. While the physical fitness test remains our universal measure of individual fitness, it is the commander's responsibility to observe and evaluate the unit's ability to perform effectively in combat. Often the commander's most effective evaluation tool is weekly physical training sessions. The commander should participate in these sessions and see that they are fully integrated into the unit physical conditioning program.

2104. INTEGRATION OF TRAINING

Training time must be used efficiently and wisely. Every opportunity to integrate physical training into other training activities should be seized. For example, foot marches under load can be integrated into the normal daily activities as a way to move from one training or work site to another. This integration reflects realism in training and should be used to maintain overall proficiency. To accomplish this requires imagination and enthusiasm on the part of the commander. The area of operations in which the training is conducted determines what can be integrated and how. If possible, training should also consider the terrain and climate of the area in which the unit will subsequently conduct training or operations. Training should include familiarization with special equipment and the application of specialized techniques to tactical principles. Activities such as gun drills and command post emplacement drills also have physical conditioning value, particularly if supervised and done competitively or measured against time standards.

2105. ACTIVITIES AT THE UNIT LEVEL

Physical readiness training is a command responsibility and is generally conducted in Marine units at the company and platoon levels. Battalion/squadron commanders must constantly supervise, making sure that all Marines are physically prepared for combat.

a. Physical readiness training should usually be conducted weekly by each battalion/squadron-level unit, using events designed for all personnel (e.g., marches under load, battalion/squadron runs, military field days, etc.).These events enable the commanders to demonstrate personal leadership, and observe and evaluate the physical condition and combat physical readiness of their Marines. This exercise period should be designed to elevate the unit's morale and emphasize unit identity through the wearing of uniforms and the carrying of battalion and company guidons. The battalion/squadron commander should make every effort to be a visible participant in the unit's physical readiness program on a weekly basis. These activities could be conditioning marches under load or a weekly calisthenics/unit run.The suggested times for these activities are early on Monday morning or late on Friday afternoon after the training week has been completed. The uniform may be physical training gear orutilities and boots.

b. Commanders should be cautious of long runs in boots and utilities. This is not to suggest that units should not run in utilities, but to remind commanders that boots are designed for marching, not running. If commanders choose boots and utilities, they should select off-road routes which will provide a cushioning effect.

c. Units should be made well aware of the commander's intent before the unit falls out for exercise. If the exercise period is conducted early in the morning, it need not necessitate an extensive shower/clean-up period afterward, so long as all Marines from the commander on down remain dirty. Commanders should inform their Marines not to fall out for these sessions in clean, pressed uniforms.

d. The physical readiness training of headquarters units is often difficult to manage. Often physical training is left for individuals or sections to develop and manage on their own. This approach can be effective, but if not supervised, it can result in a "paper" program. Headquarters unit commanders should endeavor to conduct weekly integrated training which will enable them to evaluate the physical readiness of their personnel. Often this training has to be scheduled before or after normal working hours to maximize participation of the entire unit.

2106. STEPS IN PLANNING

To implement workable and effective programs (as directed by CMC ALMAR 261/87) the planner must-

a. Determine the Type of Program Needed. Marine units are inherently different in organization and mission. The physical readiness program must be tailored to the mission and to the current physical condition of most unit personnel. Programs to meet this need are of the following types:

(1) **Developmental Programs.** Marines in a beginning or poor state of physical readiness need a program which will develop strength, endurance, physical skills, and character traits which are beneficial to successful accomplishment of military missions. Such programs should be applied progressively to rise gradually to a peak of fitness and skill.

(2) Maintenance Programs. Once Marines reach the sustaining stage of conditioning, their goal is then to maintain this level by participation in a maintenance program.

(3) Remedial Programs. The term "remedial" is usually applied to those individuals or groups who possess substandard physical fitness. For example, a remedial physical conditioning program could be applied to persons who are overweight, who fail to reach physical fitness test standards, or who have missed extended periods of conditioning due to illness, injury, extended hospitalization, or other absence. Saturday training sessions are designed as remedial sessions for all individuals who either miss the daily training or have failed to meet adequate standards. Note that personnel in light duty or no-duty status should be expectedto attend all training sessions, observing, supervising, or assisting as needed.

b. Determine the Time Required. The amount of time for training operations varies considerably. However, every unit can find time to conduct physical readiness training. Frequent (i.e.,daily) physical training of short duration-30 to 60 minutes-is preferred over occasional longer periods. Other demands for training time are so urgent that every minute of time allotted for physical readinesstraining should be used. Determine the time required per week anddivide it into daily blocks.

c. Organize for Various Group Size. It is essential to stress exercise rather than formality. Marines must complete the program where they are- -on the training field, in the motor pool, on the range, next to the classroom, in the office area, in the shop, aboard ship, or elsewhere. Although desirable, it is not always possible to assemble company-size units for physical training. Platoon-size groups are appropriate for the proper conduct of physical conditioning activities. Certain situations may require exercise programs for section- or squad-size units. This manual outlines programs for all situations and types of organizations.

d. Allow for Weather and Exercise Area. In programing and scheduling, the climate and terrain often govern the selection of activities.

(1) Weather changes cause differences in temperature, rainfall, wind chill, and snow. These changes should be anticipated as they dictate the type of program. Alternate plans should be part of the schedule. Nevertheless, normal weather changes and rain should not drive training inside.
(2) Local terrain and available exercise areas may also influence the selection of activities and the type of program which it is possible to support. Some activities can be completed in nearly any area.
e. Plan for Seasonal Change. As most physical readiness training is conducted outdoors, it is necessary to recognize seasonal change. A program should be divided into fall, winter, spring, and summer parts. Seasonal change also causes change in light. For example, an early morning program started in the summer will have ideal lightconditions, yet in the fall or winter, darkness will occur at that same hour and interfere with the conduct of the program. Develop programs in seasonal blocks, and make provisions for anticipated changes.

f. Consider Needed Facilities. An excellent program can be conducted with practically no facilities since there are exercises which require no equipment. However, a better program can be developed when supported by certain facilities and items of equipment. Proper command support, plus ingenuity, will solve this problem. (Items of equipment, when necessary to support the recommended exercises, are included in the chapters on exercise activities.)

g. Specify Appropriate Uniform. The uniform worn for exercising depends upon the season of the year, the state of the weather, and local regulations. Whenever possible, Marines should be dressed alike. Undershirts are preferred as the upper garment when the weather permits. A uniform that restricts the free movement of the body should not be worn when exercising.

h. Consider Availability of Instructors. Leaders who can lead and direct the scheduled activity must be available. Organizational units should train junior officers and noncommissioned officers down to squad or section leaders to instruct and lead the various activities.

i. Select Activity and System. With the type of program needed and the objectives in mind, select an activity package or an integrated training package for each day's scheduled physical training. At this same time, the system to be used in employment of the selected activity or activities must be determined. This selection affects equipment, areas, instructors, transportation, and other support requirements.

j. Secure Command Participation and Support. Prepare and brief the commander to assure full understanding of the objectives and administration of the program. The full participation and support of the commander will greatly improve the level of success.

k. Supervise Execution. Determine needs, publish the program, and supervise its execution as a necessary part of developing the unit's training schedule.

2107. SELECTION OF ACTIVITIES AND SYSTEMS

a. Exercise Activities. The planner must determine exercise activities which will be appropriate to include in the program. An exercise activity is a single means of exercise usually identified by the name applied; for example, running, log exercises, and obstacle course. Many exercise activi ties can become part of activity packages.

b. Activity Packages. Many of the physical activities described in this manual are arranged in prescribed sequences and are known as activity packages. An activity package is a number of exercises of the same type, assembled as a group or a set, and arranged in a specific sequence. Exercise packages are organized in such manner that not more than 15 minutes will be required to complete the execution of any package. Each type of activity is explained in laterchapters. The number of available drills, tables, or circuits; the manner of organization; and the contribution each makes to the total program are covered. Full understanding of this information will greatly assist in developing effective programs. Various designations are used to identify exercise packages; for example, conditioning exercises when arranged in a set order are known as drills, and other packaged activities are designated as tables or circuits. The following are activity packages:

(1) Activity packages for groups:

o Conditioning Drill Two
o Conditioning Drill Three
o Rifle Drill
o Log Drill
o Grass Drill
o Running Tables
o Guerrilla Tables
o Circuit-Interval Table
o Combatives Tables
o Relay Tables

(2) Strength Circuits:

o Fixed Circuit
o Movable Circuit
o Simplified Circuit (Circuit Interval Table)

(3) Activity packages for individuals:

o The 6-12 Plan
o Weight Training
o Isometric Exercise

c. Advantages of Using Packages. These of exercise packages simplifies scheduling and conducting of exercise and results in the following benefits:

(1) Schedule development is simplified as the planner assembles packages which will satisfy the training objective. There is no need to deal with selection of individual activities or to be concerned about the amount of time to be expended on each.

(2) Any 15-minute period, and in some cases less time, can be scheduled or used to perform an activity.

(3) If longer periods of time are available or if the objective demands, several packages can be assembled to provide a more complete period of activity.

(4) Marines are assured a balanced set of exercises or activities as each package is carefully arranged to reach all muscle groups.

(5) The instructor can concentrate on the conduct of a vigorous workout as the type and duration of the activity have already been determined.

d. Nonpackaged Activities. Several types of activity are not packaged. Activities in this category are conditioning marches, unit runs, obstacle courses, combat water survival swimming, teamcontests, and team sports. These activities can be scheduled in combination with packaged activities, or they may be scheduled separately. Often they require a longer period of time. Most nonpackaged activities require a 50-minute period to satisfactorily complete their objective. For conditioning marches, a few hours are required. The benefits of these activities should not be overlooked as some desirable objectives cannot be attained without them.

e. Systems of Exercise. Several methods or systems of organizing exercise and activity packages can be used. Each system is based upon a specific organization as follows:

(1) Single Activity System. The unit leader immediately assumes command of the unit at the beginning of the exercise period. The leader moves the unit to a predesignated exercise site at double time, forms the unit in a circle around the leader, groundsclothing and equipment as appropriate, and quickly moves into the exercise routine. There is usually no time to teach; therefore,the Marines must know the activity to be used. At the conclusion of a 5- to 15-minute period, the unit leader returns the unit to the instructional area at double time and releases the unit for the next scheduled activity.

(2) Progressive Activity System. All Marines (company or platoon) complete activities in the same order during the period. For example, Drill One is followed by dual combatives, and finally a 1-mile run. This system is usually progressive from a warm up activity, such as Drill One, to an activity which contributes in a major way to one of the objectives, such as aggressiveness development through combatives or cardiovascular development, such as running.

(3) Rotating Activity System. Set up the same number of activities or stations as there are platoons in the company. Each platoon rotates through each station in turn. With four platoons in a 50-minute period, about 10 minutes can be devoted to each station. With three platoons, approximately 15 minutes may be spent at each station. Activities must be of a type that can be covered in the time allotted. For example, with three platoons, Station 1 could be Conditioning Drill One; Station 2, running; and Station 3, a team contest.

(4) Circuit System. Set up a number of stations to provide various types of exercise equipment and/or items of apparatus. The idea is to keep all Marines busy and exercising vigorously for a short period of time at each station. The fixed strength circuits and the movable strength circuits are examples of this system. Station changes must be rapid, and the exercise must be started quickly after each change. Since the objective is to exercise at top speed, the motivation comes from frequent change of activity by moving to another station. Rotation by station groups continues until all Marines have covered all stations.

(5) Interval System. This system stresses the development of strength and endurance. It involves heavy work for a given distance within a specified time, alternated with lighter work and recovery, but never stopping during the workout. This procedure is repeated, and the intensity is increased gradually as physical condition improves, but always with adequate recovery. The important concept is stress, recover, stress, recover, and so on. This system is often applied through running, but other activities of a continuous nature may also be used. An example is the Circuit-Interval Table.

f. Steps in Selecting Activities and Systems. The planner should follow these steps.
(1) Evaluate the needs of one unit with emphasis upon unit mission, objective, and time available.
(2) For each day, select an activity package or a combination of activities which will contribute to the objective.
(3) Then determine the system to be used in implementing the selected activities. There are several possible choices and much opportunity for flexibility in program development. For example, the weekly program may contain various exercise packages, systems, and time periods. Additional guidance is contained in following chapters.

2108. SAMPLE WEEKLY PROGRAMS

The variety of weekly programs given shows available alternatives. Programs and schedules other than those illustrated may be assembled to provide schedules to fit any situation. Many other scheduling combinations are possible through use of the packaged activities, varied time periods, and different types of assigned duty.
a. Sample A. Figure 2-1 illustrates a 60-minute time allotment in which the single activity system is used with activity packages that vary from day to day. In this program, 5 hours of training are scheduled for the week. This program needs some type of integrated unit training to supplement it each week. A bimonthly conditioning march of 10 miles in 3 hours is suggested to round out this program. The Saturday session is designed for remedial physical training.
NOTE: A physical training activity must be conducted at a minimum of once every 72 hours. After 72 hours, muscles begin to atrophy.

Figure 2-1 is available in the illustration section at the end of the book

b. Sample B. A more comprehensive weekly program (fig. 2-2) illustrates the scheduling of various-length periods, a variety of activities, and the use of three systems during the week. In this schedule, 3.5 hours are included for the week. The following should be noted concerning this schedule:

Figure 2-2 is available in the illustration section at the end of the book

(1) On Monday, a 60-minute period is available for physical readiness training. Three 20-minute packages are scheduled. Each platoon, under its own leadership, will progress through each scheduled activity in turn.
(2) On Tuesday, Wednesday, and Thursday, only 30-minute periods are available. Here each platoon leader supervises each platoon in a single activity system.
(3) On Friday, the company commander directs training to exhibit personal leadership example and supervision. In this example, only one strength circuit and one cross-country course are available, and each will accommodate only one platoon. For this reason, Conditioning Drill One station is included. A platoon rotates to each station within the 60-minute period.

(4) The Saturday session once again is designed for remedial physical training.
(5) Bimonthly, a 10-mile/3-hour march under load will be necessary to round out this program.
NOTE: Cardiovascular/respiratory exercise to improve endurance must be of long enough duration (20 minutes or more) to induce and maintain oxygen debt.

c. Sample C. A third sample schedule (fig. 2-3) illustrates the use of the single activity system for 5.25 hours of training. This schedule includes sustaining-type activities for a unit that has passed through both the toughening and slow improvement stages of conditioning. The use of competitive activities is featured to holdinterest and provide self-motivation. A bimonthly 10-mile/3-hour march under load will be necessary to round out this program. The Saturday session once again is designed forremedial physical training.

Figure 2-3 is available in the illustration section at the end of the book

NOTE: Thc body will respond to hard work or physical training without soreness and stiffness provided the work or physical training is conducted regularly.
NOTE: Excrciscs of short duration increase strength and bulk but do little to effect endurance.
NOTE: Thc times shown in the charts arc participation times.

Section II. Remedial Physical Training

2201. NEED FOR ATTENTION

Remedial physical training is a process by which physically substandard individuals are conditioned to meet standard levels of performance. To achieve a full degree of operational readiness, it is necessary to bring all Marines up to the prescribed standard. Experience has demonstrated that some Marines have difficulty due to a poor state of fitness, obesity, or lack of motivation. Attention to these individual deficiencies will help improve unit combat readiness.

2202. IDENTIFICATION OF INDIVIDUALS

The company or battery commander identifies persons who cannot achieve the prescribed level of proficiency through the physical fitness test or as a result of observation during training. The commander notes particular weakness of body parts as indicated by failure of certain test events. These Marines are then placed in a special remedial program either at platoon, company/battery, or battalion/squadron level. Extra time is devoted to overcoming the weakness. These extra periods of conditioning may be during or after duty hours, as determined by the commanders. Other persons need reconditioning as result of hospitalization. Restoring physical fitness to damaged areas of the body is accomplished through progressively graded physical activities under professional supervision, not by the physical fitness trainers.

a. Types of Deficiencies. Physical deficiencies which can be corrected by exercise fall into several categories.

(1) Lack of Strength in One or More Body Parts. The major muscle areas concerned are the arms and shoulder girdle, back, abdomen, and legs.

(2) Lack of Overall Endurance. Usually, there is deficient muscular and cardiovascular endurance.

(3) Deficiency in Coordination and Agility. In these cases, physical skill is not developed to a satisfactory degree in activities such as crawling, running, jumping, climbing, traversing, vaulting, pushing, pulling, lifting, and carrying.

(4) Overweight or Underweight. Either condition may interfere with physical fitness and mission accomplishment. Lack of exercise is not always the cause. The cause may be malfunction of normal physiological functions or it may be poor health habits such as over- or under-eating, lack of adequate rest, or overconsumption of alcohol.

(5) Lack of Motivation. Not all persons are motivated to attain or maintain **a** desirable state of fitness. Some Marines do not understand the importance of physical fitness, some find proper exercise too difficult, and others find it inconvenient.

b. Causes of Deficiencies. There are several causes for Marines being deficient in physical fitness:

(1) Absence of exercise.

(2) Exercise which fails to develop all muscle groups and components of fitness.

(3) Exercise which is not vigorous enough or which lacks progression.
(4) Injury or illness which depletes fitness.
(5) Inadequate amounts of sleep or rest.

c. **Methods of Detection.** The commander has several means by which to detect physical deficiencies:
(1)
from those who pass. Make an analysis to determine the cause of failure as revealed by test scores.
(2) Observe Marines as they perform physical tasks-
o Marines who have difficulty during training or in physical types of work,
o Marines who have become obese and therefore experience difficulty.
(3) Be alert to those Marines who are often on sick call, returnees from hospitalization, or newly assigned.

2203. ADMINISTRATION OF REMEDIAL ACTION

a. **Group.** Attitude Marines who are singled out as being physically deficient are self-conscious and are not always convinced they need extra help. Within a deficient group, motivation may be low. These Marines must be convinced that a remedial program which is tailored to their needs will help them and will eliminate their deficiencies. Encouragement is often needed and desirable.

b. **Leadership.** The leader of this group must study individuals and know their deficiencies. The unit leader should counsel Marines individually, observe them closely as they progress through the remedial program, maintain records, and adjust the program as required.

c. **Measurement..** Whenever improvement in performance is noted, Marines should be measured by the physical fitness test either individually or by group. When testing reveals an individual to be satisfactory, that individual should be released from the remedial group. There may be exceptions to this policy in the case of Marines who are overweight or in the execution of an exercise program prescribed for some special purpose.

d. **Organization of a Remedial Group**
(1) A remedial group is usually a small group of Marines within a company or battalion. In some situations, it may be a larger group numbering 50 to 75 Marines. In the case of a larger group, divide the Marines initially into subgroups according to ability. Prescribe exercise loads commensurate with their ability. General conditioning is sufficient in the early part of the program to qualify Marines who are on the borderline. Qualification will reduce the size of the group, permitting more individual and specialized attention to those who remain.
(2) As the program progresses, regroup individuals who have like deficiencies in order that they can concentrate on their weakness. For example—

(a) One group may be weak in the arms and shoulders as revealed by failure in the chin up event. This group, with an assistant instruction charge, can work on pull-ups, rope climb, pushups, rifle or logexercises, horizontal ladder, and similar types of exercise.

(b) Another group may be weak in general endurance as revealed by the3-mile run. This group could profit from participation in conditioning drills, running, grass drill, and strength circuit.

(3) It may be necessary to form some groups to overcome weaknesses in skills, such as an inability to throw, to quickly change direction while running, to crawl rapidly, or to carry a load. Lack of coordination or lack of practice may cause these deficiencies. In this instance, instructors must provide an opportunity to practice and correct poor form and other errors as they are noted.

Chapter 3
PHYSICAL CONDITIONING ACTIVITIES
Section I. Marching Under Load

3101. GENERAL

Few physical fitness activities are as directly related to readiness for combat as foot marches under load. In addition to the obvious physical conditioning and unit cohesiveness benefits, marches under load prepare most Marines for numerous foreseeable tasks in combat. The idea that only infantry and reconnaissance units actually require training marches to prepare them for combat is clearly a mistake. In addition to the very real possibility of Marines from all types of units being used to fill combat-depleted infantry units, it is likely in maritime prepositioning force operations that aviation, combat service support, and command element personnel will have to march with weapons and equipment, from the arrival airfield to the equipment marshalling site. In American coastal cities, the harbor is nearly always 10 or more miles from the airport. It is difficult to imagine any different condition in a lesser developed country .A further combat-related benefit is that foot marching under load is the most effective way to develop leg strength in the context of a unit training program. In his classic study of troop performance in World War II, *Men Against Fire*, S.L.A. Marshall noted that the greatest single weakness of replacements of all occupational fields was leg strength. Truck drivers must push trucks which are stuck in mud, and headquarters communicators must climb hills to emplace antennas. The demands on the lower body in combat apply to all Marines, and foot marching is one activity which can help build these muscles while being conveniently integrated with the rest of the training program.

3102.TRAINING GOALS

The standard for success of a foot march is very simple to measure: did the unit arrive at the destination at the prescribed time with Marines in condition and required equipment present to accomplish the mission?A progressive program can increase a unit's readiness. It can instill pride by increasing distances and rates of march and by selecting routes over increasingly challenging terrain as the Marines become better conditioned. However, the ability to execute the mission at the conclusion of the march must remain the standard of success.

3103.MOVEMENT PLANNING

The basic considerations in planning a foot march are the mission, tactical situation, terrain and weather, and the units to participate. The success of the march will depend largely upon the thoroughness with which it is planned. A successful march is characterized by adherence to prescribed routes and time schedules, the efficient employment of the means available, and the ability of the unit to accomplish its assigned mission upon arrival at the destination.

a. Movement Orders. Movement planning culminates in the preparation and issuance of an operation order prepared in the standard, five-paragraph format. Necessary annexes are attached to furnish detailed information required for the movement. Written movement orders are rarely prepared at company level. A discussion of the annexes is contained in this chapter.

b.March Planning. March planning, as discussed here, is the planning conducted at battalion level. March planning may be organized into the following steps:

(1) **Preparation and Issuance of the Warning Order.** In order to afford subordinate units the maximum possible time to prepare for a pending move, a warning order containing all available information about the march is issued. The amount of planning time available will determine the time of issuance and the content of the warning order.

(2) Estimate of the Situation. In his estimate, the commander considers the mission, terrain, weather, time, and space factors, available routes, available transportation for the movement of equipment and/or shuttling of Marines, enemy capabilities, disposition of own forces, physical condition and training of Marines, and courses of action available to the command.

(3) Organization and Dispatch of a Reconnaissance. Party very march plan is based on as thorough and complete a ground reconnaissance as time and the situation will permit. Map and aerial reconnaissance are valuable in formulating a plan, but are not a substitute for ground reconnaissance. Route reconnaissance is accomplished by a reconnaissance party which usually consists of a reconnaissance element, an engineer element from the attached or supporting engineer unit, and a traffic control element. Unit standing operating procedures generally establish the basic composition of the reconnaissance party. It is modified as necessary to meet the requirements of a particular march. A recommended method of reporting information obtained by the route reconnaissance party is shown in figure 3-1.The minimum information required from the reconnaissance party is—

(a) Available routes and conditions.(Routes may be specified by higher headquarters.)

(b) Recommended rate of march.

(c) Selection of start point and release point, or confirmation of the suitability of start point and release point previously selected by map reconnaissance.

(d) Confirmation of location of the assembly or bivouac area.

(e) Location of critical points on the route.

(f) Distance between critical points on the route and total distance from start point to release point.

Figure 3-1 is available in the illustration section at the end of the book

(g) Location of obstacles and estimation of necessary Marines and equip- ment needed to repair and maintain routes.

(h) Number of guides required and their location on the route.

(4) Development of Detailed Movement Plan. Some of the elements included in a detailed movement plan are-

(a) Organizing the Column. To facilitate control and scheduling, units will be organized into serials and march units and given an order of march. In determining the order of march, the march planner must consider the enemy situation and the desirable order of arrival of the units at the destination. Where dispersion is required, a unit may be organized into two or more columns, each assigned a different route.

(b) Using Reconnaissance Information. Results of the route reconnaissance will be used to select the route(s); determine the start point, critical points along the route(s), and the release point; and select the rate of march.

(c) Determining March Computations. March unit pass time is based on the strength, formation, and rate of march. The pass time of the marching columns, plus necessary time distance computations, will be used to determine the completion time of the march.

(d) Drafting of Road Movement Table. Using the completed march computations, a draft road movement tabic is compiled.

(e) Checking the Plan. Using the draft road movement table and a road movement graph, the movement plan is checked to ensure that it conforms to the directive of the higher headquarters and the battalion commander's instructions.

(5) Preparation and Issuance of the Road Movement Order. After the plan has been checked and approved by the commander, an operation order is prepared and issued. The operation order may be in written form or issued orally, and is accompanied by a road movement table, overlay and/or strip map, and appropriate administrative details.

(6) Road Movement Table. A road movement table, prepared as an annex to the operation order, provides serial commanders with arrival and clearance times at critical points along the route of march. It also provides the column commander with information as to the proposed location of elements of the column at various times.

(7) Overlay/Strip Map. An overlay serves the normal purpose and should show, as a minimum, the present location of units, route of march, critical points, and the new location of units at the destination. A strip map is a schematic diagram of the route of march and shows landmarks and critical points with the distances between them. A strip map may be issued as an annex to the road movement order, in addition to or in lieu of an overlay.

c. March Computations. Prior to issuance of the operation order, the S-3 must verify time and space computations as they provide him with the necessary data for the preparation of a road movement table.

(1) Time-Distance. Time distance (TD) is determined by dividing the distance to be traveled (D) by the rate of inarch (R):

TD = D/RWhere TD = hours D = distance in kilometers R = kilometers per hour

(2) Length of ColumnThe length of column (LC) is used to determine the pass time of a column. The sum total of the following two parts determines the length of column. (See pars, (a) and (b)).The two parts are the space occupied by the Marines alone (including the distance between Marines) and the sum of the distances between the units of the column (column gap).

(a) The length of column of Marines alone is determined by multiplying the number of Marines by the appropriate factor selected from the table below. The length of column does not include distances between units. LC Marines = No. of Marines x factor.

(b) The total distance in meters between units is determined as follows:

Step One. Determine the number of gaps between serials (total serials minus one).

Step Two. Multiply the number of serial gaps from Step One by the length (in meters) between respective units.

Step Three. Determine the number of gaps between march units (total march units minus one, minus the number of serial distances).

Step Four. Multiply the number of march unit gaps from Step Three by the length (in meters) between respective units.

Step Five. Add the totals from Step Two and Step Four to get the total meters for the column gap.

EXAMPLE:A battalion foot column is organized into 12 platoon-sized march units and 3 company-sized serials. REQUIRED: total column gap distances when there are 100 meters between serials and 50 meters between march units.

(3) Pass-Time.Pass-time (PT) is the time a unit takes to pass a specified point. For foot columns, the pass-time is determined by applying the following formula:PT (minutes) = LC x FACTOR (for appropriate rate of march).

PASS-TIME FACTORS - FOOT TROOPS

.0150 for 4.0 km/h

.0187 for 3.2 km/h

.0250 for 2.4 km/h

.0375 for 1.6 km/h

EXAMPLE: Determine the pass-time of a unit whose length of column is 1,500meters and is marching at a rate of 4 km/h.PT (min) = 1,500 x .0150 (the factor for 4.0 km/h) = 22.5 mm.

(4) Completion Time. Completion time is the time of day that a march will be completed. Completion time is determined by using the following formula: Completion Time = SP time + TD + PT + Scheduled Halts (other than normal hourly halts).

EXAMPLE: A column's starting point (SP) time is 0700.The time-distance is6 hours and 40 minutes. Pass-time of the column is 30 minutes. A 35-minutelunch halt has been scheduled. What is the completion time of the march?Employing the 24-hour clock system, the formula can be applied as follows for simplified addition of the times:

HrMin SP time................0700

TD......................0640

PT......................0030

Lunch halt..............0035

Completion time.........13 105or 1445

The march will be completed atl445.

(5) Experience Tables. Based on previous movements made by a unit, data is accumulated to facilitate march planning. Such data includes approximate pass-times lor various elements of the battalion. The S-3 can utilize these data rather than computing them each time a march is scheduled. Such experience tables serve to reduce the time required to complete the computation phase of march planning. Matter appropriate to the unit standing operating procedure should be integrated therein.

d. Road Movement Graph

(1) A road movement graph is a time-distance diagram used in planning, preparing, or checking road movement tables, and for controlling marches. The graph provides a visual representation of a march plan so that conflicts and discrepancies may be prevented in the planning stage, before congestion occurs on the route. It is not usually issued as a part of the order. Road movement graphs may be applied to small units, to movements of a single column, or to a large organization scheduling separate elements, moving by various means, with different rates of march, over one or more routes.

(2) To construct a road movement graph, use the following steps: (See fig. 3-2.)

Figure 3-2 is available in the illustration section at the end of the book

Step One. Determine the number of hours available for the march. Designate the lower left corner of a sheet of graph paper as the SP time or the earliest even hour before the march is to begin. Select a scale and plot the hours available in sequence from the left on the horizontal axis.

Step Two. Determine the distance to be moved in kilometers. Indicate the SP at the lower left corner of the graph sheet and, using an appropriate scale, plot the number of kilometers on the vertical scale from the SP to the release point. Indicate the location of critical points on the vertical scale.

Step Three. At the proper distance from the SP, draw a horizontal line indicating the location of the RP. Indicate by a vertical line the hour when the movement must be completed. Plot lines representing route restriction, if any, at the proper distances and times on the graph.

Step Four. Determine the pass-times of foot and motor elements of the column.

Step Five. Starting at the SP at the specified hour, plot the movement of the head of the leading element. If the rate of march is 3 kilometers per hour, the lead element will have moved 1.5 kilometers in 30 minutes, 3 kilometers in 1 hour, etc. Plot the trace of the lead man to the RP. The last Marine will cross the SP in pass-times after the lead man. Measure this time on the graph and plot the trace of the last Marine of the column. The lines describing the head and tail of a march serial are parallel. Indicate the time subsequent serials reach the SP and plot the trace of the head and tail. Check to see that the plan complies with all restrictions and orders. If there are any violations or conflicts, the plan may be made to conform by changing the order of march, the starting time of the serials, the planned rate of march, or the organization of the column.

e. Conversion to the Metric System. The adoption of the metric system for expressing linear distances requires a simple means of conversion.

(1) To accomplish this, use the following tables:

o Multiply inches x 2.54 to obtain centimeters. Multiply centimeters x .39 to obtain inches,

o Multiply yards x .91 to obtain meters. Multiply meters x 1.1 to obtain yards,

o Multiply miles x 1.6 (or 8/5) to obtain kilometers. Multiply kilometers x .62 (or 5/8) to obtain miles.

(2) Linear distances can be accurately measured on maps which have been converted to the metric system by using a straightedge graduated in centimeters (cm).The centimeter scale can be used in lieu of the graphic scale when the scale of the map is shown, as follows:

Map Scale:	One cm Equals:
1:500,000	5,000 meters
1:250,000	2,500 meters
1:100,000	1,000 meters
1:50,000	500 meters
1:25,000	250 meters

3104. INDIVIDUAL LOAD

Since the backbone of Marine ground combat forces is the infantry, tactical mobility is largely dependent on the marching ability of the individual Marine. The load that each Marine must carry is the greatest impediment to mobility. In general, the commander must make every effort to reduce the individual load of his Marines to the absolute minimum. While the load has the greatest effect on Marines marching on foot, the value of carrying the minimum essential supplies and equipment applies for all Marine forces.

a. Fighting Load. The fighting load consists of items of clothing, equipment, weapons, and ammunition that are carried by, and essential to, the effectiveness of the combat Marine and the accomplishment of the immediate mission. The fighting load should not weigh more than 40 pounds. The commander must ensure that his Marines' loads- -and his own- -be stripped to the bare minimum. The addition of small, light-weight items in excess of the limit should not be tolerated, since cumulatively, these items will take a toll in energy. Every extra pound a Marine must carry decreases his combat effectiveness.

b. Existence Load. The existence load consists of items other than those in the fighting load that are required to sustain or protect the combat Marine, that may be necessary for increased personal and environmental protection, and that are not normally carried by the individual. The existence load is normally carried by the Marine's unit. Marching, but not engaged in combat, a Marine may be expected to carry a maximum of 50 pounds of supplies and equipment.

c. Considerations

(1) The primary consideration is not how much a Marine can carry, but how much he can carry without impaired combat effectiveness-moral or physical. The combat strength of a unit is not counted simply in numbers of Marines, but in the numbers of willing and physically able Marines.

(2) It is generally better to risk temporary inconvenience from lack of health and comfort items than to exhaust Marines due to overloading.

(3) It is a fundamental truth that men become physically exhausted more quickly when under the stress of combat. Marines must be required to carry less into battle than they are conditioned to carry in training. Marines should be conditioned for carrying weight, but should be equipped in combat for fleetness of foot.

(4) A common mistake is to base the fighting load on the gear and supplies necessary to meet every contingency. The commander cannot reasonably expect the Marines to carry enough gear for every possible eventuality. The items to be carried must be based on reasonable expectations.

(5) It is the commander's responsibility to procure the transport to carry any additional gear. As a rule of thumb, a rifle company, or unit of similar size, requires one 5-ton truck and trailer in support to carry additional gear. In cold weather, or other conditions in which the necessary personal gear increases, this requirement will increase.

(6) The commander must ensure that the supply system provides, on a dependable and timely basis, the balance of essential supplies and equipment not carried by the unit. Marines must feel confident that they will be supported with the necessary supplies and equipment.

(7) In training, commanders must implant in their Marines the pride in operating under austere conditions. To effectively demonstrate the need for this spartan mentality, the commander must personally set the standard. Marines should be trained in field craft, foraging techniques, and the use of caches and field expedients. Maximum use should be made of captured stores.

3105. MARCH TECHNIQUES AND PROCEDURES

a. Organization. A command executing a march is organized into march units, march serials, and march columns, as necessary for control.

(1) March Units. A march unit is a unit of command which moves and halts at the command of a single commander. The march unit normally corresponds to one of the smaller Marine units such as a platoon or company.

(2) March Serials. A march serial consists of one or more march units organized under the senior officer and given a numerical or alphabetical designation to facilitate control. A serial is usually a battalion, but may be a company if the battalion is marching alone.

(3) March Columns. A march column is composed of elements of a command moving over the same route. It may be composed of one or more serials. To facilitate control, a column commander is designated. A column is normally a regiment or larger unit, but may be a battalion if marching alone.

b. Order of March. The order of march is determined by the mission, terrain, and the commander's desired order of commitment of units into action.

c. Control and Coordination Measures. The commander establishes initial control of the march by designating control measures in his march order. The most frequently used measures include:

o Start point and release point.

o Time at which head or tail of column passes the start point.

o Rate of march.

o Order of march.

o Route of march.

o Location of command post.

o March communications.

d. The Pacesetter. The pace-setter is an experienced individual carrying the same load as the majority of the Marines and marching from 4 to 10 meters at the head of the column. (See fig. 3-3.)The pace-setter's primary duty is to maintain the rate of march ordered by the column commander.

(1) Setting the Pace. The pacesetter does this by establishing the pace (length of step) and cadence (steps per minute) to obtain the prescribed rate of march. He should be of medium height so normal strides will be taken. Overstriding or understriding tends to tire leg muscles quickly and affects the combat efficiency of marching Marines. The officer marching at the head of the column supervises the pacesetter to ensure that he takes normal strides and maintains a uniform cadence.

Figure 3-3 is available in the illustration section at the end of the book

(2) Reducing Accordion Effect. Accordion effect occurs when the column alternately lengthens and shortens, causing the rear elements of the column to double time to maintain distance. It is caused by changes in the rate of march by the lead elements, usually after passing a slowing obstacle. To reduce this effect as much as possible, lead elements should slow their rate of march for a sufficient time after crossing obstacles to allow the rear elements to maintain the prescribed distance without running. The rear elements can signal the head when the last man has cleared the obstacle and the rate can be resumed. Since some accordion effect is nearly inevitable, the order of march should be periodically rotated to prevent units in the rear from becoming physically exhausted at the conclusion of the march.

e. Length, Rate and Pace

(1) **Length of March.** The length of march varies depending on terrain and weather, enemy situation, and physical and mental condition of Marines. The normal length of march for a 24-hour period is from 20 to 32 kilometers (12 to 20 miles) marching from 5 to 8 hours at a rate of 4 kilometers (2.5 miles) per hour (km/hmi/h).A march in excess of 32 kilometers in a 24-hour period is considered a forced march. Well-trained units that have been progressively trained will be able to achieve a unit goal of 60 miles in a 3-day period.

(2) Rate of March. The same factors of terrain, weather, load to be carried, and condition of Marines affect the rate of march. The commander considers these and selects a rate which will place his unit at the destination in the shortest possible time in condition to accomplish the mission. Unit standing operating procedure should state the standard rate of march for that unit over normal terrain for both road and cross-country movement and in both day and night. The commander then modifies this rate if the situation requires. Normally prescribed standards are as follows:

Figure is available in the illustration section at the end of the book

(3) Pace and Cadence. The normal pace is 30 inches. A pace of 30 inches and a cadence of 106 steps per minute result in a speed of 4.8 kilometers per hour or 3 miles per hour and a rate of 4 kilometers per hour or 2.5 miles per hour if a 10-minute rest halt per hour is taken. Since the pace of each man may vary, the cadence may have to be adjusted to maintain the prescribed rate. Slope of ground and surface footing affect pace. A normal cadence is relatively simple to maintain on level or gently rolling terrain on firm footing. Mud, sand, loose gravel, and hills all greatly affect pace.

f. Halts. Halts during a day's march are taken at regular intervals to rest personnel and adjust loads.Halts are specified by standing operating procedures or march order. Such factors as midday heat or enemy action may force the adoption of long midday halts or night marches. At long halts, each unit disperses to a previously selected location near the route of march.
(1) Normal Time and Frequency. Under normal conditions, a 15-minute halt is made after the first 45 minutes of marching. Following the first halt, a 10-minute halt is made after each 50 minutes of marching. Variations of this schedule are made when a scheduled halt time occurs when passing through a built-up area or when cover and concealment are required by the tactical situation and none is available. Observation posts may be established, if required for security of the unit during halts.
(2) Actions at Halts. All units in the column should be halted at the same time. At the halt signal, Marines should move to the side of the road, staying within the immediate vicinity of their unit. Marines should remove or loosen gear and sit or lie with feet elevated. Commanders inspect Marines and equipment, and corpsmen administer medical treatment, as required.

g. Communications. The four primary means of communications used in controlling foot marches are radio, visual, sound, and messenger.
(1) Radio. Radios are normally used for communications between platoons and higher headquarters in the march column. Radio transmissions should be held to the absolute minimum necessary for control and should be at minimum power required.
(2) Visual.. Visual communications may include flashlights, luminous markers, panel sets, pyrotechnics, and hand and arm signals. When contact with the enemy is not expected, the loss of light discipline from some of these signals may be a lesser security risk than the risk of disclosure from radio transmissions.
(3) Sound. Sound communications include voice and such signaling devices as horns and whistles. Some of their uses include assembling Marines at the conclusions of halts and providing warnings of air or chemical attacks.
(4) Messengers. Messengers are particularly effective during periods of radio silence and reduced visibility. Messages should be simple and brief to preclude mistakes by the messenger.

h. Security. Both active and passive measures are used against attack by enemy aircraft and indirect fire weapons during movement. Active measures include the use of air sentries and organic and attached weapons in accordance with the unit air defense standing operating procedure. Passive measures include the use of concealed routes and assembly areas, night marches, and extended interval between elements of the column. With imagination and planning, all of these measures can be practiced in training without detracting from the physical conditioning benefit of the march. In fact, such drills as rapidly taking cover during simulated attacks and then quickly resuming the march can prove physically challenging.

i. Reconnaissance. Reconnaissance determines in advance critical points along the march route such as bridges, fords, and obstacles in order that congestion or delay may be held to a minimum and local security provided. Paragraph 3103.C describes organization of the reconnaissance party.

3106.TACTICAL MARCHES

These marches are movements of Marines and equipment not in direct ground contact with the enemy but expecting early ground contact either en route or upon arrival at the objective. Some characteristics of this type of movement include security elements to the front, flanks, and, if necessary, to the rear of the march column. Additional considerations include the selection of concealed areas on defendable terrain for start and release points and rest stops. In training, infantry units of regimental size or smaller will normally derive the greatest benefit from tactical marches. The security requirements usually dictate a slower rate of march and, therefore, reduce the physical conditioning value for other types of units which would seldom move in combat with the expectation of early ground contact with the enemy. Two key points should be remembered regarding tactical marches in a training program.

a. Security. Although security requirements often slow the rate of march, it is a valid and important training objective for infantry units to work toward reducing march times while still practicing sound tactical security. March standing operating procedures, good land navigation skills, and simple but complete march orders contribute as much to this goal as physical conditioning. The idea that proceeding slowly enhances security is seldom true. Keeping the enemy off balance by rapid movement frequently is the best guarantee of security.

b. Training. In peacetime training, it is critical that support personnel who normally are located with infantry units in combat, participate completely with infantry units in forced march training. Not only must they not slow the infantry rate of march because of inferior conditioning, but they must be thoroughly familiar with the infantry unit's standing operating procedure for security on the march. Personnel from within the ground combat element such as artillery forward observer teams and combat engineer units, can be easily identified for this training and should be habitually associated with the same infantry units. Personnel from other elements, such as low-altitude air defense teams and radio battalion detachments, are more difficult to identify and earmark for support of the same infantry unit. Their normal mode of employment is in vehicles and their units frequently provide general support vice direct support. This does not, however, lessen their requirement to be prepared to provide support to dismounted infantry over terrain or in a tactical situation not conducive to vehicles. Such training needs to take place, and planning and cooperation among the elements of the MAGTF can overcome the obstacles.

3107.ADMINISTRATIVE MARCHES

These marches are movements of Marines and equipment made when no enemy interference is expected except from aircraft or indirect fire. During these marches, units are administratively grouped for ease of control and speed of movement. The term administrative foot marches should not be taken to imply that tactical considerations are disregarded. There are simply less stringent security measures used than when ground contact with the enemy is a distinct possibility. The normal formation is the route column with one file moving on either side of the road and with negligible security. Air sentries are used and dispersion between individuals and units is practiced. Although roads usually provide the fastest route, there may be reasons to conduct these marches cross-country (particularly as part of a training program). Administrative marches are the type of foot marches most practical for training programs for non-infantry units. These marches may also play a key role in the physical conditioning program of infantry units, because of the greater speeds possible due to lessened security requirements.

3108. FORCED MARCHES

A forced march is a march which covers a greater distance than 32 kilometer (20 miles) in a 24-hour period. Normally the extra required distance should be achieved by increasing the number of hours marched in a day beyond the norm (8 hours) rather than increasing the rate of march beyond the norm (4 kilometers per hour/2.5 miles per hour). However, there will be occasions when the situation or mission demands an increase in the rate of march.

a. Maximum Recommended Distances. The maximum recommended distances for forced marches are-56 kilometers (35 miles) in 24 hours; 96 kilometers (60 miles) in 48 hours; or 128 kilometers (80 miles) in 72 hours.

b. Sample Time Breakdown. A sample time breakdown for a forced march of 52 kilometers, beginning at first light is as follows:

Figure is available in the illustration section at the end of the book

3109. NIGHT MARCHES

Night marches are characterized by closed formations, more difficult control and reconnaissance, and a slower rate of march than day marches.

a. Control. Control is increased by reducing the distance between individuals and units, and by using connecting files to maintain contact between platoons and companies. Connecting files are normally constituted from personnel from the rear most march unit.

b. Safety on Roads. Because of reduced visibility, night marches on roads used by vehicles require attention to safety procedures to prevent accidents. Commanders should not think these safety measures apply only to peacetime training. In combat, nearly all vehicles operate with blackout lights at night, making visibility even more difficult than in garrison driving. If consistent with the tactical situation, the following measures can reduce the possibility of marchers being struck by vehicles:

(1) Use off-road trails and routes as much as possible.

(2) Guards to the front and rear of march units should be marked with strips of luminous tape and carry red-filtered flashlights. These measures are generally consistent with light discipline, while still affording reasonable warning to approaching drivers.

Section II. Conditioning Drills (One, Two, and Three)

3201. GENERAL

a. Description and Objective. Conditioning drills are calisthenic exercises. Each drill contains seven exercises organized and numbered in a set pattern. Each drill takes 15 minutes to complete. The objective of conditioning drills is to exercise all major muscle areas in order to develop strength, endurance, coordination, and flexibility.

b. Area and Equipment. Any level area is satisfactory for conduct of the drills. Drills One and Two contain ground exercises. If ground conditions are unsatisfactory, Drill Three should be used as it contains no ground positions. Usually, no equipment is required;however, if the group exceeds a platoon in size, an instructor's stand is necessary.

c. Formation. The extended rectangular formation is used in this drill. (See app. A.)

d. Starting Level and Progression. The starting level is six repetitions of each exercise. An increase of one repetition for each three periods of exercise in which the drill is performed is an acceptable rate of progression. This rate is continued until 12 repetitions can be completed. To maintain, continue the drill at 12 repetitions. To progress, move to 6 repetitions at a more difficult drill which exercises the same muscle group. Progression can also be gained by moving from Drill One to Drill Two, as Drill Two is more demanding.

e. Starting Positions. Starting positions vary with the exercise and are explained in each exercise.

f. Leadership. A principal instructor demonstrates and leads the drills. The instructor must be familiar with leadership techniques peculiar to conditioning drills to include the exercises, commands, counting cadence, cumulative count, formation, method of teaching the exercises, and utilization of assistant leaders.

g. Use With Other Programs. Conditioning Drills One, Two, and Three reach all major muscles of the body. They are easy to learn and to perform, and they are simple to administer and supervise. These features, coupled with the short time required for completion, the fact that no equipment is necessary, and adaptability to most areas of execution, make these drills possible in any programs.

3202. CONDITIONING DRILL ONE

This conditioning drill is similar to the calisthenics drill commonly known throughout the Marine Corps as the "Daily Seven."The "Daily Seven" may be used as a substitute for Conditioning Drill One.

a. Exercise l:High Jumper

(1) Starting Position. Feet separated shoulder width, knees flexed, body bent forward at the waist, arms aligned with the trunk and hips, elbows locked, palms facing, fingers extended and joined, head and eyes to the front.(See fig. 3-4, A.)(Elbows remain locked throughout the exercise.)

Figure 3-4 is available in the illustration section at the end of the book.

(2) Cadence. Moderate.
(3) Movement. A four-count exercise: at the count of—

(a) ONE -- Take a slight jump into the air, swinging the arms forward and up to shoulder level.

(b) TWO -- Take a slight jump into the air and swing the arms downward and back, returning to the starting position.

(c) THREE -- Take a vigorous leap into the air, swinging the arms forward and up to an overhead position, momentarily looking skyward, on returning to the ground the knees are flexed, head and eyes return to the front.

(d) FOUR--Repeat the action of count two.

b. Exercise 2: Bend and Reach

(1) Starting Position. Feet spread more than shoulder width, arms overhead, elbows locked, palms facing, fingers extended and joined, head and eyes to the front.(See fig. 3-4, B.)
(2) Cadence. Moderate.
(3) Movement. A four-count exercise: at the count of—

(a) ONE-Bend at the knees and waist, swing the arms straight downward and reach between the legs. Touch the ground as far to the rear as possible and look to the rear. (Elbows remain locked throughout the exercise).

(b) TWO-- Recover sharply to the starting position.

(c) THREE--Repeat the action of count ONE.

(d) FOUR-- Repeat the action of count TWO.

c. Exercise 3: Pushup

(1) Starting Position. Front leaning rest position. To assume this position there is a silent one-two count:on the silent count of one, assume the squatting position, heels together, elbows locked inside the knees, hands flat on the ground directly beneath the shoulders; on the silent count of two, thrust the legs to the rear, toes and heels together, body straight from head to heels.(Seefig. 3-4, C.)
(2) Cadence. Moderate.
(3) Movement. A four-count exercise: at the count of—

(a) ONE -- Flex the elbows lowering the body until the thick portion of chest touches the ground.

(b) TWO -- Raise the body until elbows are straight and locked.

(c) THREE-- Repeat the action of count ONE.

(d) FOUR -- Repeat the action of count TWO. (On returning to position of attention, the silent one-two count is used in reverse).

d. Exercise 4: Trunk Twister

(1) Starting Position. Feet are spread more than shoulder width apart, fingers placed behind neck, thumbs pointing downward, elbows back. (See fig. 3-4, D.)(Elbows remain well back throughout the exercise).

(2) Cadence. Slow.

(3) Movement. A four-count exercise: at the count of-

(a) ONE-- Keeping the knees locked and the back straight, bend forward at the waist sharply, with a slight recovery.

(b) TWO-- Twist the trunk to the left vigorously at the waist, keeping the elbow back. The left elbow is higher than the right.

(c) THREE-- Twist vigorously to the right, so the left elbow comes under the right.

(d) FOUR-- Straighten sharply to the starting position. NOTE: Do not attempt to touch the elbows to the knees on counts two and three.

e. Exercise 5: Squat Bender

(1) Starting Position. Feet are spread less than shoulder width apart, hands on hips, thumbs in small of back, elbows back. (See fig. 3-4, E.)

(2) Cadence. Moderate.

(3) Movement. A four-count exercise: at the count of—

(a) ONE-- Assume the squatting position, maintain balance on the balls of the feet, with trunk erect thrust arms forward to shoulder level, elbows locked, palms down.

(b) TWO-- Recover to starting position. Elbows are well back.

(c) THREE-- Keeping the knees locked, bend forward at the waist, touching the ground in front of the toes.

(d) FOUR-- Vigorously recover to the starting position.

f. Exercise 6: Body Twist

(1) Starting Position. On the back, arms extended sideward on the ground, palms down. The legs are raised to a near vertical position, feet together, knees locked.(See fig. 3-4, F.)

(2) Cadence. Slow-fast.

(3) Movement. A four-count exercise: at the count of-

(a) ONE-- Lower legs slowly to the left until they touch the ground near the left hand, keeping the knees straight and shoulders on the ground.

(b) TWO-- Recover the starting position by quickly raising the legs, keep knees straight and feet together.

(c) THREE-- Repeat movement of count ONE, except the movement is to the right side.

(d) FOUR-- Recover sharply to the starting position.

g. Exercise 7: Stationary Run

(1) Starting Position. Position of attention. (See fig, 3-4, G.)

(2) Cadence. Fast.

(3) Movement.

(a) At the command of execution, start running in place at double-time cadence, lifting the left foot first time cadence. Follow the instructor as he counts two repetitions of cadence;

e.g., 1, 2, 3, 4--1, 2, 3, 4. The instructor then gives informal commands such as FOLLOW ME, running on the toes and balls of the feet, keeping the back straight, speeding up the cadence to a sprint, raising the knees high, leaning forward at the waist, and pumping the arms vigorously.

(b) To halt the exercise, the instructor will count two repetitions of cadence as the left foot strikes the ground: 1, 2, 3, 4 - 1, 2, 3, HALT.

NOTE: When counting cadence, the instructor counts only as the left foot strikes the ground. The duration of the exercise is approximately 1 1/2 minutes.

3203. CONDITIONING DRILL TWO

a. Exercise 1: Jumping Jack

(1) Starting Position. Feet separated more than shoulder width, arms overhead. (See fig. 3-5, A.)

(2) Cadence. Moderate.

(3) Movement. A four-count exercise: at the count of~

(a) ONE-- Jump to position with the feet together and assume the squatting position, swinging the arms sideward and downward, placing the hands palms down on the ground, elbows locked inside the knees.

(b) TWO-- Recover to the starting position by jumping to the side straddle and swinging the arms sideward overhead.

(c) THREE-- Repeat the action of count ONE.

(d) FOUR-- Recover to the starting position.

b. Exercise 2: Turn and Bend

(1) Starting Position. Side straddle, arms overhead. (See fig. 3-5, B.)

(2) Cadence. Moderate.

(3) Movement. A four-count exercise: at the count of-

(a) ONE-- Turn the trunk to the left and bend forward over the left thigh, attempting to touch the fingertips to the ground outside the left foot. Keep the left knee straight. On successive repetitions attempt to touch farther and farther to the side.

(b) TWO-- Recover to the starting position.

(c) THREE-- Turn the trunk to the right and bend forward over the right thigh, trying to touch the hands to the ground outside the right foot. Keep the right knee straight.

Figure 3-5 is available in the illustration section at the end of the book.

(d) FOUR-- Recover to the starting position.

c. Exercise 3: Eight-Count Pushup

(1) Starting Position. Position of attention. (See fig. 3-5, C.)

(2) Cadence. Moderate.

(3) Movement. An eight-count exercise: at the count of—

(a) ONE-- Assume the squatting position, palms on the ground directly beneath the shoulders, elbows locked inside the knees.

(b) TWO-- Thrust the legs to the rear assuming the front leaning rest position.

(c) **THREE**-- Flex the elbows until the thick portion of the chest touches the ground.

(d) **FOUR**-- Raise the body on a straight plane until the elbows are locked.

(e) **FIVE**-- Repeat the action of count THREE.

(f) **SIX**-- Repeat the action of count FOUR.

(g) **SEVEN**-- Recover to the squatting position as in count ONE (elbows locked inside the knees).

(h) **EIGHT**—Return sharply to the position of attention.

d. Exercise 4: Turn and Bounce

(1) Starting Position. Feet spread more than shoulder width apart, arms sideward at shoulder level, palms up. (See fig. 3-5, D.)

(2) Cadence. Slow.

(3) Movement. An eight-count exercise: at the count of—

(a) **ONE**-- Turn sharply to the left as far as possible, then recover slightly.

(b) **TWO**-- Again turn to the left as far as possible and recover as in ONE.

(c) **THREE**-- Repeat the action of count TWO.

(d) **FOUR**-- Recover sharply to the starting position.

(e) **FIVE**—Turn sharply to the right as far as possible, then recover slightly.

(f) **SIX**—Again turn to the right as far as possible and recover as in FIVE.

(g) **SEVEN**-Repeat the action of count SIX.

(h) **EIGHT**-Return to the starting position.

NOTE: The head and hips remain to the front throughout the exercise and the knees and elbows are locked at all times.

e. Exercise 5: Squat Stretch

(1) Starting Position. Attention. (See fig. 3-5, E.)

(2) Cadence. Moderate.

(3) Movement. A four-count exercise: at the count of—

(a) **ONE**-- Squat, placing the hands on the ground about 12 inches in front of the feet.

(b) **TWO**— Keeping the fingertips on the ground, straighten the knees completely and raise the hips.

(c) **THREE**-- Recover to position ONE.

(d) **FOUR**-- Recover to the starting position.

f. Exercise 6: Leg Circular

(1) Starting Position. On the back, arms stretched sideward, palms down, feet raised foot from ground, knees straight. (See fig 3-5,F.)

(2) Cadence. Slow.

(3) Movement A four-count exercise: at the count of—

(a) **ONE**-- Swing the legs as far as possible to the left, keeping the knees straight and the legs together.

(b) **TWO**-- Swing the extended legs over head with the thighs as close as possible to the trunk.

(c) **THREE**-- Swing the legs as far as possible to the right.

(d) **FOUR**-- Recover to the starting position.

g. Exercise 7: Stationary Run

(1) Starting Position. Position of attention. (See fig. 3-5, G.)

(2) Cadence. Fast.

(3) Movement

(a) At the command of execution, start running in place at double-time cadence, lifting the left foot first. Follow the instructor as he counts two repetitions of cadence; for example: 1, 2, 3, 4--1, 2, 3, 4. The instructor then gives informal commands such as FOLLOW ME. Running on the toes and balls of the feet, keeping the back straight, speeding up the cadence to a sprint, raising the knees high, leaning forward at the waist, and pumping the arm vigorously.

(b) To halt the exercise, the instructor will count two repetitions of cadence as the foot strikes the ground: 1, 2, 3, 4-1,2, 3, HALT.

NOTE: When counting cadence, the instructor counts only as the left foot strikes the ground. The duration of the exercise is approximately 1 1/2 minutes.

3204. CONDITIONING DRILL THREE

a. Exercise 1: Side Straddle Hop

(1) Starting Position. Position of attention. (See fig. 3-6, A.)

(2) Cadence. Moderate.

(3) Movement. A four-count exercise: at count of-

(a) ONE-- Take a slight jump into the air, moving the legs sideward (more than shoulder width apart); at the same time, swing the arms overhead (to an overhead position) clapping the palms together.

Figure 3-6 is available in the illustration section at the end of the book.

(b) TWO-- Take a slight jump into the air, swing the arms sideward and downward returning to the starting position.

(c) THREE-- Repeat the action of count ONE.

(d) FOUR-- Repeat the action of count TWO.

b. Exercise 2: Back Bender

(1) Starting Position. Standing, feet 12 inches apart, fingers placed behind the head. (See fig. 3-6, B.)

(2)Cadence. Slow.

(3) Movement. A four-count exercise: at the count of-

(a) ONE-- Bend the upper trunk backward, raising the chest high, pulling the elbows back, and looking upward. Keep the knees straight.

(b) TWO-- Recover to the starting position.

(c) THREE-- Repeat the action of count ONE.

(d) FOUR-- Recover to the starting position.

c. Exercise 3: Squat Thrust

(1) Starting Position. Position of attention. (See fig. 3-6, C.)

(2) Cadence Moderate.

(3) Movement. A four-count exercise: at the count of—

(a) ONE-- Assume the squatting position; heels together, placing the hands flat on the ground, shoulder width apart, elbows locked and inside the knees.

(b) TWO-- Thrust the legs to the rear, assuming the front leaning rest position, body in line from head to toe, heels and toes together.

(c) THREE-- Return to the squatting position as in ONE.

(d) FOUR-- Return to position of attention.

d. Exercise 4: Side Bender

(1) Starting Position. Feet are spread more than shoulder width apart, arms are raised sideward and overhead, thumbs interlocked palms to front, fingers extended and joined, elbows locked. (See fig. 3-6, D.)

(2) Cadence. Slow.

(3) Movement. An eight-count exercise: at the count of—

(a) ONE-- Bend to left as far as possible, then recover slightly.

(b) TWO-- Again bend to the left as far possible, then recover slightly.

(c) THREE-Repeat the action of count TWO.

(d) FOUR-- Recover sharply to the starting position.

(e) FIVE-- Bend to the right as far as possible, then recover slightly.

(f) SIX-- Again bend to the right as far as possible, then recover slightly.

(g) SEVEN-- Repeat the action of count SIX.

(h) EIGHT-- Recover sharply to the starting position.

NOTE:Keep the elbows and knees locked throughout the exercise. The bend should occur to the side and not the front.

e. Exercise 5: Knee Bender

(1) Starting Position. Feet are spread less than shoulder-width apart, hands on hips, thumbs in small of back, elbows back. (See fig 3-6, E.)

(2) Cadence. Moderate.

(3) Movement. A four-count exercise: on the count of—

(a) ONE-Do a knee bend, lean trunk forward at the waist, thrust arms between legs until the extended fingers touch the ground palms to the ground, hands 6 inches apart.

(b) TWO~Recover sharply to the starting position.

(c) THREE-Repeat the action of count ONE.

(d) FOUR-Repeat the action of count TWO.

f. Exercise 6: Bottoms Up

(1) Starting Position. Front leaning rest position. A silent one-two count is used as in the pushups. (See fig. 3-6, F.)

(2) Cadence. Moderate.

(3) Movement A four-count exercise: at the count of-

(a) ONE-- With the weight on the hands, and knees locked, jump forward bringing the feet as close to the hands as possible; look to the rear.

(b) TWO--Keeping the knees locked, thrust the legs backward assuming the front leaning rest position.

(c) THREE-- Repeat the action of count ONE.

(d) FOUR-- Repeat the action of count TWO.

g. Exercise 7: Stationary Run

(1) Starting Position. Position of attention. (Sec fig. 3-6, G.)

(2) Cadence. Fast.

(3) Movement

(a) At the command of execution, start running in place at double time, lifting the left foot first. Follow the instructor as he counts two repetitions of cadence; for example: 1, 2, 3, 4--1, 2,3, 4. The instructor then gives informal commands such as FOLLOW ME, running on the toes and balls of the feet, keeping the back straight, speeding up the cadence to a sprint, raising the knees high, leaning forward at the waist, and pumping the arms vigorously.

(b) To halt the exercise, the instructor will count two repetitions of cadence as the left foot strikes the ground: 1, 2, 3, 4 - 1, 2, 3, HALT.

NOTE: When counting cadence, the instructor counts only as the left foot strikes the ground. The duration of the exercise is approximately 1 1/2 minutes.

Section III. Rifle and Log Drills

3301. RIFLE DRILL

a. Description and Objective. Rifle exercises are conditioning exercises performed with a rifle. Each drill contains six exercises (fig. 3-7) and they are numbered in a set pattern. The drill takes 15 minutes to complete. The objective of rifle drills is to exercise the arms, shoulders, and back muscles in order to develop strength and endurance, particularly in the upper body. In units without rifles, log drills may be substituted.

b. Area and Equipment. Any level area is satisfactory for conducting this drill. Each Marine completes these exercises from a standing position and no ground contact is required. Each Marine will need a rifle and, if the group exceeds a platoon in size, then the instructor will need an instructor's stand.

c. Formation. The extended rectangular formation is used in this drill. (See app. A.)

d. Starting Positions. Starting positions vary with the exercises and are explained in each exercise. As in all set conditioning drills, the command used to start the exercise is STARTING POSITION, MOVE. The following directions apply to rifle drill.

(1) In those exercises which start from the rifle downward position, on the command MOVE, execute port arms as prescribed in NAVMC 2691, Drill and Ceremonies Manual, and then assume the starting position. The command to return to the men to the position of attention at the conclusion of the exercise is POSITION OF ATTENTION, MOVE.

(2) In exercises which terminate in the rifle downward position, on the command MOVE, execute the position followed by order arms as prescribed in NAVMC 2691.

(3) In the exercises which terminate in a position other than the rifle downward position, Marines first assume the rifle downward position before executing port arms and order arms.

(4) These movements are executed without command. This procedure promotes uniformity, but precision is not expected. To be effective, rifle exercises must be strenuous enough to tire the arms, but not to the point where the arms cannot move with precision.

e. Leadership. A principal instructor demonstrates and leads the drill. He must be familiar with leadership techniques for conditioning exercises and the peculiar techniques for rifle drill.

3302. EXERCISES PERFORMED WITH RIFLES

The exercises of rifle drill are outlined in the following paragraphs,

a. Exercise 1: Foreup, Behind Back

(1) Starting Position. Rifle downward, feet together.(See fig. 3-7, A.)

Figure 3-7 is available in the illustration section at the end of the book.

(3) Movement. A four-count exercise: at the count of—

(a) ONE-- Swing the arms forward and upward to the overhead position. Inhale.

(b) **TWO**-- Lower the rifle to the back of the shoulders. Exhale.
(c) **THREE**-- Recover to position ONE and inhale.
(d) **FOUR**-- Recover to the starting position and exhale.

b. Exercise 2: Lunge Side, Turn and Bend
(1) Starting Position. Rifle downward, feet together.(See fig. 3-7, B.)
(2) Cadence. Moderate.
(3) Movement. An eight-count exercise:at the count of—
(a) **ONE**-- Lunge sidewards to the left, swing the rifle forward and upward to the overhead position.
(b) **TWO**-- Turn the trunk to the left and bend forward over the left hip. At the same time, swing the rifle to a low horizontal in front of the left ankle.
(c) **THREE**-- Recover to position ONE.
(d) **FOUR**-- Recover to the starting position.
(e) **FIVE, SIX, SEVEN, and EIGHT**-- Repeat on the right side.

c. Exercise 3: Foreup, Back Bend
(1) Starting Position. Rifle downward, feet together.(See fig. 3-7, C.)
(2) Cadence. Moderate.
(3) Movement. A four-count exercise:at the count of—
(a) **ONE**-- Swing the arms forward and upward to the overhead position.
(b) **TWO**-- Bend backward, emphasizing the bend in the upper back. The face is up.Keep the knees straight.
(c) **THREE**-- Recover to position ONE.
(d) **FOUR**-- Recover to the starting position.

d. Exercise 4:Up and Forward
(1) Starting PositionRifle downward, feet together.(See fig. 3-7, D.)
(2) Cadence. Fast.
(3) Movement. A four-count exercise:at the count of—
(a) **ONE**-Swing the arms forward and upward to the overhead position.
(b) **TWO**- Swing the arms forward to shoulder level.
(c) **THREE**-Recover to position ONE.
(d) **FOUR**-Recover to the starting position.

e. Exercise 5: Foreup, Full Squat
(1) Starting Position. Rifle downward, feet in narrow stance. (See **fig.** 3-7, E.)
(2) Cadence. Moderate.
(3) Movement. A four-count exercise: at the count of—
(a) **ONE**-- Swing the arms forward and upward to the overhead position.
(b) **TWO**-- Swing the arms down to shoulder level and assume the squatting position.
(c) **THREE**-- Recover to position ONE.
(d) **FOUR**-- Recover to the starting position.

f. Exercise 6: Arms Forward, Side Bend

(1) Starting Position. Sidestraddle, regular stance, rifle forward.(See fig. 3-7, F.)
(2) Cadence. Moderate.
(3) Movement. A four-count exercise:at the count of-

  (a) ONE-- Bend the trunk to the left. Keep the knees straight.
  (b) TWO-- Recover to the starting position.
  (c) THREE-- Bend the trunk to the right. Keep the knees straight.
  (d) FOUR-- Recover to the starting position.

NOTE:Keep the rifle on the same level as the shoulders throughout the exercise.

3303. LOG DRILL

a. Description and Objective. Log exercises are conditioning exercises performed with a log. Each drill contains six exercises and they are numbered in a set pattern. The drill takes 15 minutes to complete. The objective of log drills is to develop strength and muscular endurance and, in this instance, under maximum loads. Log exercises also develop teamwork. Log may be used in lieu of conditioning drills after the Marines have become somewhat conditioned.

b. Area and Equipment. Any level area is satisfactory for conducting this drill. Each Marine completes these exercises from a standing position and no ground contact is required. If the group exceeds a platoon in size, then the instructor will need instructor's stand. Each six-person group or eight-person group will need a log. The logs should be from 6 to 8 inches in diameter. They may vary in length from 14 feet (for 6 people) to 18 feet (for 8 people).The logs should be skinned, smoother, and dried. The 14-foot logs should weigh approximately 300 pounds and the 18-foot logs, approximately 400 pounds. Rings should be painted on the logs to indicate each person's position. When not in use, the logs should be stored on a rack to keep them off the ground.

c. Formation. All the Marines assigned to the same log team should be about the same height at the shoulders. The recommended method of dividing the platoon is to have the Marines form a single file or column with short people to the front and tall people to the rear. Have the Marine assume their positions in the column according to shoulder height, not head height. When they are in position, they are given the command COUNT OFF BY SIXES (OR EIGHTS), COUNT OFF, to divide them into six- or eight-person log teams. Each team in turn, can then proceed to the log rack, shoulder a log, and carry it to the designated exercise area. The log teams form in columns in front of the instructor. With the Marine holding the log in the chest position, have them face the instructor and ground the log at least 10 yards between columns and 10 yards between log teams within the columns.

d. Starting Positions. The Marines fall in, facing the log, with their toes about 4 inches from it. The basic starting positions and commands are as follows:(See fig. 3-8.)
(1) **RIGHT-HAND STARTING POSITION, MOVE.** It the command MOVE, move the left foot 12 inches to the left, and lower the body into a flat foot squat. Keep the back straight, head up, and arms between the legs. Encircle the far side of the log with the left hand. Place the right hand underneath the log.(See fig. 3-8, A.)

(2) LEFT-HANDED STARTING POSITION, MOVE. These commands are executed in the same manner as in paragraph a. except that the left hand is underneath the log and the right hand encircles its far side. (See fig. 3-8, B.)

Figure 3-8 is available in the illustration section at the end of the book.

(3) RIGHT SHOULDER POSITION, MOVE. At the command MOVE, pull the log upward in one continuous motion to the right shoulder. At the same time, move the left foot to the rear and stand up, facing left. Balance the log on the right shoulder with both hands.(See fig. 3-8, C.)This movement cannot be performed from the left-hand starting position because of the position of the hands.

(4) LEFT SHOULDER POSITION, MOVE. These commands should be given from the left hand starting position. At the command MOVE, pull the log upward in one continuous motion, to the left shoulder. At the same time, move the right foot to the rear and stand facing right. Balance the log on the left shoulder with both hands.(See fig. 3-8, D.)This movement cannot be performed from the right-hand starting position.

(5) WAIST POSITION, MOVE. From the right hand position pull the log waist high. Keep the arms straight and fingers laced underneath the log. The body is inclined slightly to the rear and the chest is lifted and arched. (See fig. 3-8, E.)

(6) CHEST POSITION, MOVE. This command should be given after the waist position has been assumed .On the command MOVE, shift the log to a position high on the chest, bring the left arm under the log and hold the log in the bend of the arms.(See fig. 3-8, F.)Keep the upper arms parallel to the ground.

(7) To move the log from the right shoulder to the left shoulder, the command is: LEFT SHOULDER POSITION, MOVE. On the command MOVE, push the log overhead and lower it to the opposite shoulder.

(8) To return the log to the ground from any of the above positions, the command is: STARTING POSITION, MOVE. At the command MOVE, slowly lower the log to the ground. The hands and fingers must be kept from under the log.

e. Leadership. A principal instructor demonstrates and leads the drill. He must be familiar with the leadership techniques for conditioning exercises and the peculiar techniques for log drill.

3304. EXERCISE PERFORMED WITH LOGS

The exercises of log drill are outlined in the following paragraphs. Figure 3-9 graphically explains log drill.

a. Exercise l:Two-Arm Pushup

(1) Starting Position. Right or left shoulder position. Regular stance. (See fig. 3-9, A.)

(2) Cadence. Moderate.

(3) Movement. A four-count exercise:at the count of—

- **(a) ONE--** Push the log overhead until the elbows lock.
- **(b) TWO--** Lower the log to the opposite shoulder.
- **(c) THREE--** Repeat the action of count ONE.
- **(d) FOUR--** Recover to the starting position.

b. Exercise 2: Forward Bender

(1) Starting Position. Chest position. Regular stance.(See fig. 3-9, B.)

(2) Cadence. Moderate.

(3) Movement. A four-count exercise:at the count of-

(a) ONE-- Bend forward at the waist, keeping the back and legs straight.

(b) TWO-- Recover to the starting position.

(c) THREE-- Repeat the action of count ONE.

(d) FOUR-- Recover to the starting position.

c. Exercise 3: Straddle Jump

(1) Starting Position. Right or left shoulder position, feet together, fingers interlaced on top of the log.(See fig. 3-9, **C.)**

(2) Cadence. Moderate.

(3) Movement. A four-count exercise:at the count of—

(a) ONE-- Jump to a side-straddle. Pull down on the log with both hands to keep it from bouncing on the shoulder.

(b) TWO-- Recover to the starting position.

(c) THREE-- Repeat the action of count ONE.

(d) FOUR-- Recover to the starting position.

Figure 3-9 is available in the illustration section at the end of the book.

d. Exercise 4: Side Bender

(1) Starting Position. Right shoulder position, feet regular stance.(See fig. 3-9, D.)

(2) Cadence. Moderate.

(3) Movement. A four-count exercise:at the count of—

(a) ONE-- Bend sideward to the left as far as possible, bending the left knee.

(b) TWO-- Recover to the starting position.

(c) THREE-- Repeat the action of count ONE.

(d) FOUR-- Recover to the starting position.

(4) After completing the required number of repetitions, change shoulders and execute an equal number of repetitions to the other side.

e. Exercise 5: Knee Bend

(1) Starting Position. Right or left shoulder position. Narrow stance. Fingers interlocked on top of the log.(See fig. 3-9, E.)

(2) Cadence. Slow.

(3) Movement. A four-count exercise:at the count of—

(a) ONE-- Flex the knees to a quarter-squat position.

(b) TWO-- Flex the knees to a half-squat position.

(c) THREE-- Lower the body to a three-quarter squat position.(Lean slightly forward.)

(d) FOUR-- Recover to the starting position.

NOTE:Pull forward and downward on the log throughout the exercise.

f. Exercise 6: Overhead Toss

(1) Starting Position. Right or left shoulder position, regular stance. The knees are bent to a quarter-squat.(See fig. 3-9, F.)

(2) Cadence. Moderate.

(3) Movement. A four-count exercisc:at the count of-

(a) ONE-Straighten the knees and toss the log into the air approximately 12 inches overhead. Catch the log with both hands and lower it toward the opposite shoulder. As the log is caught, lower the body into a quarter-squat.

(b) TWO-- Again toss the log into the air and when caught return it to the original shoulder.

(c) THREE-- Repeat the action of count ONE.

Section IV. Grass Drills

3401. GENERAL

a. Description and Objective. Grass drills are extremely strenuous exercises and are performed at top speed for only short periods of time. No cadence is counted but the Marines continue to execute the multiple repetitions of the command until the next command is given. The grass drill consists of two drills:Drill One and Drill Two. Each drill contains six exercises. The objective of the drills is to decrease reaction time, to develop cardiovascular endurance, and to provide a vigorous workout for all major muscles.

b. Area and Equipment. Any level area suitable for ground contact and of a size to accommodate the group is adequate. No equipment is needed.

c. Formation. All movements are executed in place. The extended, rectangular formation is recommended for a platoon- or company-size unit. The circle formation is suitable for groups of squad or section size. At the beginning of an exercise program, 2 to 3 minutes of grass drills will insure a good workout.

d. Progression. Progression is gained by gradually increasing the length of time devoted to the drills. As the physical condition of the Marines improves, the periods should be gradually lengthened to 5 minutes. As the second drill is more difficult than the first, some progression can be attained by initially executing Drill One; then as the program and the Marines progress, introduce Drill Two. To extend the duration of the drill, it may be necessary to repeat the drill.

e. Starting Position

(1) The drills are started from the GO position. Other basic positions are FRONT, BACK, and STOP. (See fig. 3-10, A.)

(a) GO. Running in place (top speed): on the toes and balls of feet, knees raised high, arms pumping, body bent forward at waist.

(b) FRONT. Prone position: elbows bent (along body), palms flat on ground directly under the shoulders, legs together and straight.

(c) BACK. Supine position (flat on back): arms extended near side on ground with palms down, legs together and straight, feet toward the stand or instructor.

(d) STOP. Football lineman stance: feet spread and staggered, left arm across left thigh, right arm straight, knuckles on ground, head up, back parallel with ground.

(2) To assume the FRONT or BACK position from the STANDING, GO, or STOP position, vigorously get into the prescribed position as quickly as possible. (See fig. 3-10, B.)

(3) To change from the FRONT to the BACK position, quickly do a pushup, move the feet several short steps to the right or left, lift the arm on the side toward which the feet move, and thrust the legs vigorously to the front. (See fig. 3-10, C.)

Figure 3-10 is available in the illustration section at the end of the book.

(4) To move from the BACK to the FRONT position, sit up quickly, place both hands on the ground to the right or the left of the legs. Move the feet several short steps to the rear on the side opposite the hands. When the feet are opposite the hands, thrust the legs vigorously to the rear and lower the body to the ground.(See fig. 3-10, D.)

f. Leadership. A warm up activity of lesser intensity should proceed grass drill. During the instructional phase and conduct of these drills, the following points should be applied.

(1) The instructor executes only GO and STOP with the Marines. This allows the instructor to supervise the drill.

(2) The commands for grass drills are given in rapid succession without the usual preparatory command.

(3) To prevent confusion, the instructor should give the commands sharply to distinguish them from comments or encouragement.

(4) As soon as the Marines know the drill, they should respond to the instructor's commands and perform all exercises vigorously and as rapidly as possible. All exercises are executed continuously until the next command is given. Insist on top speed performance; anything less is not effective.

(5) The commands peculiar to each exercise are identical to the name of the exercise.

(6) Marines are not to be required to assume the position of attention once the drills are started. To halt the drill for instructions or for rest, the command UP is used. At this command, the Marines assume a relaxed standing position. Do not demand formality. At the conclusion of a fast and vigorous 5-minute grassdrill, it is physically impossible for people to stand at attention.

(7) The sequence of commands for the execution of grass drills should occur as follows:Drill One. GO, FRONT, Bouncing Ball; GO, BACK, Bicycle; GO, Full Squatter; GO, BACK, Situps; GO, FRONT, Mountain Climber; GO, FRONT, Roll Left; GO, STOP, UP.

g. Use With Other Programs. Since grass drills can be executed in a short period of time, they may be executed where only a few minutes are available for exercise or in conjunction with another type of activity. Grass drills are an excellent substitute for running when time is a factor.

3402. GRASS DRILL ONE AND TWO

a.Grass Drill One

(1) Bouncing Ball. From the FRONT position, push up, supporting the body on the hands (shoulder-width apart) and feet. (See fig. 3-11.)Keep the back and legs in line and the knees straight. Bounce up and down by a series of short, upward springs from the hands, hips, and feet simultaneously,

(2) Bicycle. From the BACK position, raise the legs and hips. Keep the elbows on the ground and support the hips with the hands. Move the legs vigorously as if pedaling a bicycle.

(3) Full Squatter. From the STOP position, assume a full knee bend, the feet on line, hands on hips. Bounce up and down in place by short, bouncing jumps.

Figure 3-11 is available in the illustration section at the end of the book.

(4) Situps. From the BACK position and with arms stretched overhead, sit up, reach forward, and touch toes. Return to the supine position.

(5) Mountain Climber. From the STOP position, place both hands on the ground directly under the shoulders. Thrust the right leg to the rear, knee straight. The left foot should be close to the left hand, the left knee outside the left arm. Shift the weight to the hands, thrust off with the rear (right) foot and bring that foot up close to the right hand, the right knee outside the right arm. At the same time, thrust the left leg vigorously to the rear, knee straight. Continue at a fast cadence, alternating the legs.

(6) Roll Left. From the BACK or FRONT position, make one complete roll in the direction commanded. On completing the roll, return to the FRONT or BACK position.

b. Grass Drill Two

(1) Legs Over. .From the BACK position and with arms stretched overhead, palms up, raise the legs upward and then swing them backward over the head until the toes touch the ground behind the head. Return legs to the starting position. (See fig. 3-12.)

(2) V-Up and Touch Toes. .From the BACK position, raise the legs with the knees straight, sit up until the trunk and legs form a V, and touch the toes with the hands. Return to the BACK position.

(3) Rocker. .In the FRONT position, clasp the hands behind the back, arch the body, holding the head back. Start rocking, using the front part of the trunk as a rocker.

(4) Bounce and **Clap Hands**. The procedure is the same as for bouncing ball, but while in the air, clap the hands. This requires a more vigorous bounce or spring. (See fig. 3-12.)

(5) Leg Spreader. .From the BACK position, raise the legs so that the heels are 10 to 12 inches from the ground, spread them apart as far as possible, then close them together. Open and close legs as rapidly as possible.

(6) Forward Roll. .For forward roll from the STOP position, place both hands on the ground, tuck the head, and do one complete forward roll, keeping the legs tucked as you roll, and come back to the STOP position.

Figure 3-12 is available in the illustration section at the end of the book.

Section V.Guerrilla Exercise

3501.GENERAL

a. Description and Objective. Guerrilla exercises are individual exercises of various basic skills that are performed rapidly while moving forward in a circle formation. There are two tables of guerrilla exercises. Each table takes 15 minutes to complete. The objective of this exercise is to increase strength and endurance, aid flexibility, and develop coordination.

b. Area and Equipment. Any level area is suitable for conducting guerrilla exercises. No ground contact is required other than the hands. There is no equipment requirement.

c. Formation

(1) The circle formation (app. A) is used for guerrilla exercises. Each platoon forms its own circle and engages in guerrilla exercises under a platoon instructor. If the platoon exceeds 30 Marines, double or concentric circles may be used.

(2) When the circle is formed, the instructor steps into the center of the circle and moves clockwise in a small circle. He commands:QUICK TIME, MARCH, 1-2-3-4. (Rapid cadence of approximately 130 counts per minute. Cadence and step are maintained between exercises.)

(3) To reform the platoon after completing guerrilla exercises, the instructor halts the Marines and places the base man or platoon guide where he wishes and commands:

(a) BASE MAN (or platoon guide),POST.

(b) FALL OUT AND FALL IN ON THE BASE MAN (or platoon guide).

d. Progression. Progression may be attained by moving from table I to table II or by shortening the quick time marching periods between exercises and performing all exercises a second time.

e. Leadership

(1) To execute the exercises, the Marines continue at quick time while the instructor simultaneously explains and demonstrates the exercise to be performed, and then commands the Marines accordingly. In each instance, the preparatory command will be the name of the exercise and, in all instances, the command of execution will ***MARCH**.. To* terminate each exercise, the command *is* ***QUICK TIME, MARCH.*** *The* Marines immediately pick up the step as the instructor counts cadence.

(2) Unless specified differently, each exercise should be continued for 20 to 40 seconds depending upon the vigor of the exercise. The leader can determine the duration of each exercise by observing its effect upon the Marines.

(3) To form for double guerrillas, the commands for pairing the Marines (who are in circle formation) are-

(a) PLATOON, HALT.

(b) FROM (designate an individual), BY TWO'S, COUNT OFF(.Examplel-2; 1-2; 1-2; etc.)

(c) EVEN NUMBERS MOVE UP BEHIND ODD NUMBERS(At this time, adjust pairs according to height and weight.)

(d) YOU ARE NOW PAIRED UP FOR DOUBLE GUERRILLAS. To change the Marines position, merely command CHANGE.)
(e) FORWARD, MARCH.

f. Place in the Program. Many Marines have not had the opportunity to perform the simple skills involved in guerrilla exercises. The conduct of these exercises is a simple matter since they can be performed easily and quickly in almost any situation. The tables of exercise are applicable to all personnel. The tables can constitute a station within a 1-hour period or be completed within a separate 15-minute period.

3502.GUERRILLA TABLES

a. Table I

(1) Double Time.(See fig. 3-13, A.)Hold arms in the thrust position. Execute a double-time run, maintaining the circle formation and the prescribed distance between your fellow Marines. Duration-1 minute.

(2) All Fours.(See fig. 3-13, B.) Face downward. Support the body with the hands and feet. Walk forward hands first.

(3) Crab Walk.(See fig. 3-13, C.) Get in the sitting position, face upward, and lift the hips. Support the body with the hands and feet. Walk forward feet first.

(4) Squat Walk.(See fig. 3-13, D.) Assume a full knee bend position. Grasp the ankles (left ankle with the left hand, right ankle with the right hand). Walk forward.

(5)Broad Jump.(See fig. 3-13, E.) Jump forward on both feet in a series of broad jumps. Swing the arms vigorously to assist the jumps.

(6)Toe-Touch Walk.(See fig. 3-13. F.) Walk forward, bending at the waist and touching one hand to the toe of the opposite foot while it is on the ground. Raise the trunk to the vertical position between steps. Keep the knees straight.

(7)Bottoms-Up Walk.(See fig. 3-13, G.) Assume the front leaning rest position and move the feet toward the hands in short steps, keeping the knees locked. When the feet are as close to the hands as possible, walk forward on the hands to the front leaning rest position.

(8) Straddle Run.(See fig. 3-13, H.) Run forward, leaping to the right from the left foot and to the left from the right foot.

(9) Fireman's Carry .(See fig. 3-13,1.)

(lO)Single-Shoulder Carry(See fig. 3-13, J.) Two men execute the carries as indicated by the diagram. No. 1 man executes one type; No. 2 man executes the other.

Figure 3-13 is available in the illustration section at the end of the book.

b. Table II

(1) Double Time.(See fig. 3-14, A.)Hold arms at the thrust position. Execute a double-time run, maintaining the circle formation and the prescribed distance between your fellow Marine.Duration-1 minute.

(2) Toe-Grasp Walk.(See fig. 3-14, B.) Bend forward and grasp toes. With knees slightly bent, walk forward.

(3) Hand-Kick Walk..(See fig. 3-14, C.) Walk forward, kicking the moving foot upward on every step. At the same time, lean forward and touch the elevated toe with the hand of the opposite arm.

(4) Pike Juinping.(See fig. 3-14, D.) Jump forward and upward from both feet, keeping the knees straight, and at the same time, swing the legs forward and touch the toes with the hands at the top of each jump.

(5) Squat Jump.(See fig. 3-14, E.) Leap forward from the squatting position, with the hands on the ground and the arms between the legs. Land on the ground with hands and legs extended. Bring up the legs to the squatting position.

(6) Steam Engine..(See fig. 3-14, F.) Lace the fingers behind the neck and walk forward in the following manner: as the left leg moves forward, raise the knee high, bend the trunk forward, and touch the outside of the right elbow to the outside of the knee. Then lower the left leg and step forward on the left foot and raise the right leg. Repeat with the right leg and left elbow.

(7) Knee-Touch Walk.(Scc fig. 3-14, G.) Walk forward, bending the knees and touching the ground on each step. The knees are bent and straightened on each step.

(8)Hobble Hopping.(See fig. 3-14, H.) Hold foot behind back with opposite hand and hop forward. On the command CHARGE, grasp the opposite foot with opposite hand and hop **forward.**

(9) Cross Carry.(See fig. 3-14,1.)(10)Saddle Back Carry. (See fig. 3-14, J.) Two Marines execute the carries as indicated in the diagram. No. 1 Marine executes onetype;No. 2 Marine executes the other.

Figure 3-14 is available in the illustration section at the end of the book.

Section VI. Running and Orienteering

3601. RUNNING

a. General

(l)Description. The general form and technique for all types of running are fairly constant. (See fig. 3-15.)The head is erect, body slightly forward without bending at the waist, and the arms are at a loose thrust position alternating from front to rear in straight planes. A cross-body arm movement wastes energy .The movement of the legs and feet will be discussed in subsequent paragraphs dealing with the different types of running. Of primary importance is the fact that in all types of running, the toes should be pointed straight ahead. Toeing out is a common error in both running and walking and should be an item of individual correction.

(2) Objective. The objective of running is to develop cardiovascular endurance. Despite the fact that Marines have developed their muscle structure and the strength of their muscle tissue, unless they have developed cardiovascular endurance to a satisfactory degree, they are not entirely physically fit or combat ready. Cardiovascular endurance (wind) depends on the efficiency of the lungs and heart. The efficiency of the lungs and heart depend on the amount of oxygen the lungs can absorb with each breath inhaled and the amount of carbon dioxide the lungs can expel. The process of absorbing oxygen and expelling carbon dioxide (cardiopulmonary process) is performed by the blood that circulates through the lungs. The condition of this process will determine the amount of effort a person can exert over a period of time. Running is one of the best activities to develop this vitally important endurance.

b. Running Skills. In the development of running skills, individuals may require instruction to improve their proficiency. Some important skills to consider are-

Figure 3-15 is available in the illustration section at the end of the book.

(1) Action of the Arms. Arm action is important. Check to see that arms are held loosely and that the action is relaxed. The faster the run, the more rapid the arm action.

(2) Breathing. Allow the individual to breathe through the mouth as the body demands a large supply of oxygen. Oxygen can be inhaled in greater quantities through the mouth.

3602. TYPES OF RUNNING

a. Double Time

(1) Description and Objective. Double timing is marching at the rate of 180 steps per minute, each step being 36 inches in length. It takes practice to double time with precision in formation. The Marines should keep in step, placing their feet flat on the ground. This, however, should not be a stamping motion, but should be done with as slight a jolt as possible. Double timing is like a jog, the difference being that in a jog the feet are lifted well off the ground and the running motion is bouncy. In double timing, the feet skim the ground and there is no bounce to the ran. Double timing is a vehicle for teaching proper running form and for the development of the cardiovascular system.

(2) Area and Equipment. This type of running can be completed over a variety of surfaces. Usually a Marine uses a field or road. There is no equipment requirement.

(3)Progression. There is no set standard for alternating quick time and double time in the early conditioning of Marines. A general rule is to begin with enough quick-time marching to ensure a thorough warming up, then double time about 100 paces. Change again to quick time until the individuals have made a reasonable recovery from the running, then double time another 100 paces. The amount of double time can be increased and the quick time decreased from week to week, until the individuals are double timing about 1800 yards. This type of training should be given at least twice a week, but by no means is it adequate as the sole means of conditioning.

(4) Leadership

(a) The instructor should be to one side of the column or group and toward the rear so there is a full view of all Marines. Inexperienced instructors have a tendency to supervise from a position too far forward.

(b) Select an individual who can maintain the proper cadence to act as the guide during double-time running.

(c) There are several ways for the instructor and group to count cadence while double timing. If not contrary to local policy, learn several methods and use them for variety.

b. Wind Sprints

(1) Description and Objective This type of running involves a series of 30- or 40-yard dashes, usually conducted in successive waves of squads. Each squad is in line and the squad leader on the right flank. Wind sprints assist indeveloping speed and cardiovascular endurance. Any flat and level area may be used which will permit the squad to form a line and run the required distance.

(2) Progression. One or two 30-yard sprints will be adequate at the beginning. As time passes, sprints can be lengthened and up to six or seven sprints may be used.

(3) Leadership

(a) At the command READY (given by squad leader), each runner assumes the sprinter's starting position. At the command GO, the squad sprints approximately 30 yards, takes 10 yards to stop, and lines up immediately with the squad leader who repeats, READY, GO, and again the squad sprints. At the conclusion of the third sprint, the squad waits until all the squads of the platoon have made three sprints. Then they all line up and the squad leaders conduct three more wind sprints in the opposite direction.

(b) Valuable time is gained by having each squad ready to go when the preceding squad has moved off its second sprint mark.

c. Cross-Country Running

(1) Description and Objective. Cross-country running is a distance run conducted on a course laid out along roads, across fields, over hills, through woods, and on any irregular ground. The cross-country ran may be utilized as a conditioner or as a competitive event; the objective is to cover the distance in the shortest possible time. The course should be 2 to 2.5 miles in length and be laid out to avoid heavy vehicular traffic. The course should be marked by directional arrows until the runners know the course. These runs build leg muscles, increase lung capacity, and develop endurance. Any local area of varied terrain is suitable.

(2) Progression. In the mass training of a large group, leaders should be stationed at the head and the rear of the column and should make every effort to keep the individuals together .After determining the abilities of the individuals in cross-country running, it is advisable to divide the unit into three groups. The poorest conditioned group starts first, and the best conditioned group, last. The starting time of the groups should be staggered so that all of them finish about the same time. In preliminary training, the running is similar to ordinary road work in that it begins with rather slow jogging, alternating with walking. The speed and distance of the run is gradually increased. As the condition of the individual improves, occasional sprints may be introduced. At first the distance run is from 1/2 to 1 mile. It is gradually increased to 2 or 2.5 miles. Well-conditioned personnel can run 2 to 2.5 miles within a 15-minute period.

(3) Leadership. Marines should not be required to take part in distance running until they have been through a progressively scheduled training program which requires a considerable amount of running. Cross-country runs should be scheduled occasionally to provide variety in the program. Cross-country running has the advantage of allowing mass participation. Interest can be stimulated by putting the runs on a competitive basis. As a single activity, short cross-country runs can be scheduled once a week, gradually increasing the distance as the physical conditioning improves.

(4)Use With Other Programs. Cross-country running can be combined with other activities such as conditioning exercises.

d. Fartlek Training

(l)Description and Function. Fartlek training is various running exercises conducted along an intense combat-like course. The running is conducted to overwork the lungs, allowing only partial recovery which is followed by another intense period of overload. This sequence is repeated for the duration of the workout. Fartlek training is a useful combat training method and general cardiovascular conditioner. The entire unit must be in uniformly good to excellent condition to qualify for this training.

(2) Progression. The following is an example of one fartlek session:

(a) Warm up by stretching 3-5 minutes.

(b) Warm up running easily 5-10 minutes.

(c) Run at a fast, steady pace for 3/4 to 1 3/4 miles (dependent on terrain).

(d) Walk/jog at a moderate pace for 5 minutes (recovery).

(e) Run easily sprinting 15 to 20 meters occasionally.

(f) Run full speed uphill for 175 to 200 meters.

(g) Warm down by running easily for 1/2 to 1 mile.

(h) Warm down by stretching 3 to 5 minutes.

(3)Leadership. To add variety, try incorporating combat movements, terrain association, a series of exercises or carrying table of equipment weapons (hit and roll). This will take some initiative in setting the course, but it allows for variety. In this manner, the fartlek course can be used as a training session or part of one. It works extremely well with highly motivated, competitive Marines.

3603.ORIENTEERING

Orienteering is land navigation over a prescribed course as a timed, competitive event. It is an excellent way to integrate land navigation training with physical conditioning and requires Marines to be skilled in both areas. Commanders adjust the training as needed to change the degree of difficulty or to emphasize either navigation or conditioning. Taken to the extreme, an orienteering exercise is conducted in full combat gear over rugged terrain against aggressors. Orienteering requires the following physical skills: endurance, speed, strength, and all-around conditioning.

Section VII. Circuit Training

3701. GENERAL

a. Description and Objective. A strength circuit consists of a series of stations where individuals in small groups exercise vigorously for a short period of time and then move (on signal) to the next station where a different form of exercise is conducted. This rotation continues until all groups move through all stations. Strength circuits contain no set or specific types of exercise stations within the circuit. The objective of circuit training is to develop strength. There are three general types of circuits.

(1) Fixed Circuit. This is a circuit in which apparatus of an immovable type (fixed into the ground) is used.

(2) Movable Circuit. This circuit consists of individual exercise apparatus which is portable and can be moved to and from the training area.

(3) Simplified Circuit. This circuit requires no equipment or apparatus.

b. Formation. The exercises are done at will, but rapid, steady, and continuous work is required of all. Each Marine's nervous and muscular system reacts differently to timed vigorous exercises. His performance should be measured on how many movements per exercise he can complete as an individual. For example, one Marine may be able to complete 5 movements, while another may be able to complete 20, and yet each is receiving the maximum benefit. All three circuits contained in this chapter are designed for platoon-sized groups. Expansion beyond this capacity requires a large amount of equipment, as each Marine in the fixed and movable types of circuits must have an item of equipment available for exercise at each station. A group larger than a platoon could be exercised through use of the simplified type of circuit; however, the group would be unwieldy and control could be a problem.

c. Place in the Program. All circuits illustrated can be completed in a 15-minute period. This feature allows the exercise of a platoonor smaller group on the circuit for a single 15-minute period, or the scheduling of the circuit as a 15-minute period within a longer period. A circuit can thus be utilized within a rotating activity system of scheduling. Choice of a circuit by the unit depends upon area, facilities, and other local factors; however, there is a circuit for every need.

3702. FIXED STRENGTH CIRCUIT

a. Description and Objective. The strength circuit is an arrangement of various types of exercise apparatus which are fixed in position. (See fig. 3-16.)Seven basic exercises are used and each exercise requires an apparatus. All apparatus of one type are positioned together to constitute a station. Each station will accommodate 10 Marines. The circuit is designed to be accomplished in 15 minutes when conducted on a time-rotation basis, as normally executed for unit training. For individual use, a Marine may complete the circuit by executing a specific number of repetitions for each exercise. In this case, the time required for completion of the circuit would vary slightly depending on the number of repetitions the Marine completed. The objective of this circuit is to provide a series of exercises which will improve and maintain the strength of the body's major muscle groups.

Figure 3-16 is available in the illustration section at the end of the book.

b. Starting Level. Marines must be thoroughly warmed up prior to participating in the circuit system. If Marines have not engaged in vigorous exercise immediately prior to starting the circuit, then an instructor should execute the following warm up exercises. These exercises should be conducted in the normal formation for set drills. Seven repetitions of each exercise will normally provide sufficient warm up. The exercises are-

(1) High jumper, Conditioning Drill 1.
(2) Bend and reach, Conditioning Drill 1.
(3) Squat bender, Conditioning Drill 1.

c. Progression. The instructor can adjust the circuit system through several methods to accommodate participants of varying physical ability. One method of adjusting an exercise is to change the method of executing the leg lift or to select a heavier weight for the barbell curls. When it becomes apparent during a unit program that the overall fitness of the group has Improved, then the exercise may be made more strenuous in two additional ways. First, the instructor can have one or more of the supplementary stations added. Second, the time spent exercising at each station can be increased in 5-second increments to a maximum of 60 seconds. Also, as Marines learn the circuit, the 45-second period for movement and instruction between stations can be eliminated, thus allowing only enough time to change stations.

d. Starting Positions. The Marine starts on any station, exercises steadily for a certain period (45 seconds initially), then moves on command to the next type of apparatus where he again exercises steadily for an equal period. The Marine continues until the required exercise is completed at each station. If it is desired to expand the number of stations in the circuit to accommodate more participants at one time, the instructor can provide four additional supplementary exercises, requiring no equipment. This will increase the amount of time required to complete the circuit.

e. Leadership. Close control of all Marines is necessary to ensure that a minimum amount of time is spent in moving them to their initial stations and in moving them between stations. One instructor can control the activity on the strength circuit. A stop watch or wrist watch with a second hand is required. When Marines arrive at the strength circuit, the instructor will form them for exercise and conduct the warm up drill. The group is then reassembled and formed into a number of files equal to the number of stations being used in the circuit. Each file is then directed to a station. As soon as all participants have reached an exercise position at a station, the instructor gives the command READY, GO. After 45 seconds of exercise, the instructor gives the command STOP, CHANGE OVER. The instructor allows Marines 45 seconds for moving to the next station and for preparing for the next exercise before the command EXERCISE is again given. In lieu of verbal commands, a whistle may be used to stop and start the exercises. For large groups, a megaphone or loudspeaker is useful.

f. Individual Conditioning Program. For the Marine working alone on the strength circuit, it would be impractical to time the exercise periods. For individual exercise, the Marine should select a number of repetitions of each exercise to complete, then rotate to the next stations after completing these repetitions. The number of repetitions selected should be at or near the maximum that the Marine is capable of doing without halting for rest.

g. Fixed Circuit Stations

(1) Primary Stations (See fig. 3-16.)

(a) Pullups/Chinups. A horizontal bar placed 8 feet above the ground is necessary for each Marine at this station. The Marine will also need a space on the bar that is 45 inches wide. On the command**EXERCISE**,the Marine grasps the bar with both palms facing either forward or to the rear, arms fully extended, and feet free of the ground and executes the exercise as described in MCO 6100.3_.The exercise is repeated as many times as possible until the command **STOP** is given. Then the Marine moves to the next exercise. If a Marine has done his maximum number of pullups/chinups prior to the command **STOP**, he will remain in the "dead" handing position until the command **STOP** and move to the next station.

(b) Twist Grip. The apparatus is a horizontal bar, free to turn, held between uprights placed 30 inches apart. The bar is 52 inches above the ground. A weight of 20 pounds is attached to the center of the bar by a light rope long enough to permit the weight to rest on the ground. The Marine stands at arm's length from the bar and grasps it with his hands on either side of the rope, palms down, thumbs under the bar. On the command **EXERCISE,** the hands are rotated so that the backs of the hands are rotated away from the body, thus winding the rope on the bar. The elbows are kept straight to ensure that the exercise is performed by the hand and forearm. When the weight is drawn up to the bar, the bar is then rotated in the opposite direction to lower the weight to the ground. This exercise is continued until the command **STOP** is given. The Marine then moves to the next station.

(c) Pulley Weights. The apparatus is a T frame with a system of pulleys that suspends a weight of about 90 pounds. The weight is attached to a light steel cable which has a drawbar attached to the other end. The Marine grasps the drawbar and sits down directly under the bar, legs extended to the front and arms extended overhead. The exercise is executed by pulling the drawbar down behind the head, then extending the arms slowly again until they are fully extended overhead. The exercise is repeated as many times as possible until the command STOP is given. The Marine then will move to the next exercise. Upon completion of the exercise, the weight is lowered slowly to the ground.

(d) Barbell Curls. A barbell is necessary for each Marine at this station. The barbell is constructed of 1 1/4-inch pipe 5 feet long, and two concrete-filled No. 10 cans. Each barbell should weigh about 40 pounds. Variance in the weight of the barbells, up to about55 pounds, will allow appropriate overload to be applied to Marines who are above average in strength or weight. The Marine grasps the bar with the palms forward and assumes a standing position with the barbell held in front of the hips, hands approximately shoulder's width apart. On the command **EXERCISE,** the Marine flexes the elbows and draws the barbell up until it touches the upper chest. The elbows remain at the sides. Breath is inhaled with the upward movement and exhaled as the barbell is lowered to the starting position. The exercise is repeated as many times as possible until the command **STOP** is given. The Marine moves to the next apparatus.

(e) Step-Up. The apparatus is a platform or ledge 18 inches high and of such size. At accommodate 10 Marines. The Marine faces the platform and on the command **EXERCISE,** steps up onto the platform, bringing his trailing foot up beside the leading foot. He then steps back down to the original position, stepping down first with the same foot he initially used in stepping up. After 10 repetitions of the exercise, he changes the order of moving the feet to use the opposite leg for stepping up. Repeat this exercise until the command **STOP** is given. The Marine then moves to the next station.

(f) Leg Lift. The apparatus is a horizontal bar constructed as described in (a). To prevent the body from swaying, a horizontal back support is added 40 inches below the horizontal bar. The arms are kept fully extended. On the command **EXERCISE,** the Marine jumps up, grasps the bar with the palms forward and the back support behind him. The exercise is executed by raising the legs to a horizontal position then slowly lowering them to the vertical position. The Marine does not flex his knees. He does not swing the legs to the rear of a vertical position to gain momentum for raising them in the next repetition of the exercise. The movement is repeated until the command **STOP** is given. The Marine then moves to the next exercise. If unable to raise his legs to a horizontal position without flexing his knees, the Marine flexes his knees and draws them up to the chest, then lowers his legs to the vertical position.

(g) Rope Climb. The rope climb is 20 to 30 feet high with five ropes suspended from a horizontal bar which forms the uppermost part of the framework. To prevent the horizontal bar from sagging and to provide safety, only five ropes are attached to it. There are two frameworks per station. The ropes are 6 feet apart. Any method may be used to climb the rope, and the Marines climb as high as possible. Marines who are proficient should climb therope several times during the time allotted. Inexperienced Marines should be cautioned to take care during descent to avoid rope burns on their hands.

(2) Supplementary Stations. The following exercises are designed to expand the basic circuit by being inserted in specific places within the system. For each supplementary station used, there should be adequate room for 10 Marines to exercise.

(a) Bent Knee Situp or Bottoms-Up. These calisthenics are designed to strengthen the abdominal muscles. These exercises will be inserted between the pullup/chinup and twist grip stations. The primary stomach exercise is the situp. In case of inclement weather or other conditions that make ground contact undesirable, the bottoms up exercise is used. In situps on the command **EXERCISE,** the Marine lies on his back (supine position) with knees flexed and both feet flat on the ground and executes the exercise (minus an assistant) as described in MCO 6100.3_.The Marine then returns to the starting position, repeating the exercise until the command **STOP** is given. The Marine then moves to the next station. In bottoms up on the commandEXERCISE,the Marine assumes the front leaning rest and executes the bottoms-up exercise as described in Conditioning Drill 3. He continues this exercise at a moderate cadence until the command **STOP** is given. The Marine then moves to the next station.

(b) Pushup. This exercise is designed to strengthen the arm- and shoulder-girdle muscles. It should be included between the twist grip and pulley weight stations. Upon the command **EXERCISE,** the Marine executes the pushup as described in Conditioning Drill 1. He continues this exercise at a moderate cadence until the command **STOP** is given. The Marine then moves to the next station.

(c) Knee Bender. This exercise is designed to build leg muscles and is included between the pulley weight and barbell curl stations. On the command **EXERCISE,** the Marine executes the knee bender as described in Conditioning Drill 3.He continues this exercise at a moderate cadence until the command **STOP** is given. The Marine then moves to the next station.

(d) Trunk Twister. This exercise strengthens the major muscles of the trunk and is included between the step-up and pullup/chinup stations. On the command EXERCISE, the Marine executes the trunk twister as described in Conditioning Drill 1. He continues this exercise at a moderate cadence until the command **STOP** is given. The Marine then moves to the next station.

3703. MOVEABLE STRENGTH CIRCUIT

a. Description and Objective. The exercises in this circuit are progressive and the course is planned to gain and hold the interest of the participating groups. The circuit consists of a series of stations, with each station designed to develop a particular group of muscles. Along with muscular development, correct posture and deep rhythmic breathing should be stressed on this circuit at all times.(See figs. 3-17 and 3-18.)

Figure 3-17 is available in the illustration section at the end of the book.

b. Equipment. The equipment is set up in files. Six files of 8 stations will accommodate a platoon

c. Formation. The platoon marches to the area where the equipment is positioned and forms a file within each lane of stations, covering down on a piece of equipment. Movements are made on the double, the important factor being that no time is wasted in getting to work.

d. Progression. Initially 40 to 45 seconds per station is adequate. As individuals become stronger the time should be increased in 5-second increments until a minute to a minute and a half is reached.

e. Leadership

(1) The leader stands in front of the barbell station and controls the rotation from this position. The leader supervises the entire group, with the assistance of several instructors who move about in the platoon correcting and encouraging the Marines.

(2) The leader starts each group but does not count cadence nor lead them through the exercises. Each individual exercises rapidly but individually.

(3) As each Marine finishes his repetitions with the barbell, he places the barbells on the ground. The leader calls, READY, followed by the command, FALL OUT ONE. All will double-time to the station directly in front of them, while the individuals on the barbell stations will do an about face to the rear station in their lane.

f. Movable Circuit Stations. The Marine can obtain best results on the movable circuit if the exercises on the various stations are given in the following manner:

(1) Station 1—The Barbell. (See fig. 3-17, 1.)The exercises at this station stress proper posture; deep, rhythmic breathing; and development of the muscles of the arms, shoulders, and upper body. An instructor teaches the proper methods of lifting before the exercises begin. The methods are to lift with the legs, to keep the back straight, and to merely grip with the hands. Two recommended exercises are given below. The instructor will use only one exercise per period .He may use either exercise. At successive periods, the other exercise is used.

(a) Exercise 1--Two-Hands Military Press. (See fig. 3-18, 1.)Grasp the barbell with both hands, knuckles up at shoulder width, and lift to the chest. Steadily press to arm's length overhead; lower to the chest resisting weight all the way. Inhale as the weight is pressed up and exhale as the weight is brought down.

(b) Exercise 2--Two-Hands Regular Curl. (See fig. 3-18, 1.) Lift the weight to the waist, with the palms of the hands out, heels together, stomach in, chest lifted and arched, shoulders back, elbows in close to the sides; inhale deeply and curl the weight to the shoulders, using the arms only, at the same time keeping the elbows close to the sides; exhale Emphasize posture and the use of the arms only. This is a very valuable exercise for the development of the biceps and the grip and should be repeated from 8 to 16 times, depending on the ability of the Marine.

(2) Station 2--The Jump Rope. (See fig. 3-18, 2.) This exercise develops strength and agility in the legs and stamina of the whole body. It makes the Marine agile on his feet and increases his footwork efficiency and timing. Each Marine should progress until able to jump rope at least 3 minutes at top speed.

(3) Station 3--The Twist Grip. (See fig. 3-18, 3.) The twist grip is an excellent exercise for the hands and forearms, and adds greatly to the Marine's ability in hand-to-hand combat. The handle is gripped and twisted, winding the rope until the weighted can is level with the height of the hands, which are held horizontal. The weight is lowered in the same manner; the Marine resists the weight all the way, occasionally stopping the twisting motion and alternately removing first one hand, then the other, from the handle. A variation of the above exercise is to wind the handle with the palms up and the arms bent and with the elbows held close in to the sides. Each Marine should maintain a good posture and keep the stomach muscles taut throughout this exercise.

(4) Station 4--The Incline Plane. (See fig. 3-18, 4,)The use of the incline plane is a very strenuous exercise and well designed for the development of the abdomen. Although 6 to 10 repetitions are sufficient for the beginners, more can be added as ability increases. Marines with hernias or recent operations will be excused from participation at this station.

(5) Station 5--The War Club. (See fig. 3-18, 5.) The war club is a simple and effective means of exercising the principal muscle groups of the body, especially those of the trunk, back, and shoulders. To gain the maximum benefit from this exercise, the Marine must keep both feet flat on the ground at all times. Throughout the exercise period, the weight is swung from arm's length as follows:

(a) As in chopping wood, first on one side, then on the other.

(b) As a batter warming up with a number of bats.

(c) In large circles, first with one hand and then with the other.

(6) Station 6--The Bicycle Ride. (See fig. 3-18, 6.) The bicycle ride is well suited to exercising many of the muscle groups of the body, particularly those of the abdomen. Vary the speed of the exercise, but keep everyone "riding" the entire period. A variation exercise may be performed by placing the legs together, raising them slowly to a height about 2 feet from the ground, and then lowering them slowly to the ground.

(7) Station 7--The Step-up. (See fig. 3-18, 7.) The step-up exercises the legs. The step-up is performed by initially stepping up with the left foot, followed by the right, then stepping down with the left foot followed by the right. Continue for 20 seconds, then change to the right foot as the lead foot for 20 seconds.

(8) Station 8--The Isometric Pull. (See fig. 3-18, 8.) Two trainees work at this station with a cable pull and perform the following exercises:

(a) Initially start with one Marine in the supine position and one Marine sitting. The sitting Marine lowers the upper body to the ground and pulls the partner up to the sitting position. The partner then performs this same action and this is continued for 20 seconds at a rapid rate.

(b) During the last 20 seconds, the same action takes place but, in this case, the Marine in the supine position resists the pull of the partner for approximately 5 seconds before allowing to be pulled up into the sitting position.

3704. CIRCUIT-INTERVAL TABLE

a. Description and Objective. The circuit-interval table is designed to develop strength and endurance within a short period of time, with no equipment requirement, through a rapid and vigorous routine of exercise. Fifteen minutes is an adequate period to execute all exercises and to secure a vigorous workout with the circuit-interval principle.

Figure 3-18 is available in the illustration section at the end of the book.

b. Formation and Starting Positions

(1) A leader forms platoons or smaller groups in an oval or circular formation with 3- to 5-yard intervals between Marines. The Marines face to the right and move forward at quick time and then double time. (See fig. 3-19, A.)After running several platoon circle laps, the leader calls out the name of an arm and shoulder exercise from the list below, orders quick time and commands, for example, PUSHUPS. On this command, all Marines immediately hit the ground and individually and rapidly begin doing pushups. No cadence is counted. (See fig. 3-19, B.) After 30 seconds of exercise, the leader commands. **QN YOUR FEET, FORWARD, MARCH** The platoon resumes the quick time cadence and the leader, when ready, gives the necessary commands for double time. The double time is continued for one or more laps and the leader calls out the name of the next exercise and the process is repeated. This continues, with running between each exercise, until every body part has been exercised.

(2) The instructor controls the running and quick time to observe the effects of the exercise upon the Marines. Cadence, step, and precision are not important to the objective and the instructor should not use them. What is important is speed and the instructor should stress this. After the exercise period is started, the Marines do not stop. This circuit method emphasizes stress and recovery, the recovery occurring during the quick time periods.

c. Activities. The leader can use the following exercises. These exercises can be repeated if necessary, during a second round. The leader can use other calisthenic exercises in the circuit-interval table.

(1) Arms and shoulders-push-ups.

(2) Stomach situps.

(3) Back-squat thrusts.

(4) Legs-bicycle (on back).

d. Progression. lhe progress is controlled by the leader, who must pace the running, quick-time movement, and exercise in such a way that Marines will receive a vigorous workout yet be able to participate throughout the 15-minute period. Marines who are in the initial stages of physical condition will not be able to double time or exercise as long as those who are better conditioned. The idea is to set a pace which can be increased during each workout, thus progressing gradually to a higher level of physical fitness.

e. Leadership. The platoon leader, platoon sergeant, or section leader can lead the group. The leader must execute the exercise with the unit to feel the effects and thereby adjust the pace.

f. Place in the Program. This activity may be scheduled whenever a short period of time is available. The only requirement is that enough space, indoors or out, be available to form the circle.

Figure 3-19 is available in the illustration section at the end of the book.

Section VIII. Basic Physical Skills and Obstacle Courses

3801. GENERAL

The purpose of this section is to list the basic military skills and the methods for their development. Many of these skills are best developed by obstacle courses but other drills are also discussed. The section explains types of obstacle courses, details of construction, and methods of negotiating the various obstacles.

3802. BASIC PHYSICAL SKILLS

a. Objective. The objective is to develop proficiency in the various military physical skills which are essential to personal safety and effective combat operations. In travel by foot over rugged terrain and in the execution of combat duties, Marines must be trained to perform basic skills, such as running, jumping, climbing, and carrying. During training, Marines will develop agility and coordination in these skills .Fast and skillful execution of these skills may mean the difference between success and failure on the battlefield.

b. Place in the Program. These skills are practiced throughout the entire program in many different activities. Many of these skills are best practiced on obstacle courses.(See par. 3803.)

c. Description. These basic skills are the minimum skills required by all Marines. The basic skills are as follows:

(1) Running. Running is used to strengthen the legs and develop cardiovascular endurance. Marines should be exposed to running in various situations: on roads, over rough ground, up and down hills, cross-country, and running over low obstacles.

(2) Jumping. In broad jumping, the take-off foot is planted firmly and the spring comes from the extension of this leg as the other leg reaches for the far side of a ditch or similar obstacle. (See fig. 3-20.) The arms are forcibly raised forward and upward to assist in propelling the body up and forward. Landing may be on one or both feet depending upon the length of the jump. In vertical jumping downward from a height, the jumper should aim his feet at the desired landing spot and he should also jump with the knees slightly bent and feet together, with the trunk inclined slightly forward. As the feet touch the ground, the shock is absorbed by bending the knees into a full squatting position. If the height is too great or the ground too hard to absorb the shock, then the jumper should forward roll or side roll thus eliminating some of the momentum.

Figure 3-20 is available in the illustration section at the end of the book.

(3) Dodging In combat situations, it is often times necessary to change directions quickly. To execute this movement while running, a lead foot is firmly planted, left foot if the direction is to the right and right foot if the direction is to the left. The opposite foot is moved toward the new direction. The knees are slightly flexed during the movement and the center of gravity is low and balanced. (See fig. 3-21.) At the time of the change of direction, the head and trunk are turned quickly in the new direction.

Figure 3-21 is available in the illustration section at the end of the book.

(4) Climbing and Surmounting. All Marines should know how to climb and surmount various types of obstacles. (See fig. 3-22.)

Figure 3-22 is available in the illustration section at the end of the book.

(a) Vertical Climbing, as in Climbing a Rope or Pole. The hands grasp the rope or pole overhead with the palms toward the face. Gripping the object, the body is pulled upward with the arms and shoulders, assisted by the feet which grip the object and assist by pushing downward. If shoulder girdle strength and body coordination are not adequate to permit alternating the hands, the arms act together in pulling upward.

(b) Climbing as in Surmounting a Wall. Ingoing over a wall, the body should be kept as close to the top as possible, since in combat operations it is important to offer as small a target as possible to the enemy. If an individual climbs a wall while carrying a rifle, both hands should be freed by slinging the rifle over the back. There are two methods commonly used for surmounting a wall of moderate height, but only one for dropping from it.The methods are as follows:

1 Run, Jump, and Vault. Approach the wall at a run, jump forward and upward at the wall and place one foot against it as high up as possible. Use the foot in contact with the wall to help push the body upward while grasping the top of the wall with the hands. Pull the body up with the arms, assisted by pressure of the foot against the wall and swing the legs over, propelling the body weight over the wall.

2 Hook and Swing. Approach the wall at a run and jump forward and upward. Hook one elbow over the wall, locking the arm in place by pulling up until the top of the wall is underneath the armpit. Grasp the top of the wall with the other hand. Draw the leg which is closer to the wall up as far toward the top as possible. Then swing the other leg over the top of the wall.The body is then carried over with a rolling motion. A variation of this leg action can be used by Marines who are unable to draw up the leg as described. While hanging with both legs fully extended, start a swinging motion with the legs together. When the legs have enough momentum, swing the outside leg over the top of the wall with a vigorous kick, then follow with the body.

3 **Dropping.** All drops from a height are executed in the same manner regardless of the method used to gain the top. One hand is placed against the far side of the wall while the other hand grasps the top. From this position, the body is rolled over the wall and vaulted away from it with the legs swinging clear. As the body passes over the wall and drops, it should at all times face the wall. This will keep the rifle and other equipment clear. Break the fall by retaining a grasp on the top of the wall as long as possible.

(c) Climbing Ladders and Cargo Nets. Rope ladders, stationary vertical ladders, and cargo nets employ the same general technique. The important element is to grasp the side supports firmly in the hands about shoulder height and place the feet on a rung which would cause the body to be fully extended. In movement upward, one hand is moved upward and a new grasp is secured and, at the same time, the opposite leg moves up a rung. As the knee straightens, the body is elevated. This process is repeated using the opposite arm and leg. Alternation continues in this manner until the climber reaches the objective.

(5) Traversing Horizontal Objects. The traversing of horizontal objects puts heavy stress on the arms and shoulder girdle area as the feet are usually suspended in the air with all of the body weight on the arms and shoulders.(See fig. 3-23.)

(a) Traversing Horizontal Ropes or Pipes. The hands grasp the horizontal support overhead with the palms facing. To propel the body forward, one hand is released and moved forward to secure a new grasp. At the same time, the opposite side of the body is swung forward (some people are able to "walk" in the air, keeping the body to the front and moving the legs in time with the arms as in walking on the ground). The other hand is then released and moved forward; this alternation is continued until the objective is reached.

Figure 3-23 is available in the illustration section at the end of the book.

(b) Traversing Horizontal Ladders. In this situation, the movement is the same as used in traversing a rope or pipe. The hands, however, are placed on the rung with the palms away from the face. Other than this difference the technique is the same.

(6) Crawling. Crawling in combat situations is an often used skill. Crawling may be high or low.(See fig. 3-24.)

Figure 3-24 is available in the illustration section at the end of the book.

(a) High Crawl.

In the high crawl, the Marine moves on hands and knees, moving one hand and the opposite knee and then continuing to move the hands in alternation with the opposite knee following the companion hand.

(b) Low Crawl. The Marine is in the prone position usually with the forearms and palms of the hand on the ground. He propels forward by bending the knee of one leg and pushing with the inside edge of the shoe. At the same time, the opposite arm moves forward and pulls to the rear. The body remains low and movement is continued by bending the opposite knee and pushing, and at the same time sliding the opposite arm forward and pulling. Alternation of hands and legs continues until the objective is reached.

(7) Throwing. Throwing may be executed from the kneeling or standing positions. The object to be thrown is held in the hand, and the throwing arm is bent at the elbow; the hand is then moved to the rear until the hand is behind the ear. The body is turned so that the lead foot and balance arm on the side toward the target are pointing at the target. (See fig. 3-25.) The balance arm is used to sight over and align the throwing hand and the target. When properly aligned, the elbow is moved rapidly forward until it is at a point just in front of the body where the arm is straightened and the wrist snapped. This whip motion propels the object to the target. Underhand throws secure momentum by the thrower bending his knees and swinging the throwing arm to the rear. As the knees are straightened, the arm is forcefully swung forward from the shoulder and the object released.

Figure 3-25 is available in the illustration section at the end of the book.

(8) Vaulting. Vaulting is employed to overcome low barriers or fences. (See fig. 3-26.)The object to be surmounted is approached at an angle. The hand on the side next to the obstacle is placed on the top of the obstacle and, with a straight arm, the body weight is pushed upward. At the same time, the leg on the side next to the obstacle is thrown upward and over the top followed by the other leg (side approach). In landing, the weight comes down on the leading leg first followed by regaining the balance on both legs. The free arm serves as a balance. A direct (front) approach can be used at which time both legs go over the object together.

(9) Man Carrying. There are three basic individual means of carrying personnel in combat situations and one of these methods may be used in carrying objects.

Figure 3-26 is available in the illustration section at the end of the book.

(a) Fireman's Carry."A" stands sideways in front of "B", "A" bends his knees and leans forward, placing one arm through "B's" crotch, grasps the wrist of "B's" arm, which is hanging over the shoulder, and then "A" runs forward.

(b) Saddle-Back Carry. With his back toward "B", "A" stands in front of "B". "B" mounts "A's" hips and clasps his arms in front of "A's" chest. "A" grasps "B's" thighs.

(c) Single-Shoulder Carry. "A" stands facing "B"."A" assumes a semi-squatting position. "B" leans forward until "B" lies across "A's" right shoulder. "A" clasps his arms around "B's" legs and straightens up, lifting "B" from the ground. "A" then runs forward. This method may also be used to carry heavy objects.

(10) Balancing. Balancing the body while walking or running on a narrow object when crossing obstacles is a skill which requires practice and confidence. Balance is required in negotiating a log placed across a stream, in crossing a narrow beam or rail, and in similar situations.(See fig. 3-27.) To perform this skill, place the feet on the object to be crossed, hold the arms to the sides at shoulder level, and fix the eyes on the object approximately 5 yards in front of the feet. Generally, it is not a good practice to look down at the feet. Walk the beam by placing first one foot and then the other in the center of the beam, thereby moving forward, using the arms to aid in maintaining balance.

Figure 3-27 is available in the illustration section at the end of the book.

(11) Falling. Injury can be avoided if Marines are taught to fall properly. They should know how to use the body momentum to their advantage during a fall rather than to try resisting that force. (See fig. 3-28.) If enough force is present, such as occurs during a fall while running or in jumping downward from a height, individuals can extend their hands to catch the weight. At the same time, duck the head and roll forward onto the feet. The key to falling without injury from the standing position is relaxation and rolling on the outside of the leg, hip, and buttocks to take the brunt of the fall.

Figure 3-28 is available in the illustration section at the end of the book.

3803. OBSTACLE COURSES

a. General. Obstacle courses are a valuable part of physical readiness training. The challenge presented by the obstacles assists in developing and testing the basic physical skills. In many combat situations, success will depend upon the Marine's ability to perform one or more of these skills, often while carrying field equipment and when fatigued. In this section, two types of obstacle courses will be discussed-Conditioning Obstacle Course and Confidence Obstacle Course.

b. Course Safety Precautions. Commanders and course instructors should take certain precautions to prevent injury on obstacle courses. A few of the precautions are:
(1) Inspect the course for faulty construction of obstacles, protruding nails, rotten logs, condition of the landing pits, and other hazards to safety.
(2) Conduct warm up exercises before the unit runs the course.
(3) Explain and demonstrate the correct techniques for negotiating all the obstacles before allowing the Marines to try them.
(4) Give Marines at least two weeks of conditioning exercises before scheduling the obstacle and/or confidence courses.
(5) Ensure that negotiation of the higher and more difficult obstacles are under the supervision of an instructor.
(6) Do not permit individuals who have neither practiced the basic skills nor run the conditioning obstacle course to participate in the confidence obstacle course.
(7) Weather conditions may cause footing or handhold surfaces to be slippery. If such is the case, postpone training on the course.

3804.CONDITIONING OBSTACLE COURSES

a. Description and Objective. The Conditioning Obstacle Course is commonly known as the Obstacle Course. This course consists of fairly low obstacles which are designed to be negotiated quickly. The obstacles serve to test various basic skills, and running the course is a test of the Marine's physical condition. After receiving instruction and an opportunity to practice the skills, Marines run the course against time.

b. Area and Equipment
(1) Complete standardization of obstacle courses should not be attempted since topographical conditions always vary. Commanders should use ingenuity in constructing a course, making good use of streams, hills, trees, rocks, and other natural obstacles. Since the course is eventually run at high speed, it should not be dangerous.

(2) The course should be wide enough for six or eight men to run simultaneously, encouraging competition. The lanes for the first several obstacles should be wider and the obstacles themselves easier than those that follow. This avoids congestion until the contestants scatter out over the course. The last two or three obstacles should not be too difficult and should not involve high climbing. This prevents injuries and falls resulting from fatigue.

(3) The total distance of the course should range from 300 to 450 yards and include from 15 to 25 obstacles. Normally the obstacles should be 20 to 30 yards apart and arranged so that those which exercise the same groups of muscles are separated.

(4) The obstacles should be substantially built. Peeled logs, 6 to 8 inches in diameter, are ideal for many of the obstacles. Sharp points and corners should be eliminated. Landing pits for jumps or vaults should be filled with sand or sawdust to prevent injuries.

(5) The course should be constructed and marked so that it is not possible to sidestep or detour obstacles. However, it is desirable to provide alternate obstacles of varying degrees of difficulty.

(6) The course should be in the shape of a horseshoe or figure eight so that the finish is close to the start and signs should be placed to indicate the course route.

c. Leadership. Before Marines run an obstacle course, they should be instructed in the proper technique of negotiating each obstacle. In each case this technique should be explained and demonstrated in detail, with emphasis on avoiding injury. Every individual should be given an opportunity to practice on each obstacle until he becomes reasonably proficient at negotiating it. Before the course is run against time, it is advisable for individuals to make several runs at a slower pace. During such practice or trial runs, the instructor should observe the performances and make appropriate corrections. Marines should never be permitted to run the course for time until they have practiced on all obstacles. The best method of timing the runners is to have the timer stand at the finish and call out the minutes and seconds as each individual finishes. If several watches are available, each wave may be timed separately. If only one watch is available, the different waves should be started at regular intervals, such as every 30 seconds. If an individual fails to negotiate an obstacle, a previously determined penalty should be exacted.

d.Types of Obstacles

(1) Jumping-Type Obstacles. These obstacles may be ditches which are cleared with one leap, trenches which the individuals can jump into, heights which require jumping downward, or hurdles which an individual can jump over. (See fig. 3-29.)

Figure 3-29 is available in the illustration section at the end of the book.

(2) Dodging-Type Obstacles. Obstacles of this type are usually mazes consisting of posts set in the ground at irregular intervals. (See fig. 3-30.)The intervals between posts should be rather narrow so that the Marines must pick their way carefully through and around them. Lane guides may be constructed to guide the Marines to dodge and change direction. Obstacles may be put into a maze pattern to cause the Marines to change direction.

Figure 3-30 is available in the illustration section at the end of the book.

(3) Vertical Climbing and Surmounting Type Obstacles. These obstacles may be climbing ropes, either plain or knotted and 1-5 inches in diameter; cargo nets or walls 7 or 8 feet high; or vertical poles 6 to 8 inches in diameter and 38 feet high.(Seefig. 3-31.)

Figure 3-31 is available in the illustration section at the end of the book.

(4) Horizontal Traversing-Type Obstacles. Horizontal obstacles may be pipes, beams, ladders, or ropes. (See fig. 3-32.)

(5) Crawling Type Obstacles. Obstacles which require crawling may be constructed of large pipe sections, low rails, and wire. (See fig. 3-33.)

Figure 3-32 is available in the illustration section at the end of the book.

(6) Vaulting-Type Obstacles. Obstacles of 3 to 3.5 feet in height such as low walls or fences may be used as a vaulting obstacle.(See fig. 3-34.)

(7)Balancing-Type Obstacles. Beams, logs, and planks may be used as balancing-type obstacles.(See fig. 3-35.)These items may be used to span water obstacles and dry ditches, or raised off the ground somewhat to simulate natural depressions.

Figure 3-33 is available in the illustration section at the end of the book.

3805.CONFIDENCE OBSTACLE COURSES

a. Description and Objective. The Confidence Obstacle Course is commonly known as the Confidence Course. This course is composed of higher and more difficult obstacles than those used in the conditioning course. The confidence obstacle course is designed to give Marines confidence in their mental and physical capacities and to cultivate their spirit of daring. They are encouraged but not compelled to negotiate this course and the course is not run against time. The negotiation of a confidence course, however, is strenuous enough to be an excellent physical conditioner. Marines should NEVER attempt to take the obstacles at high speed and should not compete for speed. The obstacles vary from fairly easy to extremely difficult. Some are of considerable height to accustom Marines to climbing such heights without fear. Considerable emphasis is placed on obstacles that train and test an individual's balance.

Figure 3-34 is available in the illustration section at the end of the book.

b. Area and Equipment

(1) The confidence course accommodates four platoons, one platoon at each group of six obstacles. The course should be made up of about 24 obstacles, numbered and marked as follows:! to 6, white numbers on red background; 7 to 12, black numbers on a white background;

13 to 18, white numbers on a blue background; and 19 to 24, white numbers on a black background.

(2)A few simple pieces of equipment should be provided for individuals who do not have the strength, courage, or ability to negotiate the obstacles.

Figure 3-35 is available in the illustration section at the end of the book.

c. Formation. The obstacles should be divided into groups of six, and each group is designated by a different color. Each platoon starts at a different color. Individuals are separated into groups of 8 to 12 at each obstacle. At the starting signal from the company commander, they proceed numerically through. Anyone may skip an obstacle who is afraid to try. Individuals proceed from obstacle to obstacle until time is called, then assemble as ordered.

d. Leadership. If the Marines are new to the confidence course, an instructor will demonstrate or will give a brief orientation at each obstacle, including an explanation and demonstration of a method of negotiating it. Marines are encouraged to try the various obstacles, but they are not compelled to do so. No compulsion is to be used. The manner of negotiating any obstacle is left to the discretion of the Marine. However, the instructor assists anyone who experiences difficulty. Instructors must supervise closely at all times to prevent injuries, as some of the obstacles are quite high. Also, some of the obstacles should not be used when slippery or wet. The example of instructors and especially selected demonstrators will serve to inspire the individuals to greater effort.

e. Negotiating the Obstacles. Although personnel need not conform to any one method of negotiating the obstacles, there should be some uniformity in the approach to them. A general method of negotiating the obstacles is indicated below.

(1) Red Group This group contains the first six obstacles. (See fig. 3-36.)

(a) The Belly Buster Individuals may vault, jump, or climb over. Warn them that the log is not stationary.

(b) Reverse Climb. Climb the reverse incline and go down the other side to the ground.

(c) The Weaver. Move from one end of the obstacle to the other by weaving the body under one bar and over the next.

(d) Hip-Hip. Step over each bar, either alternating legs or using same lead leg each time.

(e) Balancing Logs. Step up on log, retaining the balance, walk or run along log.

(f) Island Hopper. Jump from one log to another until the obstacle is negotiated.

(2)White Group. This group is composed of the second six obstacles. (See fig. 3-37.)

(a) Tough Nut. Step over each X in the lane.

(b) Slide for Life. Climb the tower, grasp the rope firmly, and swing the legs upward. Hold the rope with the legs to distribute the weight between them and the arms. Braking the slide with the feet and legs, proceed down the rope. Warn Marines that there is danger of getting rope burns on their hands. When the rope is slippery or wet, this can be a dangerous obstacle.

(c) **Low Belly Over.** .Mount the low log and jump onto the high log, both arms grasping over the top of the log, the stomach area in contact with it. Swing the legs over the log and lower the body to the ground.

(d) **Belly Crawl** Move forward under the wire, belly down, to the end of the obstacle.

(e) **The Dirty Name** Mount the low log and jump to or reach the higher logs in succession, then jump or drop to the ground. Warn the Marines about the height of the final log.

(f) **The Tarzan** Mount the lower log and walk the length of it and each successive, higher log until reaching the horizontal ladder. Grasp two rungs of the ladder and swing the body into the air. Negotiate the length of the ladder by releasing one hand at a time and swing forward, grasping a more distant rung.

(3) **Blue Group.** This group is formed by the third group of six obstacles. (See fig. 3-38.)

(a) **High Step over**. Step over each log, alternating the lead foot or using the same lead foot.

(b) **Swinger.** Climb onto the swinging log and over to the ground on the opposite side.

(c) **Low Wire.** Move under the wire on the back, using the hands to raise the wire to clear the body.

(d) **Swing,** Stop and Jump. Gain momentum with a short run, grasp the rope, and swing the body forward to the top of the wall. Release the rope while standing on the wall and jump to the ground.

(e) **Six Vaults.** Vault over the logs, using one or both hands.

(f) **Easy Balancer.** Walk up one inclined log and down the one on the other side to the ground.

(4) **Black Group.** The last group is formed by the final six obstacles. (See fig. 3- 39.)

(a) **Incline Wall.** Approach the underside of the wall, jump up and grasp the top and pull the body up and over. Slide or jump down the incline to the ground.

(b) **Skyscraper**. Jump or climb to the first floor, climb up the corner posts or assist each other to any desired floor. Descend to the ground in any desired manner.

(c) **Jump and Land**. Climb up the ladder to the platform and jump to the ground.

(d) **Confidence Climb**. Climb the inclined ladder to the vertical ladder. Go to the top of the vertical ladder, then down the other side to the ground.

(e) **Belly Robber.** Step on the lower log and assume the prone position on the horizontal logs. Crawl over the logs to the opposite end of the obstacle.

(f) **The Tough One** Climb the rope or pole on the higher end of the obstacle, then go down the ladder and across the log platform. Climb over or between the logs at the end and go down the rope or pole to the ground. Vault over the final log.

Figure 3-36 is available in the illustration section at the end of the book.

Figure 3-37 is available in the illustration section at the end of the book.

Figure 3-38 is available in the illustration section at the end of the book.

Figure 3-39 is available in the illustration section at the end of the book.

Section IX Individual Exercise Programs

3901.GENERAL

Marines at times will **be** stationed on independent duty and consequently will be responsible for their own physical fitness program. This section will assist them in understanding the need for exercise and will aid in the planning and execution of an individual exercise program. Exercise activities included are the Bench Conditioner, 6-12 Plan, Weight Training, and Isometric Contraction. Keeping physically fit is a problem that faces every Marine. Even though we are frequently engaged in training that requires some physical effort, in many cases, it is not enough to prepare us to meet the intense physical demands of combat. Attaining a satisfactory level of physical readiness is not an insurmountable objective. Available time appears to be the most difficult obstacle to the development of physical readiness. In most cases, regular physical training programs are centralized, requiring the individual to temporarily leave the work area. The problems involved in setting an hour aside two or three times each are numerous. However, most of us can devote a few minutes each day to physical fitness with little, if any, impact on our daily work schedule, especially if it does not require us to leave our work area.

a. Type of Program. There are many good physical fitness programs available to the individual or group. Regardless of the type or duration, to be effective, the program must contain exercises that are strenuous and are challenging to the individual. Space will not permit the inclusion of all available means of individual exercise. The programs selected for this chapter have met the requirement of minimum space and minimum time.

b. Need to Augment Program. These programs are quite strenuous and will develop a satisfactory level of physical readiness. However, if the Marine desires additional development of endurance, it is recommended that he supplement these programs with a 15- to 30-minute period of wind sprints and double timing on an alternating daily basis.

c. Progressive Training. If Marines are performing duties which require little or no physical activity, they must plan a physical conditioning program that assures a moderate beginning, moderate but steady progression, and sufficient warmup before starting the vigorous exercise. To avoid organic of bodily harm, a Marine activity without adequate warmup. He should conduct conditioning programs on a daily basis over an extended period of time, never on an unduly accelerated or crash basis.

3902.THE BENCH CONDITIONER

a. Description and Objective. The bench conditioning program uses a modified bench to employ both isotonic (moving) and isometric (stationary) exercises are as the nucleus of the program. The exercises are designed to develop strength and endurance in all the major muscle groups of the body. The principles of progression, overload, and balance are employed when the exercises are performed properly.

b. Area and Equipment. The conditioning apparatus can be constructed in any unit motor pool with welding equipment found in most salvage yards.(See figs. 3-40, 3-41.)Additionally, there are available any number of commercially produced apparatus available. The important thing is that in utilizing this equipment the following exercise routine be adhered to.

c. Starting Level and Progression.. The program consists of two tables, each with 10 exercises. The Marine controls progression by required repetitions or, in some cases, by application of maximum effort. Each table can be completed within 15 minutes.

d. Starting Position. To start the program, the Marine begins with the Table I and executes each exercise for the required number of repetitions as indicated. The Marine controls the starting level and progression. When the Marine executes the maximum repetitions for Table I within a 15-minute exercise period, he progresses to Table II. To maintain this level of development, the Marine should also execute the maximum repetitions for Table II within a 15-minute exercise period. The Marine should keep substitution of exercises to a minimum. However, it he completes a full 15 minutes of strenuous exercise and exercises all muscles, then there should be no appreciable difference in the overall development.

e. Bench Conditioning, Table I

(1) Exercise 1: Side-Straddle Hop. This is a two-count warmup exercise done at moderate cadence. The starting position is the position of attention. On count ONE jump slightly into the air, swinging the arms out to the sides and up to a vertical position, hands touching.(See fig. 3-40, A.)At the same time, spread the feet wider than shoulder-width apart. On count TWO, using a slight flexing of the knees and ankles, jump slightly into the air and return to the starting position by swinging the arms back down to the sides. Twenty repetitions of this exercise is the standard dosage throughout the program.

(2) Exercise 2: Hand Walk. Remove the lower horizontal bar. Adjust the upper horizontal bar so that it is high enough to permit a "dead hanging" position, with the feet off the ground.(See fig. 3-40, B.j From the "dead hanging" position, release one hand and drop the arm to the side of the body. Then raise that arm and regrasp the horizontal bar. Release the bar with the other hand and drop that arm to the side. Repeat this as many times as possible.

(3) Exercise 3: Situps. Lie down with the fingers interlocked and placed behind the head. Hook the toes under the foot braces. Raise the trunk and upper body to an upright sitting position, twisting it to the left and then forward and downward until the right elbow touches the left knee.(See fig. 3-40, C.)Lower the body to the starting position .Sit up again but twist the body to the opposite direction as before, touching the left elbow to the right knee. Again lower the body to the starting position. The starting dosage is 20 situps. The Marine should continue the progression until he has attained 40 situps.

Figure 3-40 is available in the illustration section at the end of the book.

(4) Exercise 4: Double Step-upStarting at one end of the bench, step up onto the bench, and walk across it. Step down from the other end; turn around and repeat the process to return to the starting point.(See fig. 3-40, D.)Each return to the starting point constitutes a repetition. The starting dosage is 20 repetitions. Maximum dosage is 35 repetitions. This exercise should be done at a rapid cadence.

(5) Exercise 5: Isometric Bar Lift. Adjust the lower bar so that it is slightly higher than the beltline. Placing the feet on the footplates at the base of the frame, grasp the lower bar so that the hands are spread shoulder-width apart. Assume a crouched position and lift with maximum effort using the arms, back and legs.(See fig. 3-40, E.)Starting dosage is 4 repetitions of a stress time of 5 seconds followed by a 5-second rest prior to the next repetition. The Marine obtains progression by lengthening stress periods to 6 and later 7 seconds. Do not increase the number of repetitions.

(6) Exercise 6: Knee Lift. Adjust the upper bar to the same height used in Exercise 2.Adjust the lower bar so that is stops rearward movement of the hips when the "dead hanging" position is assumed.(See fig. 3-40, F.) Keeping the arms extended, flex the legs and raise the knees as high as possible. Hold this position for 5 seconds, then return to the starting position. After 2 seconds in the starting position, raise the knees again. Each return to the starting position constitutes one repetition. The dosage is five repetitions. Progression is the same as in Exercise 5.

(7) Exercise 7: Isometric Pull. Adjust the lower horizontal bar to a position where it is slightly higher than the beltline. Grasp the handles and pull outward.(See fig. 3-40, G.)Apply maximum effort and hold for approximately 5 seconds. Relax for 5 seconds between repetitions; perform four repetitions. Moving the body closer to or farther away from the bar will change the stress from the upper arms to the forearms. Progression is the same as in Exercise 5.

(8) Exercise 8: Isometric Compression. Maintain the position as in exercise 7.(See fig. 3-40, H.)Grasping the handles in the same manner, press in **with** maximum effort and hold for approximately 5 seconds. Relax for 5 seconds between repetitions. Progression is the same as in Exercise 5.

(9) Exercise 9: Isometric Press. Remove the lower horizontal bar. Adjust the upper horizontal bar until it is about 6 inches lower than the extended arms can reach. Stepping on the footplates at the bottom of the frame, grasp the bar with both hands and push up.(See fig. 3-40,1.)Keep both the legs and arms slightly flexed and the back straight. Apply maximum effort for 5 seconds then relax for 5 seconds. Complete four repetitions. Progression is the same as in Exercise 5.

(10) Exercise 10: Pushups. Grasping the foot braces with both hands, assume the front leaning rest position.(See fig. 3-40, J.)Keeping the back and legs straight, lower the body until the chest is lower than the **hands,** then return to the starting position. The Marine should complete the maximum possible number of repetitions.

f. Bench Conditioning, Table II. There is no limit on the maximum number of repetitions attainable in Exercises 3, 4, and 6 of Table II. The only limit imposed is that the entire program of 10 exercises outlinted in either table should not exceed 15 minutes.

(1) Exercise 1: Side-Straddle Hop. This is a two-count warm up exercise done at a moderate cadence. The starting position is the position of attention. On count ONE, jump slightly into the air, swinging the arms out to the sides and up to a vertical position, hands touching.(See fig. 3-41, A.)At the same time, spread the feet wider than shoulder-width apart. On count TWO, using a slight flexing of the knees and ankles, jump slightly into the air and return to the starting position by swinging the arms back down to the sides. Twenty repetitions of this exercise is the standard dosage throughout the program.

(2) Exercise 2: Pull-up. Adjust the horizontal bar so that is is high enough to permit a "dead hanging" position with the feet off the ground. Grasp the bar with both hands, palms facing forward. By flexing the arms, raise the body to a position where the chin is higher than the bar.(See fig. 3-41, B.)Then lower the body to the "dead hanging" position. Repeat as many times as possible.

(3) Exercise 3: Bench situps. Sit on the bench and hook the feet under the foot braces. With the fingers interlocked behind the head, lean back until the head touches the floor.(See fig. 3-41, C.)Return to the starting position. The starting dosage is 15 situps.

(4) Exercise 4: Step-Up. Face the bench and step up on it with one foot, bringing the trailing foot up next to the leading foot. Step back down again, leading with the same foot used first in stepping up.(See fig. 3-41, D.)Perlorm half of the total repetitions, then change the sequence of moving the feet to use the other leg in stepping up, and repeat the same amount of exercise. The starting level is a total of 40 step-up. **This exercise should be done at a rapid cadence.**

(5) Exercise 5: Isometric Bar Lift\djust the lower bar so that it is slightly higher than the beltline. Placing the feet on the footplates at the base of the frame, grasp the lower bar so that the hands are spread shoulder-width apart. Assume a crouched position and lift with maximum effort using the arms, back, and legs.(See fig. 3-41, E.)Starting dosage is four repetitions of a stress time of 8 seconds followed by a 5-second rest prior to the next repetition. Progression is obtained by lengthening stress periods to 10 seconds. Do not increase the number of repetitions.

Figure 3-41 is available in the illustration section at the end of the book.

(6)Exercise 6: Leg Lift. Adjust the bars and assume the starting position as shown in figure 3-41, F. Keeping arms and legs extended, raise the legs to a horizontal position and hold in that position for 2 seconds. Then lower the legs slowly to the starting position. Five repetitions is the starting level.

(7) Exercise 7: Isometric Pull. Adjust the lower horizontal bar so that it is slightly higher than the beltline. Grasp the handles and pull outward.(See fig. 3-41, G.)Apply maximum effort and hold for approximately 8 seconds. Relax for 5 seconds between repetitions; perform four repetitions. Moving the body closer to or farther away from the bar will change the stress from the upper arms to the forearms. The Marine obtains progression by lengthening the stress period to 10 seconds.

(8) Exercise 8; Isometric Compression. (See fig. 3-41, H.)Maintain the position as in exercise 7.Grasping the handles in the same manner, press in with maximum effort and hold for approximately 8 seconds. Relax for 5 seconds between repetitions; perform four repetitions. The Marine obtains progression by lengthening the stress period to 10 seconds.

(9) Exercise 9: Isometric Pres. Remove the lower horizontal bar. Adjust the upper horizontal bar until it is about 6 inches lower than the extended arms can reach. Stepping on the footplates at the bottom of the frame, grasp the bar with both hands and push up.(See fig. 3-41,1.)Keep both the legs and arms slightly flexed and the back straight. Apply maximum effort for 8 seconds, then relax for 5 seconds. Complete four repetitions. The Marine obtains progression by lengthening the stress period to 10 seconds.

(10) Exercise 10: Inclined Pushup. Assume the front leaning rest position with the feet on the bench.(See fig. 3-41, J.)Keeping the back and legs straight, lower the body until the nose touches the ground. By extending the arms, raise the body to the starting position. Repeat as many times as possible.

3903. THE 6-12 PLAN

a. Description and objective. The 6-12 plan of physical fitness has been developed to assist in regulating quantity and progression and to provide a convenient set of exercises. This is a basic program and will take 18 weeks to complete if you follow the moderate progression as to the time prescribed for each level of achievement. The time can be shortened as explained below. This plan consists of six basic exercises a day which can be completed in 12 minutes. There are six tables of six exercises each, thus allowing you to progress from table to table. The plan is progressive, fits any age group, contains balance and variety, and applies the principle of overload in a safe and gradual manner. Begin at Table I, Progression Guide, with the number of repetitions as indicated by age.

b. Formation. If just starting an exercise program, do not rush through the first table. Remember, individuals should remain at each level for about a week before moving upward. The time allotment stated for each exercise at the bottom of the tables is a guide; some people may take more and some less time on the individual exercises. At the end of a one-week period, when the individual can comfortable perform the six exercises in 12 minutes, move on to the next level. To a certain degree, the individual must be the judge of his ability to progress form level to level and table to table. If starting with a certain degree of fitness, some of the beginning tables may present little challenge.

c. Starting Level **and Progression.** There are three levels of achievement for each age group, indicated as A, B, and C. Start at the C level for the appropriate age group. At the end of a one-week period, or when the individual can do all exercises at that level within 12 minutes, progress to the B level. At the end of the second week, or when the individual can accomplish that level within 12 minutes, progress to the A level. At the conclusion of the third week or when the individual can achieve the A level within the time limitation, move on to table II.(See Tables I through VI, Progression Guide.)

d. Maintenance Level. Attempt to work through all six tables. If this proves to be too difficult, then maintain exercise at the—

(1) A-level on Table IV, Progression Guide it in the 45 to 49, 50 to 59, or over 60 age group.

(2) A-level on Table V, Progression Guide if in the 17 to 29, 30 to 39, or 40 to 44 age group.

e. Precautions. To achieve the maximum benefit, perform each exercise exactly as specified. Read the descriptions and study the illustrations. Do not slight the movements. Use a sensible approach and follow these four points as they apply before starting or during your exercise program.

(1) If you have the slightest doubt about your ability to participate in this exercise program, consult a physician.

(2) Stop immediately if you notice unusual breathlessness or chest pain while taking part in these exercises. If these conditions persist, consult a physician.

The following Tables of Progression Guides and Exercises are the progression of the 6-12 Plan program.

Table I. Progression Guide is available in the illustration section at the end of the book.

Table I. Exercises.is available in the illustration section at the end of the book.

Table II. Progression Guide.is available in the illustration section at the end of the book.

Table II. Exercises.is available in the illustration section at the end of the book.

Table III. Progression Guide.is available in the illustration section at the end of the book.

Table III. Exercises.is available in the illustration section at the end of the book.

Table IV. Progression Guide.is available in the illustration section at the end of the book.

Table IV. Exercises.is available in the illustration section at the end of the book.

Table V. Progression Guide.is available in the illustration section at the end of the book.

Table V. Exercises.is available in the illustration section at the end of the book.

Table VI. Progression Guide.is available in the illustration section at the end of the book.

Table VI. Exercises.is available in the illustration section at the end of the book.

(3) Unless you have exercised regularly and know yourself to be in good physical condition, start at table I with the C level appropriate to your age.
(4) If you are out of shape, admit that fact to yourself. Hide your pride; after all, you are in the privacy of your own quarters. Set your goal for the longer, steadier pull toward fitness. Resist the urge to pass over the lower numbered tables to find a table that will test your fitness. You are not trying to test, but rather to develop.

3904.WEIGHT TRAINING

Weight or barbell training should not be confused with the more common types of weight lifting used as a competitive sport. Weight lifting is designed to develop specific muscle groups so that the individual is capable of lifting a large amount during a single lift. In contrast, weight training is the systematic development of all the major muscle groups by the use of calisthenics reinforced with weight to provide resistance.

3905.WEIGHT LIFTING

a. Description and Objective. The weight lifting program is progressive and applies the principle of overload in a safe, gradual manner. The exercises of the table can be completed in 15 minutes. The objective of the exercises is to develop strength and muscular endurance, and muscle tone of the five major muscle groups:legs, arms, back, trunk, and shoulder girdle.(See fig. 3-42.)

b. Warmup. A warm up exercise is important to prepare the body for the more vigorous exercises that are to follow. Ten repetitions of the high jumper exercise are excellent for a warm up period. Also, an exercise which will require fast body movement is needed.To provide such exercise, 3 to 5 minutes of rope skipping is recommended to increase the individual's development potential.

c. Formation. Marines should take care in the completion of these weight lifting exercises. They should ensure that their back is straight during the lifting phase of all exercises. When exercises require assuming the standing position with the weight, they should always grasp the weight while in a squatting position and then rise to a standing position.

d. Progression. Each exercise has a starting number of repetitions and specified pounds of weight. After each 4th or 5th day of exercise, the Marine should increase the repetitions by one until he has reached the maximum of 10 repetitions. At this time, the Marine should increase the weight by 5 pounds. Then he should repeat the process again with the initial number or repetitions.

e. Weight Training

(1) Exercise 1:Squat(See fig. 3-42, A.)Starting level—six repetitions, 50 pounds (commonly called the flatfoot deep knee bend). Place the bar upon the shoulders. Stand with feet about 18 inches apart. Keeping the feet flat, lower the body into the low squat position. Stand erect and repeat. Exhale while lowering into the squat position and inhale while standing. This constitutes one repetition.

(2) Exercise 2:Waist Bender(.See fig. 3-42, B.)Starting level-6 repetitions, 40 pounds. Assume the standing position with the bar across the shoulders and the feet a shoulder-width apart .Bend forward at the waist until the upper body is parallel to the ground; return to the starting position. Each return to the upright position constitutes one repetition.

(3) **Exercise 3:Curl** (See fig. 3-42, C.)Starting level—6 repetitions, 40 pounds. Grasp the barbell with the palms facing to the rear and assume the standing position with the feet a should-width apart. With the barbell held in front of the hips, flex the elbows and lift the weight until the bar touches the upper chest. Lower the barbell back to the hip level position. Inhale deeply with the upward movement and exhale on the downward movement. Each time that the bar touches the chest will constitute one repetition.

(4) **Exercise 4:Side Bender**(.See fig. 3-42, D.)Starting level—6 repetitions per side, 40 pounds. Assume the standing position, with the bar across the shoulders, with feet a shoulder-width apart. Bend to the left as far as possible and return to the starting position. Repeat six times and then execute the same procedure to the right for six repetitions.

(5) **Exercise 5:Standing Press**(See fig. 3-42, E.)Starting level—6 repetitions, 45 pounds. Grasp the bar with the palms facing forward and assume the starting position. Curl the weight to the upper chest position. Inhale deeply and press the bar upward to an overhead position. Exhale while lowering the bar to the chest position. Each time that the bar is pressed upward constitutes one repetition.

(6) **Exercise 6:Upward Row.** .see fig. 3-42, F.)Starting level—6 repetitions, 40 pounds. Grasp the bar, hands close together, palms to the rear, and assume the standing position. Starting with the bar held in front of the hips, flexing the elbows and the shoulder girdle muscles, lift the bar straight up to an overhead position. Inhale deeply while lifting the bar. Exhale while lowering the bar to the hip position. Each time that the bar returns to the hips constitutes one repetition.

(7) **Exercise 7:Shoulder Curl.** See fig. 3-42, G.)Starting level—6 repetitions, 25 pounds. Grasp the bar, palms down, and assume the standing position. Keeping the elbows locked, curl the bar, pivoting the arms at the shoulders until the bar is in an overhead position and as far to the rear as possible. Return the bar in the same manner to the hip position. Each time that the bar returns to the hip position constitutes one repetition.

Figure 3-42 is available in the illustration section at the end of the book.

3906.ISOMETRIC EXERCISING

a. Description and Objective. Isometric exercising is the application of maximum effort during an exercise period. The isometric principle is to apply force gradually over a 5- to 10- second period until the maximum effort is applied. Relaxation follows for approximately 5 seconds and then force is applied again. The Marine continues this process at the prescribed level for each exercise for a period of 15 minutes or less. The objective of isometric exercises is to create muscle growth. Isometric exercises are the fastest means of creating muscle growth. They are founded on the fact that a muscle will grow only so fast regardless of the type or duration of the activity. The principle of overload—that the muscles develop commensurate with demand—reinforces that fact. However, isometrics will not develop cardiovascular or muscular endurance. Consequently, Marines who choose to use isometrics in their exercise routine must also include running or some other type of aerobic activity.

b. **Area and Equipment** isometric exercises may be designed to be performed with or without equipment.

c. Use With Other Programs. The isometric exercises presented in this section will provide a variety from which to choose. In addition to the exercises contained in this section, the application of isometric force is used in some of the exercises in section II.

d.Door Frame Exercises. The following exercises are designed for use with a standard door frame found in all offices or barracks.(See fig. 3-43.)

(1) Exercise l:Arm Press. Stand in the doorway with the legs straight, knees locked. Using the arm muscles, press hard upward against the top of the door frame. Repeat for three repetitions applying gradual effort to maximum contraction.

(2) Exercise 2:Leg Press. Stand in the doorway with the hands on the top of the door frame, elbows locked .With knees bent, press hard with the leg muscles .Repeat for three repetitions beginning with a gradual effort and increasing to maximum contraction. A low platform may be necessary to reach the top of the door frame and still maintain a bent knee position.

(3) Exercise 3:Side Press. Extend both arms to the side of the doorway. Palms are shoulder high, facing outward. With both arms, press hard against the sides of the door frame. Repeat for three repetitions. Begin gradually and increase to maximum contraction.

(4) Exercise 4:Lateral Raise. Extend both arms to the sides of the doorway, arms down, palms facing inward. With the back of the hands, press hard against the sides of the door frame .Repeat for three repetitions. Begin with a gradual effort and increase to maximum contraction.

Figure 3-43 is available in the illustration section at the end of the book.

(5) Exercise 5:Neck Press?lace the forehead against the door frame, hands clasped behind the back Using the neck muscles, press hard against the door frame. Repeat for three repetitions, then reverse position so that the back of the head is resting on the door frame. Again do three repetitions .Begin gradually with both exercises and increase to maximum contraction.

(6) Exercise 6:Door Pull Stand facing the edge of the open door and grasp the doorknobs. Pull outward with both arms (if doorknobs are not available grasp the edge of the door).Applying outward pressure, move the body toward and away from the door .Repeat the three repetitions. Begin with gradual effort and increase to maximum contraction.

Chapter 4
COMBAT WATER SURVIVAL
Section I. Marine Corps Water Survival Program

4101.MILITARY SWIMMING

Military swimming emphasizes strokes that result in staying power rather than speed. Marines need to swim easily, quietly, and with adequate vision. Strokes should be used that allow the Marine to carry basic combat equipment and to tow or push a wounded buddy while keeping the face out of the water to allow breathing. The best strokes to use are the side stroke and the breast stroke. Marines must be proficient in floating, drown-proofing, and in using their basic combat equipment to make flotation devices. They must be drilled on using the Kapok and Mae West buoyancy compensators and on emergency egress from helicopters, landing craft, and amphibious assault vehicles.

4102.PROGRAM DEVELOPMENT

Commanders of the Landing Force Training Commands (Atlantic/Pacific) are tasked with developing and maintaining the individual water survival and swimming training program. They are responsible for training all Marine Corps water safety and survival instructors and integrating this training in to their combat readiness training program. All Marines are deployable and as soldiers of the sea must constantly practice the skills necessary to achieve mission accomplishment. Commanders must be imaginative in program development, integrating basic skills development which enhances the total mission accomplishment. An example is teaching waterproofing of the Marine's backpack so that it can be used as a flotation device in streams or rivers. Once learned, this skill would be practiced first in a pool and then in a stream or a river. Routine inspections during deployments would ensure that backpacks could be used as flotation devices.

4103.QUAL STDS AND TRAINING GUIDELINES

The inherent nature of Marine Corps operations and training requires that Marines achieve an ability to survive in water. Water survival and swimming training is designed to reduce fear of water, instill self-confidence, and develop a Marine's ability to survive in water. As a minimum, every Marine should be qualified as swimmer, third-class (S3).This qualification should be met during recruit training or Officer Candidate School. Once qualified, a Marine need not be requalified unless requalification at a specific level (S3 or higher) is required by unit mission, military occupational specialty or duty assignment, or for other reasons determined by the commander. The following classifications and standards apply to Marine Corps water survival and swimming training. Abbreviations will be used for service, medical, and training record entries citing MCO 1510.29A, Individual Water Survival and Swimming Training, as authority.

a. Qualification standards and test procedures are described in paragraph 4104.

b. Swimmer classification and abbreviations are:

(1) Unqualified - UQ.

(2) Swimmer, third-class (S3).Minimum aquatic skill level for all Marines.

(3) Swimmer, second-class (S2)-

(a) Minimum standard for all naval aviators, naval flight officers, and aircrewmen with written waiver from the commanding officer.

(b) Requirements for military occupational specialty 1803 (assault amphibian vehicle officer), 1833 (assault amphibian vehicle crewman), 1302 (engineer officer), 1371 (combat engineer), 1381 (shore party specialist), and 1379 (engineer operations chief).

(4) Swimmer, first-class (S1)-

(a) Naval aviator, naval flight officer, and aircrewman qualification requirement.

(b) Requirements for military occupational specialty 0321 (reconnaissance man), 8652 (reconnaissance man, parachute jump qualified), 8653 (reconnaissance man, self-contained underwater breathing apparatus qualified), 8654 (reconnaissance man, parachute/self-contained underwater breathing apparatus qualified), 9952 (self-contained underwater breathing apparatus Marine (officer/enlisted)), and 9953 (parachutist/self-contained underwater breathing apparatus Marine (officer/enlisted)).

(5) Water Survival Qualified (WSQ).Ultimate water survival/swimming goal for Marines.

c.Instructor and assistant instructor classifications are-

(1) **American Red Cross Advanced Lifesaver.** Individuals who are currently certified as American Red Cross advanced lifesavers may assist qualified instructors during Marine Corps water survival/swimming training and testing. Certification is valid for 3 years.

(2) American Red Cross Water Safety Instruction. Individuals who are currently certified as American Red Cross water safety instructors may conduct swimming training and qualify individuals as 3d, 2d, and 1st class swimmers. Certification is valid for 2 years.

(3) United States Marine Corps Water Safety and Survival Instruction. Those personnel certified as water safety and survival instructors by one of the landing force training commands may conduct Marine Corps water survival and swimming training and American Red Cross swimming and water safety training. These instructors may also participate as assistant instructors during the conduct of landing force training command-sponsored water safety and survival instructor courses. Certification is valid for2 years.

(4) American Red Cross Water Safety Instructor Traine. Individuals qualified who have been certified as instructor trainers by the American Red Cross may train American Red Cross water safety and basic swimming instructors. Certification is valid for 2 years.

(5) United States Marine Corps Water Safety and Survival Instructor Traine. Individuals qualified who are certified as instructor trainers through special training courses conducted jointly by the American Red Cross and the Marine Corps, under the direction of the Commandant of the Marine Corps, may train water safety and survival instructors and American Red Cross water safety and basic swimming instructors. Certification is valid for 2 years.

d. Training Guidelines Instructors engaged in training Marines as swimmers will use the current edition of the American Red Cross swimming and water safety courses instructors' manuals as the authoritative textbooks for swimming procedures, strokes, breaks, and holds.

4104.WATER SURV/SWIM QUAL STD AND TEST PROCS

a. The following qualification standards and test procedures are applicable to Marine Corps water survival and swimming training. The utility uniform without cover, boots/oxfords or socks will be worn for all swimming tests through swimmer, first class. The utility uniform with boots/oxfords will be worn as prescribed for the water survival qualification.

(1) Swimmer, Third-Class (S3) To qualify as a swimmer, third-class, a Marine must enter the water feet first from a minimum height of 5 feet and remain afloat for 5 minutes. During this time the Marine must swim 50 yards using any stroke or combination of strokes. This test should be taken by all Marines as early as possible in their initial training. Those unable to pass this test will be classified as UQ and should be given instruction in fundamental swimming skills. Those who barely meet requirements are swimmers who need help and should receive additional instruction in fundamental swimming skills, whenever possible.

(2) Swimmer, Second-Class (S2.) To qualify, a Marine must enter the water feet first from a minimum height of 10 feet and remain afloat for 10 minutes. The Marine must swim 100 yards and use the three basic survival swimming strokes (side stroke, backstroke, and breast stroke) for a minimum distance of 25 yards each.

(3) Swimmer, First-Class (S1)A prerequisite to qualification as swimmer, first-class, is successful completion of the test for swimmer, second-class. To qualify as a swimmer, first-class, a Marine must be able to do each of the following:

(a) Approach a person of approximately the same size while in the water, demonstrate a release, get that person in a carry position, and tow the "distressed swimmer" 25 yards.

(b) Enter water feet first and immediately swim underwater for 25 yards. Swimmer is to break the surface twice for breathing during this distance at intervals of approximately 25 feet.

(c) Remove trousers in water, inflate for support, and remain motionless for a minimum of 1 minute.

(d) Swim 220 yards using any survival swimming stroke or any combination of survival strokes.

(4) Water Survival Qualification (WSQ)A prerequisite to water survival qualification is successful completion of the test for swimmer, first-class. To be water survival-qualified, a Mannemust:

(a) Enter water by jumping from a height of 10 feet.

(b) Stay afloat in full utility uniform (less cover) for 1 hour. Boots/oxfords will be removed after 5 minutes in the water but will be retained by the swimmer.

(c) Traverse 75 yards in deep water in full utility uniform (less cover) with boots/oxfords.

b. All water survival and swimming training will include instruction in the following:

(1) Employment of safety lines or other comparable expedients to assist in swift stream/river crossings.

(2) Adverse physiological effects caused by cold water (hypothermia) and the precautionary measures to be taken prior to exposure to such an environment.

(3) Employment of a standard and expedient flotation devices in a water survival situation.

(4) Removal of individual field equipment after unscheduled water entry. Entry will be made from a minimum height of 10 feet with a complete set of field equipment, properly packed as an expedient flotation device and waterproofed. The ALICE pack will then be removed, the rifle laid across the top of the pack, the helmet remaining or put back on the individual's head, and the individual will traverse 25 yards to the side of the pool or water's edge.

c. The additional water survival qualification (flight physiology training) and requalification required of personnel assigned to flight status will be conducted in accordance with OPNAVINST 3710.7_,Promulgation of NATOPS General Flight and Operating Instructions. Swimming qualification (first-class, SI) is a prerequisite to water survival (flight physiology) training and is a one-time requirement.

Section II. Water Sports and Competitive Activities

4201.GENERAL

The commander, when developing the combat readiness program, needs to look carefully at the mission. If the mission requires Marines who must perform mission-essential water-related activities of long duration, he may want to look to water sports/competitive activities as a method of developing endurance. Endurance and confidence can only be developed by extended time in the water. While defined training can accomplish this goal, the goal can be enhanced through training which is integrated with the physical conditioning program. This approach will provide variety and enthusiasm. Two activities which will accomplish this goal are water relays and water polo. The important consideration is that the unit achieve total participation by all members of the unit.

4202.WATER RELAYS

Development of water relays will generally conform to the guidelines in chapter 5, section IV. When contemplating relays, the commander should review chapter 5 and adapt the principles to the size of the unit and the facility to be used.

4203.WATER POLO

The official rules provided in this paragraph will need to be adapted to the unit's unique situation, its size, and the size of the facility. Commanders should make every effort to have all members of the unit participate.

a. NCAA Rules This style of play is based on two popular sports: swimming and basketball. Playing is done on the surface of the water by teams of seven players each. The size of the playing area is 25 yards by 14 yards, which is the size of the typical high school or college indoor pool. The deeper the water, the better. If the entire playing area is deep-6 feet or more—this is ideal, although most indoor pools have a shallow end. At each end of the pool is a goal. In deep water, the goal is 10 feet across and 3 feet high. In shallow water, where the goalie can stand on and jump from the bottom, the goal is 10 feet across and 5 feet high. One of the seven players on each team is the goalie. He is the only player on the team who can stand on or jump from the bottom (if it is shallow enough to permit this), catch and pass the ball with both hands at the same time, or hit the ball with a clenched fist. However, the goalie may not swim across or pass the ball across the middle of the pool.

b.Equipment. The goalie on the visiting team must wear a white cap with the number 1 clearly marked on it; his teammates, who are designated as guards and forwards, must wear white caps numbered 2 through 7, with the substitutes wearing white caps numbered 8 through 21.The goalie on the home team must wear a dark-colored cap with the number 22 clearly marked on it; his teammates designated as guards and forwards, must wear similar dark-colored caps numbered 29 through 42.The guards and forwards on each team may swim freely up and down the pool as they see fit, interchanging positions as often as they wish. They may not stand on or jump from the bottom, touch the ball with more than one hand at a time when catching or passing or shooting it, or enter inside the opposing team's 2-yard line unless preceded by the ball. The ball is similar to a soccer ball, except that it is yellow and covered with a waterproofing substance that makes it easier to handle with one hand.

c. **Playing Time.** In intercollegiate competition the game consists of four 7-minule quarters, the teams changing ends after every period of play. In high school competition the game consists of four-,five-, or six-minute quarters, depending on the ages of the participants. Here again the teams change ends after every period. There is a very brief interval between each quarter. At the start of each period the teams line up at their respective ends, and with ablast of the whistle, the head referee throws the ball into the middle of the pool. The fastest swimmers on each team then race to gain possession.

d. Officials. The head referee walks along one side of the pool watching for infractions. He is aided by the assistant referee, who patrols the opposite side of the pool. The referees are both equipped with whistles and two official's flags. These flags should be 12 inches square and should correspond in color to the caps worn by the respective teams. Whenever a referee sees an infraction, he blows his whistle and signals with his flag indicating that the player nearest to the ball with the appropriate colored flag will be given possession of the ball. Whenever the whistle is blown, all players should react by looking immediately to see which color flag is being waived which will indicate which team should be moving into the defense. Other officials are a timer and a scorer. As in all sports, it is important that the officials be skilled and competent.

e. **Technical Fouls.** The following are some of the common technical fouls: starting before the referee blows his whistle to open play; holding onto or pushing off from the side of the pool during play; taking hold of the ball underwater when tackled by an opposing player; swimming inside an opponent's 2-yard line unless preceded by the ball; touching the ball with both hands at the same time (goalie excepted); standing, walking on or jumping from the bottom when taking an active part in the game. When a technical foul occurs, thereferee blows the whistle and with his flag awards possession. The player on the team awarded possession nearest the point of infraction then has 5 seconds to put the ball into play by passing to a teammate or by dropping the ball into the water and swimming after it.

f. Personal Fouls. The following are some of the more common personal fouls: committing any of the technical fouls mentioned above for the purpose of scoring or preventing a goal; holding, ducking, pulling, pushing off from, swimming over, or impeding the arm or leg movement of any opponent who is not touching the ball; and splashing water in the face of an opponent. When a personal foul occurs, the referee will follow the same procedure as with a technical foul. While doing this, the referee will also call out clearly and loudly the number of the player who was guilty of the infraction and a personal foul will be marked against that player by the official scorer. When a player accumulates five personal fouls, he fouls out of the game and must be replaced by a substitute.

g. Penalty Shot. A penalty shot can be awarded by either of the officials or the scorer when:

o An offensive player inside the opponents' 4-yard line not touching the ball is held, ducked, pulled-back, kicked, or struck.

o A team has accumulated a total of ten personal fouls. When the former occurs, the head or assistant referee should immediately blow the whistle and by holding a two-flagged stick in a vertical position above his head, signal that a penalty shot has been awarded. When the latter occurs, the scorer should use a buzzer located at the scorer's tabic to signal that one team has accumulated ten personal fouls, thereby entitling the other team to a penalty shot. A penalty shot is taken from the 4-yard line in front of the goal. All players except the defending goalie must leave the 4-yard line until the shot is taken. No player can be within I yard of the shooter. After ascertaining that the shooting player is on the 4-yard line and the goalie is onThegoal line, the referee will ask the shooter to lift up the ball. When he docs, the referee will give a sharp blast from the whistle. At the whistle, the shooter must shoot without delay and without any faking at the goal. The goalie may try to block the shot. If he does, or the shot is otherwise missed, it is immediately in play and action continues.

h. When a Goal is Scored. When a goal is scored, either from a shot taken by a player out in the field or by penalty shot, the defending goalie must then pass the ball to the nearest referee. The referee will then pass the ball back to the goalie, who puts it into play by a pass to one of his teammates as soon as possible.

Section III. The Battle Swimming Test

4301.GENERAL

The battle fitness test is the evaluation contained in the Guidance for Basic Warrior Training (BWT) as part of the Basic Warrior Training Concept Plan.

4302.THE BATTLE SWIMMING TEST

The battle swim test consist of the following:

a. Be able to climb a three meter high board, while wearing a field uniform, carrying a M-16, web gear, and four canteens. Jump off blindfolded, recover and swim 100 meters with the M-16.

b. Repeat the same process, without the blindfold, and upon entering the water a second time, recover and retrieve all submerged gear.

c. Remain afloat and tread water for five minutes.

d. Demonstrate a knowledge of how to cross a stream using field expedient techniques and a single strand rope bridge.

e. Successfully complete helicopter egress training.

Chapter 5
COMPETITIVE CONDITIONING ACTIVITIES
Section I. Organization of Competitive Activities

5101.LEADERSHIP OF COMPETITIVE ACTIVITIES

a. Place in the Program. Competitive conditioning activities consist of dual combatives, relays, team contests, and team sports in which individuals or teams compete against an opponent to win. Competition is one of the best ways of maintaining interest in the physical training program during the sustaining stage. Organized competition provides enjoyable, vigorous physical activity that has proved to be one of the best supplements to conditioning drill activities. The benefits of competition are the development of aggressiveness, unit pride and identity, teamwork, and the will to win. These activities help develop cardiovascular and muscular endurance, strength, and coordination. Scheduling competitive activities in an orderly and progressive manner is desirable. The progression should be from relays to dual combatives, to team contests, and finally to team athletics. These activities should be conducted as part of the program after a basic level of conditioning has been developed, usually during the slow-improvement stage of conditioning. Muscles and joints should be strengthened by preconditioning to withstand the strain placed upon them by sudden stops and turns, body contact, bearing of weight, and falls. Competitive activities, however, should not be allowed to dominate the physical training program.

b. Area and Equipment. Some of the competitive activities included in this chapter require specific types of areas and equipment; others do not. The area requirement can usually be satisfied on available training fields. When items of equipment are required or specific courts or field layouts are to be marked off, such information is included.

c. Leadership. The principal factor for success as a leader of competitive activity is an energetic, dynamic, enthusiastic approach. The leader's attitude is reflected by the group, so the commander must carry on the activity in a snappy and vigorous manner. Confidence on the part of the commander will create an impression of decisiveness and certainty. Confidence grows out of experience and a thorough knowledge of the activity. Mastery of subject is the first step in developing confidence, assurance, and poise.

(1) The following suggestions are offered for leaders of competitive combatives, contests, and sports:

(a) Get the activity underway quickly by selecting and teaching only the essentials.

(b) Use rules to add to the enjoyment of the activity and do not allow them to interfere with the spirit of competition.

(c) Stop the activity before interest begins to lag.

(d) In team contests, clearly distinguish sides by attempting to maintain unit identity.

(e) Always insist on fair play, enforcing the rules impartially.

(2) The following procedure is recommended for presenting a competitive activity.

(a) Name the activity.

(b) Briefly explain the objective of the activity and give only the pertinent rules.
(c) Have a demonstration at slow speed and answer questions.
(d) Organize groups into teams and appoint captains.
(e) Arrange teams in the proper starting positions.
(f) Conduct the activity.

d. Competitive Units. Units for competition should be the same organizations in which Marines train: battalions, companies, platoons. In most situations, the unit is the squad.

e. Provisions for Instruction. One of the most effective methods of maintaining interest and participation in competitive activities is to provide instruction in activities with which most Marines are unacquainted. Such instruction can be conducted during regular physical training periods. Careful planning is required to keep all individuals continuously engaged in vigorous activity. The materials in the following sections should be used as a guide for instructional purposes.

f. Officiating and Control. Every effort should be made to provide good officials for all competitive activities. Poor officials quickly cause dissatisfaction among participants in team activities and create a situation which can turn an organized activity into a brawl. Each company should have or develop several competent and qualified officials available for games on company and platoon levels.

5102.MILITARY FIELD MEETS

When units reach the latter part of the slow-improvement and sustaining stage of training, interest in the program may lag. A change in course content can arouse the desire to participate. An event which does not require a high degree of skill yet demands strenuous activity is ideal for the military field meet. This is a series of team contests conducted on a station-to-station basis during a given period of time. Team contests carried on simultaneously provide essential training, vigorous exercise, and stimulating competition in an atmosphere that is enjoyable for all. Because of the healthy rivalry that a field meet arouses, it is an excellent form of inter-unit competition.

a. Objective and Advantages. The objective of the military field meet is to provide activity for everyone in the participating units. Activities are chosen that will develop aggressiveness, teamwork, a will to win, and competitive spirit, and that will stimulate interest and build esprit de corps. The military field meet can **be** included as part of the physical training program or as part of the off-duty recreational program. It is a form of contest that can be conducted in nearly all circumstances because it can be easily modified, requires a minimum amount of equipment, and can be readily organized.

b. Level of Competition. The military field meet is flexible. It can be adjusted to large or small groups. A company-size unit is the most desirable, but it may also be administered within a larger unit. If it is conducted within a company, the participating units will be squads. If it is held within a larger unit, platoons compose the teams. A larger-size unit requires more extensive organization and administration.

c. Selection of Events. In organizing a military field meet, select events that are simple and easy to administer. All rules and regulations should be clearly understood by everyone. No event should require previous practice. In selecting the events, consider the interest and capabilities of the Marines and available equipment and facilities. Select events which will require that all members of the units participate. Events should not be dominated by the athletes; rather, MOS-related events which have significance in producing skills and combat readiness should be used.

d.Equipment and Facilities. The site must be large enough to permit events to be grouped about a central control point. Use available facilities such as parade decks, Softball fields, or tracks. A public address system is desirable at the control point for the initial orientation of teams and for subsequent announcements of time lapses, cumulative scores, and final standings. A tally board is necessary so that all units can constantly monitor how the events are progressing.

e.Administrators. Efficient administration of the military field meet depends on the referees, judges, and scorers. For these positions, choose individuals within the units who have had athletic or officiating experience. Prior to the day of the event, all administrators should be briefed and assigned a specific task in order that they may become familiar with the rules and organization of the contests they will conduct. The following officials should be available:

(1) A primary instructor or supervisor who is in charge of the control center and who is responsible for the successful operation of the athletic carnival. The supervisor must have an assistant to act as a timer and scorer.

(2) One assistant instructor in charge of each event. This person should be-

(a) Familiar with the rules of the event.

(b) Effective as a leader to ensure proper supervision and control over the participating teams.

(c) Enthusiastic to provide proper motivation.

(d) Self-confident of the ability to judge infractions of the rules. The assistant instructor must be fair in judgment and penalize without hesitation when infractions occur.

(3) Runners between event stations and the control point. These individuals collect and deliver scores.

f. Team Organization for Competition. The size of the teams is determined by the level on which the field meet is organized. Maintaining the integrity of the unit promotes esprit de corps, but this does not preclude grouping two squads into one team.

g.Conduct of Events

(1) The assistant instructor at each station takes charge of the group and gives a brief explanation of the major rules of the event. The assistant instructor speaks clearly and distinctly from a position to be seen and heard by all. A short demonstration is desirable if it will help clarify the event.

(2) The assistant instructor should make certain that teams can be clearly distinguished by the use of, for example, T-shirts and fatigue jackets, or caps and no cap, or colored jerseys. The assistant instructor should teach a whistle response (teams stop play immediately upon hearing whistle), get the event started as quickly as possible, and make any necessary corrections as the contest progresses. The rale of good officiating is to use a minimum of calls, yet maintain control of the contest. Penalize when necessary, but refrain from disqualifying contestants or teams.

(3) Keep the activity moving as rapidly as possible. When the central control point sounds the whistle to stop the play, all competition ends immediately. The assistant instructor then assembles the group, forwards the team scores to the central point, and, upon a signal from the central control point, rotates the teams to their next station. It is essential that rotation and orientation of teams be quick and orderly to reduce time spent between contests.

(4) Upon completion of the final event, assistant instructors move their teams to the central control point for the announcement of winning teams, presentation of awards (if any), and final critique.

h. Scoring System. The system for determining the winner of the military field meet should be simple and efficient. At the completion of competition, the scorer totals the points that each team has scored in all contests. The scorer then subtracts the number of points scored against a team (penalty points for nonparticipation,etc.) from the total. The resulting scores are placed in a column with the highest score at the top and the lowest at the bottom. The team with the highest total is the winner. This type of scoring system encourages a team to ensure total unit participation.

Section II.Combatives

5201.DESCRIPTTON AND OBJECTIVE

Combatives are strenuous, short competitive contests in which two individuals attempt to overcome each other in a bout of skill and strength. These contests help to develop Marines' resourcefulness, confidence, strength, agility, coordination, and will to win. Any level ground area can be used. However, extremely hard ground should be avoided as some of the combatives require ground contact. A whistle is needed to control the bouts during the competition since voice commands may go unheeded. The extended rectangular formation is used for dual combatives. Combatives are conducted on an informal basis. Marines are allowed to remain at ease between activities and are allowed to brush themselves off after being on the ground. To get the most out of combatives, individuals must be urged to overcome their opponents as quickly as possible as would be required in combat.

a. Benefits for Marine Training. The possibility of close contact with an enemy in combat faces Marines at all times. They must be trained to react aggressively and violently. Combatives may be used as an introduction to such hand-to-hand contact and should be followed by hand-to-hand combat training. Marines enjoy competition, and this type of activity is a welcome change from formal conditioning activities.

b. Instructor Responsibility

(1) The instructor tells the individuals that all combatives begin and end on the whistle signal. The instructor demonstrates each activity before having it performed, explaining it in simple terms.

(2) After stopping one activity, the instructor gets everyone in place for the next.

(3) The instructor must closely supervise combatives to insure that contestants do not use unfair or unsportsmanlike tactics. To avoid unnecessary injury, instructors must see that the bouts are closely controlled and opponents equally paired. Adjustments should be made in apparent cases of mismatched abilities.

5202.COMBATIVES TABLES

There are three tables of combatives, each of which can be completed in 15 minutes. These tables become progressively more difficult from lower to higher numbers.

a.Combatives Table I

(1) Open-Hand Slap Boxing(See fig. 5-1, A. individuals assume a boxer's stance, palms open, fingers extended and joined. Each contestant tries to slap the opponent about the head and upper body with the open hand. This is a good warm- up activity.

(2) Wrist Tug-of-War(See fig. 5-1, B.)Two individuals sit on the ground with the soles of their feet in contact. Each grasps the opponent's wrists so that the hands are directly over the feet. At the whistle, each individual tries to pull the opponent from sitting to standing position.

(3) Arm-Lock Wrestling.(See fig. 5-1, C.)Two individuals sit back-to-back with legs spread and arms locked at the elbows. Each right arm is inside the opponent's left. Each tries to force the opponent's left arm or shoulder to the ground. The individual who first wins three bouts is the winner.

(4) Bulling(See fig. 5-1, D.)Two individuals assume the Westmoreland wrestling hold, each grasping the opponent's neck with the right hand and the opponent's right elbow with the left hand. Each tries to force the other to move one foot by pushing, pulling, or otherwise manipulating. The individual who first wins two boutsis the winner.

(5) Indian Wrestling(See fig. 5-1, E.)Two individuals lie on the ground, side by side, with their heads in opposite directions. They link right elbows. On the instructor's signal or by mutual agreement, each individual raises the right leg approximately straight and far enough to engage the opponent's heel. To start the contest, each person usually raises the leg three times rhythmically and on the third time engages the opponent's heel. Each tries to roll the other over backward. The right leg is used for three bouts, then the left leg for three bouts.

Figure 5-1 is available in the illustration section at the end of the book.

b. Combatives Table II

(1) Wrist Bending.(See fig. 5-2, A.) Opponents pair off, face each other, raise their arms forward, and, with palms forward, interlock their fingers. At the starting signal, each individual attempts to bend the opponent's wrist. The arms are kept up and forward and are not allowed to swing around and down to the sides. The individual who first wins two bouts is the winner.

(2) Back-to-Back Push.(See fig. 5-2, B.)Two individuals stand back-to-back with elbows locked. Each right arm is inside the opponent's left arm. At the starting signal, each pushes backward, trying to move the opponent. Opponents are not allowed to lift and carry each other; only pushing is permitted. The one who pushes the opponent the farthest wins the bout. The individual who first wins two bouts is the winner.

(3) Hop and Pull Hand.(See fig. 5-2, C.) The individuals are matched in pairs. Each grasps the opponent's right hand and, hopping on the right foot, attempts to pull the opponent off balance. Contestants automatically lose if they touch their free hand or their lifted foot to the ground. For successive bouts, they alternate hands and feet.

(4) Westmoreland Wrestling. (See fig. 5-2, D.)Each contestant grasps the back of the opponent's neck with the right hand and the opponent's right elbow with the left hand. In this position, each attempts to pull, push, or force the opponent to touch the ground with any part of the body other than the feet. The individual who first wins two bouts is the winner.

(5) Crab Fight.(See fig. 5-2, E.)Two individuals sit on the ground facing in opposite directions with their hands on the ground behind them. At the whistle, they raise their hips and push with their shoulders and bodies, each trying to make the other's hips touch the ground. The individual who first wins two bouts is the winner.

Figure 5-2 is available in the illustration section at the end of the book.

c. Combatives Table III

(1) Hand Wrestling.(See fig. 5-3, A.) Two individuals stand facing each other. Their right feet are forward and braced side by side. They grasp right hands for the first bout, left hands for the second bout. Each pulls, pushes, makes sideward movements, and otherwise maneuvers to force the opponent to move one or both feet from the original position. The contestant who first wins two bouts is the winner.

(2) Back-to-Back Tug.(See fig, 5-3, B.) Two individuals stand back-to-back with both arms linked at the elbows. Each has the right arm inside the opponent's left arm. At the starting signal, each attempts to pull the opponent. Lifting and carrying are permitted. The contestants must maintain their original direction and keep their arms linked. After a predetermined time, the player pulled or carried the farthest is the loser.

(3) Wrestling to Lift off Feet.(See fig. 5-3, C.) Contestants face each other. Each places the right arm under the left arm of the opponent and around the body. The left arm is over the opponent's right shoulder. Each tries to lift the other off the ground. The individual who first wins two bouts is the winner.

(4) Arm Pull Between Legs.(See fig. 5-3, D.) Two individuals are paired off, back-to-back. Each bends forward and, extending the right arm between the legs, grasps the opponent's right wrist. At the starting signal, each person attempts to pull the opponent After a predetermined time, the player who has pulled the opponent the farthest is the winner of the bout. The one who first wins two bouts is the winner. Repeat with the left hand and then both hands.

(5) Rooster Fight.(See fig.5-3, E.) Each contestant grasps the left foot with the right hand from behind, and the right arm with left hand. Each hops on the right foot, and, by shoulder-butting the opponent or by feinting and sudden evasions, forces the opponent to let go of the foot or arm. The contestant who first wins two bouts is winner.

Figure 5-3 is available in the illustration section at the end of the book.

Section III .Relays

5301. DESCRIPTION AND OBJECTIVE

Relays are races in which each member of a team runs one leg of the race. The team effort decides the winner. Relays provide stimulating competition and contribute to the conditioning of personnel. They also develop aggressiveness, team spirit, and the will to win. Relays should be dispersed throughout the program for short periods of time to provide a change of activity.

a. Team Organization. Relays are conducted most efficiently in platoon-size groups. Teams of equal size must be organized. Competitive spirit is encouraged and teams are organized better by basing teams on units such as squads, crews, or sections. Team captains should be designated. Extra Marines may be used as officials. The number of individuals on a relay team should be limited to squad size. If larger teams are used, the runners will spend too much time awaiting their turns and too little time actually participating. Two to six teams are ideal for relay competition. It is difficult to keep track of winners when more teams compete.

b. Administration of Relays

(1) The time spent on any one relay should be relatively short. If one team achieves a substantial lead in a long relay, the competitive spirit and enthusiastic participation of the other teams may decrease. Several short relays are generally better than one long relay.

(2) To maintain competitive spirit throughout a number of relays, determine the teams that win, place, and show in each relay and their total points for all relays. This can be done by awarding points to all teams on the basis of position at the finish of each relay. The team with the greatest number of points is the winner of the entire set of relays.

(3) Difficulties commonly encountered in conducting relays may be avoided by the following procedures:

(a) The last player in a relay race should be conspicuously identified. For example, the last runner can tie a handkerchief around the head or arm, take off the shirt, put on a hat or take it off, or use some other means.

(b) Another way to keep track of the progress of the race is to have each player sit or squat as soon as each is finished.

(c) Judges at the starting line can keep the runners from starting too soon.

(d) To prevent contestants from turning before they run the full distance, they should be required to run around a peg, pole, or assistant instructor.

(e) Batons, handkerchiefs, tent pegs, or other objects should be passed from one runner to the next when relays are run on a circular track.

(4) Before a unit's first participation in a relay, inform participants of the rules and scoring system. Violation of the rules should not result in disqualification. Instead, impose point penalties. A point penalty is imposed by subtracting a point from the team total at the conclusion of the relay.

(5) Careful administration will prevent most violations. For efficient conduct of relays, follow this procedure:

(a) Announce the name of the relay.

(b) Form the individuals in relay position.

(c) Briefly explain the relay and the rules for running it.
(d) Demonstrate.
(e) Have a definite finish line, and ensure that all know where it is.
(f) Answer questions.
(g) Conduct the relay.
(h) Determine winner and award points.

5302. RELAY TABLES

a. Events. The following relays are grouped into a table of activities. Each relay table can be completed in 15 minutes. Thus, relays can be used as a sole activity or as a part of a longer period. Each table is planned for a platoon-size group (30-60Marines). Adequate warmup for participants is provided by conducting several repetitions of exercises 1 and 2 of a conditioning drill. The recommended relay tables require an area 40 by 60 meters in size. Each table provides a variety of activity. The tables are progressive in the overload applied and should be scheduled in numerical order although not necessarily on successive days.

b. Relay Table I
(1) **60-Meter Lane Relay** (See fig. 5-4, A.) Each team is assembled in single file behind the starting line. On signal, the first individual of each team runs to the turn-around line 30 meters away, then runs back and touches the next teammate waiting at the starting line. The winning team is the first team to get its last member across the finish line. If an individual starts before being touched by the preceding runner, the team may be penalized.
(2) Wheelbarrow Race.(See fig. 5-4, B.) The players of each team pair off and line up in single file behind the starting line. The first individual walks on the hands while the partner grasps the ankles. They advance to the distance line (25 meters) behind which they exchange positions and return to the starting line. The rear individual must always hold the partner's ankles. After the first pair returns across the starting line, the next pair starts.

Figure 5-4 is available in the illustration section at the end of the book.

(3) Squad Front Relay .(See fig. 5-4, C.) The teams form in a line along the starting line with a 10-foot interval between teams. The members of each team lock elbows so that they are linked together. At the starting signal, the teams run to the distance line (20meters) where the left flank individual acts as a pivot. The team swings around on the pivot and returns to the base line. If a team breaks its links, it must reform before continuing. The first team to completely cross the base line intact is the winner.
(4) Crab-Walk Race.(See fig. 5-4, D.) The players of each team line up in single file. The first person of each team assumes the crab-walk position with feet forward on the starting line, hands on the ground behind it. At the starting signal, contestants move forward to the distance line (10 meters). They touch the line with their feet and then return to the starting line in the reverse position with the head and hands leading. The second person may not start until the first individual touches the finish line.

c.Relay Table II

(1) 100-Meter Lane Relay.(See fig. 5-5, A.)This relay is conducted exactly as the first relay of table 1 except that the start and turn-around line are 50 meters apart. This relay provides progression in sprinting.

(2) Frog-Jump Relay. (See fig. 5-5, B.) Each team lines up in single file. The first individual assumes a squatting position on the starting line. At the starting signal, the contestant progresses to the distance line (15 meters) and back by leaping forward, catching the weight on the hands, and bringing up the legs to the squat position for the next leap forward.

(3) Simple Relay. (See fig. 5-5, C.)Each team lines up in single file. Place a marker on the distance line (20 meters) in front of each team. Each team member, behind the first one, grasps the belt of the individual ahead. At the starting signal, each team runs as a unit to the marker, circles it, and returns to the starting line. The first team to completely cross the starting line intact is thewinner.

(4) Fireman's Carry Relay.(See fig. 5-5, D.)The players of each team line up in pairs, one behind the other. One individual in each pair carries the partner to the distance line (30 meters), using the fireman's carry. At the distance line, individuals exchange places and return to the starting line. As a variation, the person to be carried lies on the ground. The one carrying lifts the partner to the proper position. This relay may be performed with the other carries described in guerrilla exercises in chapter 3.

d.Relay Table III

(1) 200-Meter Circle Relay .(See fig. 5-6, A.)A course is laid out in a circular, rectangular,or oval pattern that is 200 meters around. Each team provides one runner on the starting line. On signal, the runner races around the 200-meter track and touches the next teammate, waiting at the starting line, who runs the same course. Each team member runs one lap of the course.

Figure 5-5 is available in the illustration section at the end of the book.

(2) Bear-and-Crab Race.(See fig. 5-6, B.) Each team lines up in single file. At the signal to start, the first contestant in each column assumes the bear-walk position (hands on ground in front), walks to the distance line (15 meters), and then runs back to the starting line. The returning contestant touches off the second individual and goes to the rear of the line. The second contestant assumes the crab-walk position (hands on ground behind) and crab-walks with feet leading to the distance line. This contestant also runs back to the starting line and touches off the next individual, who walks bear fashion. The rest of the members of each team alternate in this manner. The relay ends when the first person is back at the head of the line.

(3) Pilot Relay.(See fig. 5-6, C.) The players are grouped in threes, arms interlocked at the elbows, and outside players facing backwards. The middle person runs forward; the two outside individuals run backward. They run to the turning point (15meters), where they start back, this time with the middle person running backward and the two outside individuals running forward. The next set of three players starts when the first set crosses the starting line.

(4) Saddle-Back Relay.(See fig. 5-6, D.) Mark two parallel lines 15 meters apart. Each team selects a rider. The remaining members of each team count off. The even-numbered players from each team form in single files behind one line and the odd-numbered players from each team form in single files behind the other line directly across from their teammates. At the starting signal, the rider mounts the back of the first player of the team who carries the rider across the other line. There the rider changes mounts to the first player in the second line without touching the ground. This person carries the rider to the next player waiting in the first line. The relay continues until all of the mounts have carried the rider. A rider who falls off must mount again at the point of the fall. A rider who falls in changing mounts must get back on the original mount before making the change.

e. Relay Table IV

(1) 100-Meter Circle Relay (See fig. 5-7, A.)A course is laid out in either a circular, rectangular, or oval pattern that is 200 meters around. Each team is divided in half with each half positioned at starting lines on opposite sides of the track. Each runner races halfway around the track and touches a teammate who completes the lap. Each runner then waits in file at the halfway point until touched by the runner and then completes the second half of the lap. The first team to return all runners to their original starting line is the winner.

Figure 5-6 is available in the illustration section at the end of the book.

(2) In-and-Out Relay.(See fig. 5-7, B.) Each team lines up in a file with players 2 meters apart. At the starting signal, the first player runs back through the column in a zigzag fashion. He alternates, going to the right of one teammate and to the left of the next. Upon completing the run, the player lines up 2 meters behind the last one in the column. As soon as the first runner has passed the second person, the latter starts to run. This continues until all the players have realigned their original order. The team that finishes first is the winner. It may be desirable to have this relay continue until all individuals have run through their entire team two or three times in succession.

(3) Circle Race.(See fig. 5-7, **C.)** Each team forms a circle and holds hands with all individuals facing out except one who faces in and is the "driver."At the starting signal, the teams race to the distance line (20 meters) and back, keeping the circle intact. All the individuals in the circle must completely cross the distance line. The "driver" gives directions and orders. When the circle breaks, it must be reformed before it can continue. The first team completely over the starting line is the winner.

(4) Horse-and-Rider Relay.(See fig. 5-7, D.)Each team lines up in a single file. At the signal to start, the second player in each column leaps upon the back of the first who carries the rider across the distance line (30 meters). At the distance line, the rider dismounts and runs back to the starting line. There the rider picks up the third individual in the column, and carries this player to the distance line where the first player has remained. This continues until the last person is carried across the distance line.

Figure 5-7 is available in the illustration section at the end of the book.

Section IV. Team Contests and Athletics

5401. DESCRIPTION AND OBJECTIVE

Team contests are competitive activities in which Marines as a team compete with another team to win. They are guided by simple rules and organization. The function of team contests is to provide competition and an opportunity for body contact and to contribute to the development of physical readiness. In competing and working together as a team, individuals develop aggressiveness, the will to win, and teamwork. Team athletics deserve a prominent place in the physical training program because they contribute to increased combat efficiency. Because of the competitive nature of athletics and their natural appeal, individuals take part with enthusiasm. Athletic teams formed at intramural and higher levels are a strong unifying influence and provide one of the best means of developing esprit de corps.

a. Area and Equipment A level training field is sometimes the only area required. Many contests need no equipment. In contests requiring equipment, the need is for standard items such as logs, balls, nets, goals, and similar types of equipment. Specific requirements for area and equipment are listed with each contest description.

b. Progression. Team contests of a strenuous nature should be introduced after a basic period of conditioning has been completed and individuals are in the slow improvement stage of conditioning. Progression can take place from the less active to the more vigorous contests, and then from the non-contact to the contact or combative-type contests.

c. Necessity for Preconditioning Individuals must undergo conditioning prior to participation in athletics. Muscles, organs, joints, and ligaments not accustomed to stress and strain from sudden stops and starts, falls, body contact, rapid turns, prolonged running, and other rigors of athletic competition are subject to injury. Although athletics should not be introduced until players are physically prepared, there is still opportunity to engage in competition through lead-up contests. Individuals learn many of the skills required for athletics while participating in team contests.

d. Benefits. Athletics are beneficial primarily to sustain interest in the program and to maintain an achieved level of physical fitness. Athletics are a supplement, not a substitute for other types of conditioning activities which should continue. All of the desirable traits of physical fitness cannot be developed through athletics, yet their contribution is significant. For athletics to make a proper contribution to physical conditioning, the selected sports must bevigorous. The team contests recommended here are carefully selected for their simplicity, aggressiveness and applicability to squad- teaching teamwork. They are generally strenuous and involve the whole team throughout the entire duration of the contest.

5402. TEAM CONTESTS

a. Pushball

(1) **Players.** 10 to 50 players on a side. (See fig. 5-8.)

(2) Equipment A large pushball 5 to 6 feet in diameter.

(3) Area. A field 240 to 300 feet long, 120 to 150 feet wide. Mark a center line in the middle of the field parallel to the end lines. Mark a line 45 feet on either side of this center line and parallel to it, extending it across the width of the field. Mark another parallel line 15 feet from each end line, extending it across the width of the field.

(4) Game. Four 10-minute quarters are played. Give 2-minute rests between quarters and 5-minute rests between halves. The object of the game is to propel the ball over the opponent's goal line by pushing, rolling, passing, carrying, or any other way except kicking.

(a) The ball is placed on the center line with the opposing captains 3 feet from the ball. The rest of the players are 45 feet from the ball, on their half of the field. On the referee's starting whistle, the captains immediately play the ball with their respective teams coming to their assistance.

(b) At quarter time, the ball remains dead for 2 minutes at the spot where it was when the quarter ended. At half time, the teams exchange goals. The play is then started as it was in the beginning.

Figure 5-8 is available in the illustration section at the end of the book.

(c) Players may use any means of interfering with an opponent's progress except striking and clipping. Clipping is throwing the body across the back of an opponent's leg, while the opponent is running or standing. Legal use of force may be applied to all opponents whether or not they are playing the ball. For striking an opponent, the offender is removed from the game. The team penalty is half the distance to their goal. The penalty for clipping is the same.

(d) When any part of the ball goes out of bounds, it is a dead ball. The teams line up at right angles to the side lines and 3 feet apart at the point where the ball went out. The referee then tosses the ball between the teams.

(e) When, for any reason, the ball remains in one spot for more than 10 seconds, the referee declares the ball dead. The ball is then put into play as it is for an out-of-bounds situation.

(5) Scoring. A goal is scored when the ball, or any part of it, is propelled across the opponent's end line. A goal counts 5 points. The team scoring a goal has the privilege of trying for a point after the goal. To try for this extra point, the ball is placed on the opponent's 5-yard (or 15-foot) line. The teams line up on either side of this line separated by the width of the ball. Before the whistle blows, one player on each team may place hands on the ball. On the referee's signal, the ball is put into play for 1 minute. If any part of the ball is driven across the goal line in this 1-minute period, the offensive team scores 1 point. The defending team may not score during the opponent's try for the extra point.

b. Line Rush

(1) Players. Any number up to 50 on each side. (See fig. 5-9.)

(2) Equipment. None.

(3) Area. A field, 75 by 100 feet.

(4) Game. One team lines up behind one goal line and the other in midfield. On the starting signal, the team standing behind the goal line seeks to cross the goal line at the opposite end of the field within 30 seconds. The team in the center seeks to prevent it by catching and holding the runners. At the end of 30 seconds, the teams change.

(5) Scoring. Count the number of individuals who have crossed the far goal at the end of 30 seconds. After each team has had from three to five tries, the scores are added and the winner declared. A player scores 1 point when any part of the body is across the goal line.

Figure 5-9 is available in the illustration section at the end of the book.

c. Human Tug-of-War

(1) Players. 10 to 20 on a team.(See fig. 5-10.)

(2) Equipment. None.

(3) Area.40 to 60 feet.

(4) Game. Draw a line in the center of the area. Divide the players into two equal teams. Place them in single file on opposite sides of the center line facing each other. Each individual places the arms around the waist of the teammate in front. The two leaders of the opposing teams grasp each other around the waist. On signal, the teams try to pull each other over the center line within 30 seconds.

(5) Scoring. The team pulled across the center line loses. If neither team is pulled over the center line, but one team breaks its file, that team loses the match.

(6) Variation. Use a 3/4- or 1-inch rope. Space the leading players on each team 10 feet apart, each holding one end of the rope. The team pulled across the center line loses.

Figure 5-10 is available in the illustration section at the end of the book.

d. Master of the Ring

(1) Players. Any number. (See fig. 5-11.)

(2) Equipment. None.

(3) Area. A clearly marked circle large enough to contain all the players.

(4) Game. All the players stand inside the circle. At the signal, all players attempt to throw each other out of the circle. All tactics are fair except unnecessary roughness. When any part of the body touches across the line, the player is out and leaves the circle at once. Several officials are needed to spot the players who cross the line.

(5) Scoring The player who remains in the circle when all the others are out is the master of the ring.

(6) Variation 1. The players are divided into two equal teams. Each team is clearly marked. On signal, each team tries to throw the opponents out of the circle. The winning team is the team that eliminates all opponents from the circle.

(7) Variation 2. The players are divided into two equal teams. Each team sends only one individual into the circle. When one person has been forced out of the circle, the losing side sends in only one player. The team which eliminates all opponents is the winner.

(8) Variation 3.A pit, approximately 4-feet deep, is used rather than a circle on level ground. The contest may then be an individual or team activity.

Figure 5-11 is available in the illustration section at the end of the book.

e. Log Pivot Circle. Space teams far enough apart so each team can pivot in a circle without colliding with any other team. Each team holds a log in the bend of the arms in front of the chest. (See fig.5-12.) At the command **CIRCLE RIGHT, MOVE**,the left-flank individual holds the pivot while the log is carried around 360 degrees in a counter- clockwise motion, back to the original position. This movement may also be performed to the left by facing in the other direction and pivoting clockwise(**CIRCLE RIGHT, MOVE**).Other commands may be used such as: **CIRCLE HALF-RIGHT, CIRCLE HALF- LEFT** and so on. The first team to complete the prescribed movement is the winner.

Figure 5-12 is available in the illustration section at the end of the book.

f. Log-Rolling Race. Each team tries to roll its log a measured distance by pushing it with the hands and driving the body forward with the legs. (See fig. 5-13.)The first team to get the entire length of the log across the finish line wins.

Figure 5-13 is available in the illustration section at the end of the book.

g. Prone Pushing Contest.. Two teams lie prone, facing each other with a log between them. (See fig. 5-14.)Both teams place their hands against the log, keeping their arms straight. Then, by driving with the legs, each team attempts to push the other a measured distance to the rear.

Figure 5-14 is available in the illustration section at the end of the book.

h. Shuttle Relay Race. Each team in this race divides into two parts, A and B. Members of part A run 50 yards with a log held under their right arms. (See fig 5-15.)At the distance line, they give the log to their teammates in part B who bring it back to the starting line. The team pair finishing first is the winner.

Figure 5-15 is available in the illustration section at the end of the book.

5403.CROSS-COUNTRY AND DISTANCE RUNNING

a. Value of Running. Long-distance running gives some benefits that cannot be obtained in the same degree from any other sport. It builds powerful leg muscles, increases lung capacity, and develops endurance. For these reasons, cross-country and distance running should be included in the physical training program. These sports require only a few miles of open space which is generally available. Short cross-country runs and middle-distance runs can be used to supplement other activities, particularly team sports or the sports that develop precision or agility rather than endurance. Short cross-country runs can be scheduled once a week, gradually increasing the distance as the physical condition of the runners improves.

b. Cross-Country Runs. A cross-country run is a distance run held on a course laid out along roads, across fields, over hills, through woods, and on any irregular ground. A flat cinder or dirt track is not a suitable surface for cross-country running. Some runs are as long as 6 miles while others are as short as 3 or 4 miles. If cross-country running is being used to supplement other activities, the 3-mile course is long enough for most. Cross-country runs should be used only after individuals reach the sustaining stage of conditioning. This type of running should then be scheduled occasionally to provide variety. Cross-country running has the advantage of allowing mass participation. Interest can be stimulated by putting the runs on a competitive basis.

c. Practice Methods. Conditioning is essential to distance and cross-country running. Championship distance running depends on stamina, which can be developed only through constant training. An individual of only average ability can become an outstanding distance runner by steady and careful training. Hiking is an excellent method for getting into condition before the season opens. Long walks build up leg muscles. During the first month of the season, training should be gradual, starting with short distances, and increasing day by day. At first the legs will become stiff, but the stiffness gradually disappears if running is practiced for a while every day. To prevent strain, it is essential to limber up thoroughly each day before running.

(1) In training a large group, leaders should be stationed at the head and the rear of the column. They should make every effort to keep the runners together. After determining the abilities of the unit in cross-country running, it is advisable to divide them into three groups. The poorest-conditioned group is started first, the best-conditioned group, last. The starting time of the groups should be staggered so that all of them come in at about the same time.

(2) In preliminary training, running is similar to ordinary road work in that it begins with rather slow jogging, alternating with walking. The speed and distance of the run is gradually increased. As conditioning improves, occasional sprints may be introduced. Atfirst, the distance run is from one-half to 1 mile. It gradually increases to 2 or 3 miles. On completing runs, require runners to walk for 3 or 4 minutes before stopping to permit a gradual cooling off.

d. Facilities and Equipment. There should be at least one timer with a stopwatch (preferably three) for timing the runners. A course 3 or 5 miles long should be measured and marked by one of the three following methods:

(1) Fasten directional arrows to the top of tall posts and place them at every point where the course turns. Such signs should also be placed at every other point where there may be doubt as to the direction of travel.

(2) Place a lime line on the ground over the entire course.

(3) Place flags, clearly visible to the runners.

(a) A red flag indicates a left turn.

(b) A white flag indicates a right turn.

(c) A blue flag indicates the course is straight ahead.

e. Rules/Scoring. A cross-country team consists of seven individuals, unless otherwise agreed. In dual meets, a maximum of 12 on one team may be entered, but 7 or fewer are scored. First place scores 1 point, second place 2, third place 3, and so on. All who finish the course are ranked and tallied in this manner. The team score is then determined by totaling the points scored by the first five of each team to finish. The team scoring the least number of points is the winner. If fewer than five (or the number determined prior to the race) finish, the places of all members of that team are disregarded. If two or more teams score the same number of points, the event is called a tie.

5404.TEAM ATHLETICS

Team athletics add variety to the combat readiness program. At no time should they be allowed to replace intramural activities or to provide practice time for intramural or base teams. If team athletics are integrated into the program, the goal should be total participation and competition. Therefore, the games selected can be played by the skilled or the unskilled. They require limited equipment, yet will teach competitiveness, aggressiveness, and teamwork while they improve conditioning.

a.Soccer. Soccer is one of the best athletic activities for developing endurance, agility, leg strength, and great skill fulness in using the legs. The most popular sport in the world, soccer is the national game of many European, Asian, and Central and South American countries. In recent years, it has become popular in the United States. A soccer ball is the only equipment needed for the game, and Marines can learn to play it easily. Players do not need much skill to participate, but the amount of skill they can develop is almost unlimited.

(1) Place in the Program. Soccer should be introduced during the latter part of the slow improvement stage of physical training. It can be used as a competitive activity in the sustaining stage. It is primarily a spring or fall sport. Any level field is suitable for competition. The boundaries for the soccer field are similar to the dimensions for a football field. (See fig. 5-16.) Goal posts are essential, but they are easily constructed and usually temporary so that they may be removed when not in use.

Figure 5-16 is available in the illustration section at the end of the book.

(2) Basic Skills

(a) Passing. Passing with the feet is the primary means of moving the ball. Short passes are easier to control. Emphasis should be continually placed on the skill of passing.

(b) Dribbling. The ball is dribbled by a series of kicks with the inside or outside of the foot. Do not kick with the toe. Keep the head over the ball when kicking and propel it only a short distance at a time. Keep it close to the feet. If a player lets a ball move far from the feet, an opposing player can easily take it away.

(c) Instep Kicking. The instep kick is the basic soccer kick. The toe does not come in contact with the ball. The toe is pointed downward, and the instep (the area under the shoe laces) is applied to the ball with a vigorous snap from the knee.

(d) Inside-of-the-Foot Kicking. The ball is kicked with the inside of the foot, and the leg is swung from the hip. The toe is turned outward, and the sole of the foot is parallel with the ground as the foot strikes the ball. This kick is used for short passes and for dribbling.

(e) Foot Trapping. The foot trap is the method of stopping the ball by trapping it between the ground and the foot. Place the sole of the foot on top of the ball at the instant it touches the ground, keeping the foot relaxed. This is an effective way to stop a fast-moving ball.

(f) Shin Trapping. .The shin trap is a method of stopping the ball with the shins. Stand just forward of the spot where the ball should strike the ground and allow it to strike the shins in flight or on the bounce. Use either one or both legs from the knee down so as to absorb the bounce.

(g) Body Trapping. .Intercept the ball with any part of the upper body except the arms and hands. Keep the body relaxed and inclined toward the ball. To keep the ball from bouncing, move backwards from it as it strikes the body. This drops the ball at the feet in position for dribbling or passing.

(h) Heading. Heading is a technique for changing the direction of the flight of a ball by butting it with the head. Tense the neck muscles and jump up to meet the ball. Butt the ball with the forehead at about the hairline to reverse its direction. Use the side of the head to deflect it to the side.

(3) Offensive and Defensive Positions. Forwards usually play on the offensive half of the field. Fullbacks usually play on the defensive half of the field. Halfbacks are the backbone of the team. They move forward on the offense and back on defense. The goalkeeper almost always remains within a few feet of the goal.

(4) Abridged Rules

(a) A soccer team is composed of 11 players. (See fig. 5-16.)

(b) The player propels the ball by kicking it with the feet or any part of the legs, by butting it with the head, and by hitting it with any portion of the body except the arms or hands.

(c) The goalkeeper is the only individual allowed to place hands on the ball, but only in the goalkeeper's area. The term "hands" includes the whole arm from the point of the shoulder down.

(d) A goal is made by causing the ball to cross completely the section of the goal line lying between the uprights and under the cross bar.

(e) Each goal scores 1 point for the team scoring the goal.

(f) The penalty for a foul committed anywhere on the playing field (except by the defensive team in its penalty area) is a free kick awarded to the opposing team.

(g) All opponents must be at least 10 yards from the ball when a free kick is taken.

(h) The penalty for a foul committed by the defensive team in its penalty area is a penalty kick.

(i) A penalty kick is a free kick at the goal from the spot 12 yards directly in front of the goal. The only players allowed within the penalty area at the time of the kick are the kicker and the defending goalkeeper.

(j) An official game consists of two 30-minute halves,

(k) Teams change goals at the end of every quarter.

(l) In the event of a tie, an extra quarter may played. This may be followed by a sudden death period. If the tie still exists, the game can be decided by penalty kick.

(m) After a team has propelled the ball across a side line and out of play, the ball is put back into play by a throw-in from the side line by a member of the opposing team. The ball is thrown in from the point where it crossed the side line as it went out of bounds. The ball must be thrown in with both hands using an over-the-head motion, keeping both feet on the ground.

(n) When the offensive team propels the ball across the defensive team's goal line but not in the goal, the defensive team is awarded a goal kick-a free kick taken from within its own goal area. The ball must come out of the penalty area to be in play.

(o) When the defensive team causes the ball to go behind its own goal line, excluding the portion between the goal posts, the offensive team is awarded a corner kick—a free kick taken by a member of the offensive team at the quarter circle, at the cornerflag post nearest to where the ball went behind the goal line. The flag post must not be removed.

(p) The game is started and, after a goal has been scored, is resumed by placing the ball in the center of the midfield line. Players must be on their side of the line until the ball is kicked. The ball must be kicked forward and must move at least 2 feet to be legal. The first kicker may not touch the ball twice in succession at the kick off. The opposing team must be 10 yards from the ball until it moves.

b.Speedball. Speedball offers vigorous and varied action with plenty of scoring opportunities. It is easy to learn and provides spontaneous fun. Little equipment is needed-a ball is all that is absolutely necessary. Speedball combines the kicking, trapping, and intercepting elements of soccer; the passing game of basketball; and the punting, drop kicking, and scoring pass of football.

(1) Place in the Program.Speedball, like soccer, should be introduced into the physical training program during the latter part of the toughening stage and used as a competitive activity in the sustaining stage. Speedball may be played any time weather permits but is primarily a spring or fall activity.

(2) Method of Play.The game is played as follows:

(a) Two teams of 11 each play the game under official rules, but any number of players may successfully constitute a team.An inflated leather ball, usually a soccer ball, is used. The playing field is a football field with a football goal post at each end. (See fig. 5-17.)

(b) The game starts with a soccer-type kickoff. The kicking team tries to retain possession of the ball and advance it toward the opposite goal by passing or kicking it. Running with the ball is not allowed, with the result that there is no tackling or interference. After the ball touches the ground, it cannot be picked up with the hands or caught on the bounce but must be played as in soccer until it is raised into the air directly from a kick. Then the hands may be used.

(c) When the ball goes out of bounds over the sidelines, it is given to a player of the opposing team and is put into play with a basketball throw- in. When it goes over the end line without a score, it is given to a player of the opposing team who may either pass or kick it onto the field.

(d) When two opposing players are contesting possession of a held ball, the official tosses the ball up between them as in basketball.

(e) Points are scored by kicking the ball under the crossbar of the goal post, drop kicking the ball over the crossbar, or completing a forward pass into the end zone for a touchdown.

Figure 5-17 is available in the illustration section at the end of the book.

(3) Basic Skills

(a) Shared Skills. Skills include kicking, passing, heading, and trapping (from soccer); punting, dropkicking and forward passing (from football); and passing, receiving, and pivoting (from basketball).

(b) Kickups. The kickup is a play in which a player lifts the ball into the air with the feet so as to legally play the ball with the hands. The kickup generally makes the transition from ground play to aerial play. The technique of making the play depends upon whether the ball is rolling or stationary. If the ball is rolling or bouncing toward the player, the foot is held on the ground with the toe down until the ball rolls onto the foot. Then the foot is raised, projecting the ball upward. If the ball is stationary, the player rolls it backward with one foot. Then the player places a foot where the ball will roll onto it. The player then raises the foot, projecting the ball as before. If a ball is rolling away, the player should stop it with a foot and play it as a stationary ball. The player can also raise the ball by standing over it with a foot on either side. The player presses the feet against the ball and jumps into the air, propelling the ball into the hands.

(4) Offensive and Defensive Play. The positions of players in speedball are much the same as in soccer. However, some of the positions have different names. Each team has 11 players. The forward line is composed of five players, the right end, right forward, center, left forward, and left end. The second line consists of right halfback, fullback, and left halfback. In the next line is the right guard and left guard. The player who defends the goal is the goal guard. The strategy employed in speedball during offensive play is very similar to that of soccer. There are two types of defensive formations in speedball: man-for-man and position defense. Man-for-man defense is recommended for beginning players.

(5) Abridged Rules

(a) Field. A speedball field is 100 yards long with additional 10-yard end zones on each end and 55 yards wide, the same as a regulation football field.

(b) Players. Eleven players are on a team. The goal guard has no special privileges.

(c) Time. Ten-minute quarters. Two minutes between quarters. Ten minutes between halves. Five minutes for extra overtime periods.(Begin the first overtime by a jump ball at center, same goals; change goals in the event of a second overtime period.)

(d) Winner of Toss. The winner of the toss has the choice of kicking, receiving, or defending a specific goal.

(e) Kickoff. A kick off starts the first and second half of a game and starts play after a score. The kick off is made from any place on the 50-yard line. The kicking team must be behind the ball when it is kicked. The receiving team must stay back of its restraining line (10 yards distance) until the ball is kicked. The ball must go toward the opposing team's goal before it can be played. Kick off out of bounds goes to the opponent at that spot. A kick off touched by the receiving team before going out of bounds, with no impetus added, still belongs to the receiving team. A kick off, in possession and control of the receiving team and then fumbled out of bounds, belongs to the kicking team at that spot.

(f) Starting Second and Fourth Quarters. The ball is given to the team that had possession at the end of the previous quarter, from out of bounds, as in basketball.

(g) Second Half. The first-half receiving team kicks off at the beginning of the second half.

(h) Fly and Ground Ball. The most characteristic playing rule of speedball is the difference between a fly ball (or aerial ball) and a ground ball. A player is not permitted to touch a ground ball with the hands and must play it as in soccer. A fly ball is one that has risen into the air directly from the foot of a player (for example, a punt, drop kick, place kick, or kickup).Such a ball may be caught with the hands provided the catch is made before the ball strikes the ground again.

(i) Kickup and Overhead Dribble. A kickup is a ball that is so kicked by a player that he can catch it himself. A bounce from the ground may not be touched with the hand because it has touched the ground since being kicked. This rule prohibits the ordinary basketball dribble, but one overhead dribble (throwing the ball into the air and advancing to catch it before it hits the ground) is permitted.

(j) Out of Bounds. If a team causes the ball to go out of bounds over the sidelines, a free throw-in (any style) is given to the opposing team. When the ball goes over the end line without scoring, it is given to the opponents who may pass or kick from out of bounds at that point.

(k) Tie Ball. In case two players are contesting the possession of a held ball, even in the end zone, a tie ball is declared. The ball is tossed up between them.

(6) Scoring

(a) Field Goal(3 points).A soccer-type kick, in which a ground ball is kicked under the crossbar and between the goal post from the field of play or end zone, is a field goal. (A punt going straight through is not a field goal for it is not a ground ball. The ball must hit the ground first.) A drop kick from the field of play that goes under a crossbar does not count as a field goal. A drop kick from the end zone that goes under the crossbar counts as a field goal;if it goes over the crossbar, it is ruled as a touch back, and the other team takes possession of the ball.

(b) Dropkick(2 points). A scoring drop kick must be made from the field of play and go over the crossbar and between the uprights. The ball must hit the ground before it is kicked (usually with the instep).

(c) End goal(1 point). This is a ground ball which receives its impetus (kicked or legally propelled by the body) from any player, offensive or defensive, in the end zone and which passes over the end line but not between the goal posts.

(d) Penalty Kick(1 point).A ball kicked from the penalty mark that goes between the goal posts and under the crossbar is a penalty kick. The penalty mark is placed directly in front of the goal at the center of the goal line.

(e) Touchdown(1 point). A touchdown is a forward pass from the field of play completed in the end zone. The player must be entirely in the end zone. If the player is on the goal line or has one foot in the field of play and the other in the end zone, the ball is declared out of bounds. If a forward pass is missed, the ball continues in play, but it must be returned to the field of play before another forward pass or drop kick may be made.

(7) Substitutions. Substitutions may be made any time when the ball is not in play. A player who is withdrawn may not return during that same period.

(8) Timeout. Three legal timeouts of 2 minutes each are permitted each team during the game.

(9) Fouls

(a) Personal Foul. Kicking, tripping, charging, pushing, holding, blocking, and unnecessary roughness of any kind, such as running into an opponent from behind, are personal fouls. Kicking at a fly ball and thereby kicking an opponent is a personal foul. Four personal fouls disqualify a player from the game.

(b) Technical Foul. Illegal substitution, more than three timeouts in a game, unsportsmanlike conduct, and unnecessarily delaying the game are technical fouls.

(c) Violation. Traveling with the ball, touching a ground ball with the hands or arms, double overhead dribble, violating tie ball, and kicking or kneeing a fly ball before catching it are violations.

(d) Penalties. The offended player shall attempt the kick.

(e) Officiation of Fouls

1 Fouls in the field of play allow no follow-up while fouls in the end zone always allow follow-up.

2 On penalty kicks, with no follow-up, only the kicker and goalie are involved.

3 On penalty kicks, with a follow-up, the kicking side is behind the ball and the defending side behind the end line or in the field of play. No one is allowed in the end zone or between the goal posts except the goal guard. The kicker must make an actual attempt at goal and cannot play the ball again until after another player plays it.

c. Volleyball. Volleyball is a popular sport. The game entails much physical activity, yet it is not strenuous. It is, therefore, a game for young and old alike, for beginners and for skilled players. It may be played indoors or outdoors on any type of terrain. As an informal activity, volleyball can be played by any number; as an organized activity, it provides, as few other sports do, a game for 12 to play in a limited area. While volleyball requires no great skill to play, it permits a high degree of proficiency. Players naturally get more enjoyment when they know the game and play it well. For this reason, instruction in the basic skills should be provided.

(1) Place in the Program Volleyball may be used occasionally as a competitive activity during the sustaining stage. Though it is a year- round sport, it should be included in the physical training program only when a more strenuous activity is impractical. Volleyball is an excellent off-duty activity.

(2) Instruction. Usually during the first three or four classes, a 10- to 15- minute period of instruction, followed by scrimmage, is enough to teach the basic skills, rules, and techniques of volleyball. More time, if available, can be given to teaching basic skills, but emphasize competitive play rather than formal instruction. Divide the unit into 6-person teams, organized on the basis of ability .Teams should be as nearly equal as possible.

(3) Basic Skills

(a) Passing the High Ball. The chest pass is the most effective method of playing the ball. To receive the ball, the knees are fixed, and the body is tilted forward. The elbows are raised sideward to a point in line with the shoulders. The wrists are extended in line with the forearm. The arms, wrists, and hands are rotated inward. To pass the ball, the hands are chest-high, thumbs pointing inward. The fingers are flexed, forming a cup, allowing them to contact the ball. On contact with the ball, the wrists are snapped while the fingers and elbows are pushed upward, sending the ball upward. A high ball is much easier to handle than a low one.

(b) Passing the Low Ball. A ball that is lower than the waist is one of the easiest to hit, but it is also a frequent cause of fouls (holding or carrying the ball). The best position for handling a low ball is to have the knees flexed and arms flexed at the elbows and rotated so the thumbs are pointing outward, the palms up. When the fingers contact the ball, the entire body reacts in a lifting motion. The arms and hands swing upward in a scooping action. It is important that the fingers, not the palms, contact the ball, and that the ball is batted, not thrown.

(c) Underhand Serve. Take a position behind the back line facing the net, left foot forward, holding the ball in the palm of the left hand. The left knee is flexed; the right knee is straight. Swing the right arm back and at the same time move the left hand (holding the ball) across the body in line with the right hip. Then swing the right arm forward hitting the ball off of the left hand with the palm of the right hand, raising the hips and arching the back in the same motion. Be certain to swing the right arm in a straight line, or the ball will be difficult to control. When the opposition is in formation, the server should try to place the ball in the right or left back area, and not near the net.

(d) Setup. A setup is a ball hit into the air near the net by one player, so that a teammate may hit or "spike" it sharply downward into the opponent's court. The chest pass is the best pass to use. The ball is sent approximately 10 feet into the air toward the spiker so it will descend from 4 to 20 inches from the net.

(e) Spiking. The spike is a leap into the air and a sharp downward hitting of the ball into the opponent's court. A spiker must be able to spring easily from the floor, judge the movement of the ball, and strike it with a downward movement of the arm. To jump from the floor, step off with one foot and jump with the other. Stand with the right or left side to the net, facing the setup player. Much depends upon the setup player to place the ball in the proper position. The spiker jumps into the air and strikes the ball above its center so as to drive it downward. A snapping movement of the arm and wrist will drive the ball forward and downward with power and control. Aim for a weak spot in the opponent's defense.

(f) Blocking. The block is a defense technique used to prevent a spiker from driving the ball across the net. It is an attempt by one or more defensive players at the net to block a hard-hit shot by using the force of the ball to send it immediately back into the opponent's court. In an effective block, forwards on the defensive team spring into the air at the time of the spike, placing both hands and arms in the expected path of the ball. An effective block tends to upset the offense and presents another element for the spiker to worry about. To be effective, the blocker must anticipate the path of the ball and time the block with the spike.

(4) Abridged Rules

(a) The volleyball court is 30 feet wide by 60 feet long. (See fig. 5-18.)

(b) The top of the net is 8 feet high.

(c) A volleyball team consists of 6 players.

(d) A match consists of the best two out of three games.

(e) The first team scoring 15 points wins the game, provided that they have 2 points more than their opponents.

(f) A deuce game is a game in which both teams score 14 points. The game is continued until one team obtains a 2-point advantage over the other.

(g) Only the serving team can score. If the serving team commits a fault, it loses the serve to the opposing team.

(h) The team members who receive the served ball rotate one position in a clockwise direction.

(i) The ball is put into play by serving from behind the back line.

(j) A served ball touching the net results in the loss of the serve. At any other time during play, a ball touching the net is still in play.

Figure 5-18 is available in the illustration section at the end of the book.

(k) The ball is out of play when it touches the ground or goes outside one of the boundary lines.

(l) All line balls are good.

(m) The players must hit or bat the ball; they may not throw, lift, or grasp it.

(n) A player may not touch the ball with any part of the body below the knees.

(o) A player may not play the ball twice in succession. In receiving a hard-driven spike, a defensive player may make several contacts with the ball even if they are not simultaneous. All such contacts, however, must constitute one continuous play, and all must be above the knees.

(p) The ball may be touched no more than three times on one side of the net before being returned across the net to the opposing team.

(q) A player must not touch or reach across the net.

(r) A player may touch the line under the net, but may not cross it.

Chapter 6
EVALUATION OF PERFORMANCE DURING TRAINING

6001.OBJECTIVES OF COMBAT READINESS

The objectives of the Marine Corps physical fitness program are-

a. To contribute to the health and well-being of every Marine through regular exercise and health education.

b. To develop Marines who are physically capable of performing their duties in garrison and in combat.

c. To develop Marines with a reserve level of physical fitness that will help them win in combat.

d. To provide a medium for developing the self-confidence of the individual Marine and thereby enhancing overall discipline, morale, esprit de corps, unit efficiency, and the desire to excel within the Marine Corps.

6002. THE PHYSICAL FITNESS TEST

The physical fitness test (PFT) has distracted attention from the proper goal of the physical fitness program. Simply stated, the goal of the physical fitness program is the success of Marines in combat. Unit commanders must not allow the PFT to become an end in itself. The PFT should be used within the commander's program of combat readiness training as a standard, an index of the physical fitness of individual Marines at a given time and place. The PFT is not difficult, and it must be kept in perspective. All athletes should be first class, and first class is attainable for all Marines. The conduct of the PFT is described in Marine Corps Order 6100.3_.

6003. OTHER METHODS OF EVALUATION

The commander should use other ways of evaluation to provide variety and imagination. Intangible objectives, such as confidence and aggressiveness, are hard to measure. The use of either inspection or observation is necessary to evaluate objectives of this nature. The commander has several other methods available when evaluating the physical condition of the command. The additional methods of inspection, observation, medical examination, and testing are summarized as follows:

a. Formal Inspections. Formal inspections, using inspection officers and standardized rating criteria, may help evaluate unit physical fitness.

b. Daily Observation. Routine observation of physical performance and appearance can serve as an indicator of an individual's or unit's physical readiness. However, mere observation is not a totally reliable or accurate means of evaluation.

c. **Physical Examination.** Medical examination may detect any individual disability or detrimental physical condition. It may guide in application of remedial, therapeutic, or limited exercise programs.

d. Other Physical Fitness Tests. Some other tests are described in detail in this manual for information purposes. Commanders are cautioned, however, not to place too great an emphasis on testing.

6004. RESPONSIBILITY

The commander is responsible for the physical fitness of the command and for the measurement and evaluation of its physical readiness. The goal of combat readiness training remains to prepare Marines and units for the physical demands of combat. The bulk of the commander's combat readiness training program should be such activities as obstacle and confidence courses, endurance courses, and progressive load-bearing marches which will prepare Marines for combat. Group activities which are competitive and combative in nature are advantageous and add imagination and enthusiasm to the program. Activities such as orienteering, which is not only physically demanding but also competitive and which teaches essential subject skills, are excellent and motivate all concerned. A commander's goal should be to provide an integrated training program that is so aggressive and demanding of Marines and leaders alike that a PFT administered at any time will reflect that each and every Marine is both physically and mentally prepared for the demands of combat.

6005. COMMANDER'S PHYSICAL FITNESS PROGRAM

The Commandant of the Marine Corps has directed that the physical fitness test shall be the universal measure of adequate individual physical fitness. The focus of a commander's physical fitness training program, however, should not be to prepare Marines to pass the physical fitness test, but to prepare them for the physical rigors of combat.

a. Administration. The physical fitness program requires each Marine to spend a minimum of 3 hours per week in physical fitness training, to be tested semiannually, and to obtain a minimum standard of third class. The attainment of a higher score is a laudable individual goal which should be encouraged but should not become a unit objective. Overemphasis could be detrimental to training required to develop the complete Marine. All commanders should exercise sound judgment as well as positive, aggressive leadership in striving toward the training objective.

b. Variety of Activities. The program must include the following elements:

(1) Physical Readiness Training. physical readiness training is a complete physical training program which develops and maintains the strength, endurance, and physical skills needed to sustain the individual during combat.

(2) Remedial Physical Conditioning. Remedial physical conditioning is a process by which physically substandard Marines are conditioned to meet prescribed standards.

(3) Competitive Conditioning Activities. Competitive conditioning activities consist of teams or individuals competing against an opponent to win. This involves a combination of sports and military skills designed to foster competitive attitudes and develop unit pride and esprit de corps.

(4) Occupational Conditioning. Marching long distances, dry net training, etc., should not be ignored as a means to fulfill the time requirement.

c.Time. In order to develop the desired level of physical fitness, unit commanders are advised that a minimum program of 3 hours per week is mandatory. The program must encompass at least three exercise periods weekly. Each period should include calisthenics, running, or other forms of vigorous activity .The minimum desirable length for each exercise period is 60 minutes. Shorter periods may be authorized by the unit commander if dictated by the local training situation. Unit commanders are encouraged to use the normal working day to satisfy this training requirement. However, commanders are authorized to conduct required physical fitness training during off-duty hours when the mission, workload, personnel status or other significant factors preclude fulfilling it during duty hours.

d.Action. Marine Corps Order 6100.3_ requires that commanders-
o Establish and maintain an effective physical fitness program which ensures that all Marines maintain an acceptable level of physical fitness.
o Ensure that all individuals are medically qualified prior to participating in the physical fitness program.
o Establish a minimum physical fitness program of at least 3 hours per week. Physical training may be authorized on an individual basis at the discretion of the unit commander.
o Ensure that all Marines participate in physical fitness conditioning activities commensurate with their medical qualifications and limitations.
o Conduct physical fitness testing for all Marines.
o Place all Marines who fail to pass the PFT on a daily command-supervised remedial physical conditioning program until they pass the test.
o Ensure that results of physical fitness testing are entered on section A of the fitness report. Comments are placed in section C in connection with attainment of superior physical performance or a medical excuse from the PFT.

6006. OTHER FORMAL TESTS

Commanders must realize that the best index of combat readiness is personal observation of performance on conditioning marches of 15 miles while under load. This manual presents three additional indicators of unit fitness. These indices are the physical readiness test, which was formerly the Marine Corps standard and is now routinely administered at Officer Candidate School; the airborne trainee physical fitness test which many Marines who attend U.S. Army formal schools are required to take; and the battle fitness test, the ultimate test of the combat readiness of Marines. Commanders may wish to include these tests or portions thereof within their combat readiness programs.

6007. PHYSICAL READINESS TEST

a. Purpose. The purpose of this test is to measure the Marine's ability to meet the minimum standard of physical readiness for duty in the field. This test is currently administered to all male officer candidates at Officer Candidate School, MCCDC, Quantico, VA. Female candidates do not take this test.

b. Test Area. Administration of this test requires an area suitable for crawling, jumping, and running 160 yards to include a trench that is 8 feet wide.

c. Equipment. The equipment required for this test includes—
o A platform, rail on a ledge, or a box 18 inches high,
o A 20-foot climbing rope,
o A stopwatch.

d. Conduct of Test. Adequate timers and other supervisory personnel must be provided by the company staff. All events should be conducted in a single session of one morning or afternoon. The events may be run in any sequence.

e. Uniform. The uniform for the physical readiness test is as follows: full utilities, cartridge belt, belt suspender straps, two canteens (full of water), rifle, and helmet. Depending on the heat condition, the PT shirt may be worn in place of the utility jacket. All equipment will remain secure throughout the test. Except for emergencies, canteens are not to be emptied.

f. Events

(1) Event 1: Climbing Uphill (Step-ups)This event simulates marching uphill at a rapid and steady pace. The Marine stands in front of the step. On the command GO, the Marine places either foot on the step and steps up. Hand pressure on the knee may be used if desired. The Marine then stands erect and steps backward and down, on one foot at a time. The same leg may be used for each repetition or the legs may be alternated. For 100 points, Marines will complete 60 repetitions in the maximum allotted time of 1 minute and 20 seconds (80 seconds).

*(2) Event 2: Rope Climb. T*his event resembles entering and leaving a hovering helicopter, using ropes in house-to-house fighting, and ascending and descending landing nets. The Marine assumes a sitting position on the ground at the bottom of the rope with hands at the highest point that can be reached. On the command GO, the Marine jumps up and climbs to the top. The Marine must touch the 20-foot mark with one hand, then descend without sliding or dropping. Marines must reach the 20-foot mark in the maximum allotted time of 30 seconds or less.

(3) Event 3: Evacuation. This event simulates reaching and evacuating a wounded Marine under fire. The Marine assumes the prone position at a distance of 50 yards from a casualty who is of approximately equal weight. On the command GO, the Marine springs up and covers the distance in a sprint. The Marine lifts the casualty to a fireman's carry and returns to the starting point. Assistance by the casualty is permitted. The tested individual must carry all equipment; the casualty has none. Marines must negotiate the course in a maximum time of 47 seconds or less.

(4) Event 4.'Advance by Fire and Maneuven. This event simulates advancing and assaulting during an attack. The Marine assumes the prone position at the starting point. On the command GO, the Marine then alternately crawls, runs in a zigzag manner, rolls, jumps, and assumes different firing positions for the next 130yards. Before reaching the objective, the Marine must "hit the deck," roll over, and assume the firing position three times. While assuming the firing position, the Marine must pull back the charging handle, take aim, and squeeze the trigger before rising from the ground. After rising from the ground for the third time,the Marine continues running and jumps to clear the 8-foot trench in one leap. The Marine then finishes the course in

the maximum allotted time of 2 minutes. The Marine is allowed multiple attempts only within the specified time limit.

(5) Event 5: Forced March. This event is a combat run of 3 miles without halts and against time. The Marine must double-time. Marines must complete the 3-mile course in a maximum allotted time of 30 minutes or less.

g. Event Standards

(1) To successfully pass the physical readiness test, the Marine must pass each event with a minimum of 80 points for each. Points are then assessed for each event in accordance with the scoring tables. (See fig. 6-1.)

(2) If a Marine fails one event, the highest possible score that the Marine may receive for the entire test is 79 percent. (See example 1 in fig. 6-2.) If two events are failed, the highest possible score for the entire test is 69 percent, regardless of the total points accumulated. (See example 2 in fig. 6-2.) If three events are failed, the highest possible score is 59 percent. Failure in four or more events will result in a score of zero for the entire test.

(3) To convert the total points for all five satisfactorily passed events to a final percentage score, total the accumulated points for all five events and divide by five. (See example 3 in fig. 6-2.)

Figure 6-1 is available in the illustration section at the end of the book.

Figure 6-2 is available in the illustration section at the end of the book.

6008. AIRBORNE TRAINEE PHYSICAL FITNESS TEST

a. Use and Composition of the Test.

(1) Use. The airborne trainee physical fitness test is a means of determining the physical ability of the applicant for acceptance to and retention in the airborne course of instruction.

(2) Test Events The test battery consists of five events as follows: chinups, knee bender, pushups, situps, and an endurance run.

b. Method of Scoring and Standards

(1) Scoring. The Marine will be scored by a trained scorer who is thoroughly familiar with the minimum standards for the test events. The Marine will be scored on a pass or fail basis. The performance on each event may be recorded on the reverse side of DA Form 705 (Physical Fitness Testing Record).The examinee records the personal information on the face of the card as specified on the first, second, and third lines. The face of the card is identified by writing diagonally across the lower half of the card "Airborne Trainee PPT—See reverse side."Using line 7 on the reverse side of the scorecard, the scorer enters the test title in the block entitled "(Other (Specify))" and completes the required information on the remainder of that line. The "Remarks" section can be used to record the test event titles and the applicant's performance on each test event.

(2) Standards. To successfully pass the test, the examinee must reach the standard in each test event. The standards follow:

(a) Chinups

(b) Knee bender 80(2-minute period)

(c) Pushups 45(2-minute period)

(d) Situps45(2-minute period)

(e) Endurance runs:2 miles in 15:54 minutes or less in athletic gear and 4 miles in 32 minutes in utilities and running shoes.

c. Uniform for Testing

(1) Examinees. The prescribed uniform for test participation is boots and the work uniform of the season. No headdress is worn. When climatic conditions permit, jackets or outer shirts may be removed.

(2) Officials. Scorers and other test officials should be uniformly and distinctively dressed for contrast with Marines being tested.

d. Test Administration Procedure

(1) Preparation for Test. The administration of the test to a large group makes it mandatory that the test effort be organized and efficiently operated. All testing is not completed with large test groups; on certain occasions, individuals and small groups are tested. Care must be exercised to administer the test uniformly and to standardize the conduct of all elements of the test. Regardless of the size of the test group, the following elements of sound test administration should be included:

(a) An orientation to include the purpose, method of administration, scoring of the test, preparation of the score card, and required standards.

(b) A correct demonstration of each event to ensure that there is no misunderstanding of the proper form and required standards.

(c) Completion of all five test events in one test period with all Marines taking the events in the same sequence.

(d) Adequate rest periods between the test events to allow for recovery before the next event.

(2) Method of Administration with a Large Group. With a minimum of 12 lanes per test event, 14 officials can administer the test battery to 150 or 200 Marines in 2 hours. The officials are designated as follows: one officer in charge, one demonstrator,and 12 scorers. If more or fewer Marines are to be tested, a greater or lesser number of officials will be required. The following procedure is recommended:

(a) Conduct an orientation and ensure the examinees have properly completed their score cards.

(b) Assign Marines to lanes and caution them to remain in the same lane order throughout the test.

(c) Explain and demonstrate the chinup event, administer it, and score it. Then proceed to the knee bender and pushup events, and administer them in the same manner.

(d) Grant a 5- to 10-minute rest period after the pushup event. Advise against excessive consumption of water during the break period.

(e) Explain and demonstrate the situp event, administer it, and score it as prescribed. Then move to the run area, explain the running event, and complete it.

(f) Retain the scorecards at the completion of the running event.

(3) Method of Administration with a Small Group. A similar procedure is followed for the testing of individuals and small groups. The informality usually associated with small groups must not conflict with sound test administration. With fewer examinees, fewer officials are required.

e. Description and Explanation of Test Events

(1) Test Event l:Chinups

(a) Purpose. This event is devised to test arm and shoulder flexor strength.

(b)Equipment There is one horizontal bar per lane, made of plumber's pipe or a gymnasium horizontal bar 1 1/2 inches in outside diameter. The bar must be rigidly supported at a height of 8 feet above the ground. The upright supports must be 5 feet apart. There must be a movable stand at each bar for short Marines to stand on to reach the bar.

(c) Officials There is one scorer per lane.

(d) Organization. The examinees, holding their scorecards, stand in order behind the restraining line in their respective lanes. The scorers take each scorecard when the Marine is called forward for the test.

(e) Starting Position The bar is grasped with the palms turned toward the face, the thumbs underneath the bar. The body is fully extended in a "dead" hanging position with the arms straight and the feet above the ground.

(f) Movement. Pull the body directly upward until the chin is placed over the bar. Lower the body until the elbows are completely straight and the body is again in the "dead" hanging position. Repeat as many times as required.

(g) Instructions. Explain and demonstrate the fully extended "dead" hanging position with the proper grasp. Show that the chin is placed over the bar at the top of the movement and that the arms are fully extended, the elbows completely straight, at the bottom of the movement (the hanging position). Explain that the body must be kept from swinging and that it is permissible to raise the legs and flex the hips when pulling up, but that any kicking, bicycling, or jerking motion with the trunk or legs is not acceptable. Inform the examinees that no penalty is exacted for hanging on the bar to rest in the bottom position but that this is not to their advantage. Tell them that half-completed chinups are not counted, and that the scorer will repeat the number of the last correct chinup when incorrect execution is detected.

(h) Administration and Scoring. Caution the examinees to assume the "dead" hanging position and wait for the scorer's command to begin. The scorer is at the examinee's left with a clear view of the bar. If the examinee begins to swing widely, the scorer should stop the swinging by extending the left arm across the front of the examinee's body, being sure not to hinder the execution of the chinup. The scorer counts aloud the number of chinups correctly executed. When a chinup is not correctly executed, the scorer repeats the number of the last correct one. The scorer records the number of correct chinups on the scorecard and returns the card to the examinee.

(2) Test Event 2:Knee Bender

(a) Purpose. This event measures the strength and endurance of the leg muscles.

(b) Equipment. None.

(c) Officials There is one scorer per lane.

(d) Organization. Marines stand in numerical order behind the restraining line in their respective lanes. The scorer takes each scorecard when the Marine is called forward for the test.

(e) Starting Position. The feet are spread less than shoulder width apart, hands on hips, thumbs in the small of the back, elbows back.

(f) Movement. Do a knee bend and at the same time bend slightly forward at the waist and thrust the arms between the legs until the extended fingers touch the ground. The hands are about 6 inches apart. The bend is approximately a three-quarters bend. From this knee bend position, recover to the starting position by moving the body upward, straightening the knees, and returning the hands to the hips. Repeat as many times as required.

(g) Instructions. Explain and demonstrate the correct starting position. Be certain examinees understand the correct knee bend and that only the tips of the fingers touch the ground. Tell them the scorer will repeat the number of the last correct knee bender when incorrect execution is detected. Some of the common errors are failure to correctly bend the knees, failure to touch the ground, and failure to assume the completely erect position after the bend has been executed.

(h) Administration and Scoring. The scorer stands to one side so as to see that the knees are properly bent and the fingers touch the ground as prescribed. From this position, the scorer can view the examinee to see that a properly erect position is assumed after each knee bend. The scorer counts aloud the number of correctly executed knee bends. When a knee bend is done incorrectly, the scorer repeats the number of the last correct one. The scorer records the number of correct knee bends on the scorecard and returns it to the examinee.

(3) Test Event 3:Pushups

(a) Purpose. Pushups measure arm and shoulder extensor strength.

(b) Equipment. None.

(c) Official. There is one scorer per lane.

(d) Organization. Marines stand behind the restraining line in their respective lanes until the scorer calls on them to perform. The scorer takes each scorecard when the Marine comes forward.

(e) Starting Position. The front-leaning rest position is the starting position. The body is straight from head to heels, palms are flat on the ground directly underneath the shoulders, and elbows are straight and locked. The body weight is supported on the hands and toes throughout the event.

(f) Movement. Bending only the elbows, lower the body in one straight plane until the chest touches the scorer's hand. Straightening and locking the elbows, raise the body in one straight plane, returning to the original front-leaning rest position. Repeat as many times as required, keeping the body in a straight line from head to heels.

(g) Instructions. Explain and demonstrate that the arms are straight at the beginning and completion of the movement and that the chest must touch the scorer's hand, but the stomach and thighs must not touch the ground. Also explain that the whole body must be maintained in a straight line as it is lowered and raised; that is, there is to be no breaking at the hips or shoulders so that any body part is lowered or raised in advance of the other or as a separate segment .Dipping or rolling through the shoulders is illegal, as is lowering or raising the body with one arm or shoulder at any time. Resting is not permitted during repetitions. Instruct

Marines that the scorer will repeat the number of the last correct pushup when incorrect execution is detected.

(h) Administration and Scoring. It is recommended that Marines assume a prone position while placing their feet and hands in the proper positions. This permits them to rest while the scorer gets into position and, at the same time, provides a feel of the body in a straight plane from head to heels. The scorer lies on the right hip and side to the right of the examinee. This gives the scorer a clear view of the examinee's body to see any errors. The palm of the scorer's right hand rests flat on the ground underneath the lowest part of the examinee's chest. By keeping the right forearm flat on the ground at an angle in front of the examinee's right arm, the scorer's position will not prevent the examinee from lowering the body completely. The scorer's left hand is free to test the straightening of the elbow at the completion of the movements and to point out body segments being lowered or raised separately. When in position and ready, the scorer has the examinee assume the starting position and begin doing pushups. The scorer counts aloud the repetitions done correctly and repeats the number of the last correct pushup if one is done incorrectly. There is no penalty if the contour of the examinee's body causes the hips to protrude slightly out of line, provided that the whole body is raised and lowered simultaneously. The scorer enters the number of repetitions on the scorecard and returns it to the examinee.

(4) Test Event 4: Situps

(a) Purpose. This event primarily measures abdominal strength.

(b) Equipment. None.

(c) Official. There is one scorer per lane.

(d) Organization Marines stand behind the restraining line in their respective lanes until the scorer calls on them to perform. The scorer takes each scorecard when the Marine comes forward.

(e) Starting Position. The examinee lies flat on the back with knees flexed, both feet flat on the ground. The correct angle of the thighs to the ground is 45 degrees. If the heels are too near the buttocks, the applicant will not be able to sit up. Legs are spread shoulder width apart. The examinee interlaces fingers and places them behind the head in contact with the ground. The feet are not held by another person.

(f) Movement. Bend forward at the waist and raise the upper body until the head is directly over the knees. Heels are not to leave the ground. Elbows remain in the same plane to the head and body throughout the event. The upper body is slowly lowered to the starting position until the head touches the ground. Repetitions are done at a slow cadence with no rest periods.

(g) Instructions. Explain and demonstrate the correct starting position and the proper execution of the situps to be sure that Marines understand the movement. Warn them that their knees must remain flexed during each situp, the heels cannot leave the ground at any time, and they may not roll up on one side and push up with one elbow. Tell them they must do the repetitions at a slow cadence with no rest periods. Instruct Marines that the scorer will repeat the number of the last correct situp when incorrect execution is detected.

(h) Administration and Scoring. When the performer is in position and ready, the scorer has the performer assume the starting position and begin doing situps. The scorer counts aloud the correct executions. When a situp is improperly done, the scorer repeats the number of the last correct one. No situp is credited if the hands are unclasped from behind the head, if the back is used to bounce up from the ground (which means the shoulders would not touch the ground), or if one shoulder or elbow is used to push up. The examinee is not penalized if the heels slide forward slightly so long as the knees remain flexed and the heels maintain contact with the ground. The scorer enters the number of repetitions on the scorecard and returns it to the examinee.

(5) Test Event 5:Endurance Run

(a) Purpose. This event measures cardiovascular endurance.

(b) Equipment One stopwatch or watch with a sweep second hand.

(c) Area. A large training field on which a quarter-mile track has been staked out or a level road over flat terrain may be used as a running surface. A 1-mile route is designated with wooden stakes marking the start point, finish point, and one-quarter mile intervals.

(d) Officials For large groups, a scorer times the event and controls the conduct of the run, and a guide runs with the group and sets the pace.

(e) Organization The run is conducted with groups of Marines in a column formation. Company-sized units may run at the same time with platoons serving as running groups. The scorer issues the command to assume double-time.

(f) Starting Position. Marines are assembled in proper column formation (column of twos, threes, or fours, as appropriate to the size of the group) with short men to the front. When all is ready, the column is moved forward a short distance before the running period is started.

(g) Movement. At the command DOUBLE-TIME, MARCH, the examinees retain their places in the column formation and execute the command. Length of steps is about 40 inches. The scorer has the group execute the run. The formation is maintained during therun.

(h) Instructions. Marines are instructed to maintain formation while running and are informed that the guide will set the proper pace. They will be instructed in the command used to control the column in the execution of the test. The scorer should announce the 4-minute, 2-minute, 1-minute, and 30-second remaining time intervals.

(i) Administration and Scoring. The event may be administered as previously prescribed to **a** large group, to several Marines, or to an individual. An individual examinee usually does not require a guide or pacer. If the event is administered on a training field, the scorer may stand in the center as the runner(s) circles about the field. This method of administration relieves the scorer of running with each group to be tested. Scoring is based on successful completion of the run as prescribed. The scorer should announce the remaining times as prescribed in (h).

6009. BATTLE FITNESS TEST

The battle fitness test is a physical fitness evaluation which is currently under development as part of the basic warrior training concept plan.

Chapter 7
THE HUMAN BODY

7001.GENERAL

To implement an effective physical conditioning program, a leader must understand how the human body functions and how exercise affects the body .This chapter provides only an introductory discussion of anatomy and functioning.

a. Body Functioning During the Stages of Conditioning. For more on the stages of conditioning, see chapter 1.

(1) During the toughening stage, the waste products of muscle activity (lactic acids) collect more rapidly than the blood can remove them. This acid waste builds up in the muscle tissue and irritates the nerves in the muscle fiber, causing pain and stiffness. As the exercise program continues, more blood is carried through the muscle, removing the waste materials more rapidly and eventually causing the soreness to disappear.

(2) During the slow improvement stage, the blood circulation in the muscles increases, and the body as a whole becomes a more efficient machine. The improvement is rapid in the first few weeks, but as a higher level of skill and conditioning is reached, the improvement becomes less noticeable. The body reaches its maximum level of performance after 6 to 10 weeks and should then be maintained at this peak.

(3) During the sustaining stage, it is possible to maintain this state of conditioning through 15 to 20 minutes of exercise a day, but the exercise must be quite strenuous.

b. Diet. Regular exercise increases the appetite. If the desire for greater amounts of food is satisfied by a balanced diet, the body benefits.

(1) There are two main types of foods: body-building and energy- producing.

(a) Body-building foods consist of proteins, which build up tissue and repair wear and tear.

(b) Energy-producing foods include carbohydrates and fats. Carbohydrates provide a quick source of energy, while fats act as a reserve store of energy.

(2) In addition, food contains vitamins, mineral salts, and water. During hot weather and strenuous training periods, the body requires greater fluid intake.

(3) Diet should be supplemented with proper rest to allow the digestive system to digest the lood.

(4) Occasionally, especially during early stages of conditioning, strenuous exercise may cause vomiting. Although vomiting is not a frequent occurrence, it is not usually a cause for concern.

7002. SYSTEMS OF THE BODY

The systems of the body include the skeletal, muscular, circulatory, respiratory, endocrine, digestive, genitourinary, and nervous systems. Each has a different function, but all must work in cooperation with one another to insure a sound body. Of these systems, the first four are the most affected by exercise and are discussed in succeeding paragraphs.

7003. THE SKELETON

a. Bones. The skeleton (see fig. 7-1) is composed of about 206 bones, which-

o Provide structure for the body and support for the attachment of muscles,

o Protect vital organs, such as the brain, lungs, and heart,

o Manufacture red blood cells, which carry oxygen through the body.

Figure 7-1 is available in the illustration section at the end of the book.

b. Joints. A joint is a place of union between two or more bones. Joints can be-

o Immovable, such as in the face and head.

o Slightly movable, such as in the vertebrae or sternum.

o Freely movable, such as in the shoulder, hip, knee, ankle, and elbow. The bones in a freely movable joint are held in place by ligaments. Freely movable joints are of the greatest importance in physical training because exercise improves their mobility and stability.

c. Cartilage and Ligaments. The joints in the body, where bones connect, are supported by cartilage and ligaments.

o Cartilage is a tough, elastic, translucent tissue that acts as a shock absorber or buffer between bones. Examples are the discs between the vertebrae, the tissue attaching the ribs to the sternum, and buffers in the knee joints. Damaged cartilage does not heal.

o Ligaments are connective tissue that bind bones together. They contract and extend but are not elastic. Consequently, ligaments can be torn or strained. Damaged ligaments will heal to some extent.

7004. THE MUSCLES

a. General. Muscles are tissue, or an organ composed of tissue, which contract or extend to produce movement. Muscles are of three classifications: involuntary, voluntary, and cardiac. The voluntary and cardiac muscles are the most affected by physical training; physical training has little effect on involuntary muscles.(Seefigs. 7-2, 7-3, 7-4.)

o People have no control over the involuntary muscles, an example of which is the diaphragm,

o Voluntary muscles are the larger skeletal muscles which cause movement and which the individual can control.

o Cardiac muscle is found only in the heart and for all practical purposes is an involuntary muscle. Physical training has a significant effect on the fitness of cardiac muscles.

b. Muscle Structure. Muscle is composed of cells which are grouped into bundles called fibers. Groups of fibers form muscles. These bundles of fibers are held in place by sarcolemma, a thin, sheath-like material that surrounds the muscle bundles and secretes a fluid that lubricates the muscle tissue. The fused ends of the sarcolemma form the tendons which attach the muscles to the bones.

c. Attachment of Muscles. The arrangement of muscles on the skeleton provides the proper angle of pull to make movement possible. Voluntary muscles are usually attached to the skeleton in two places.

d.Action of Muscles. To produce motion and to do work a muscle usually shortens its fibers. The movement may be flexion such as bending the arm at the elbow, or extension, such as the straightening the arm. Some muscles raise the arms or legs, others depress the raised limb. Some muscles have the primary function of rotating the trunk from side to side. In addition, muscles act as stabilizers as well as prime movers.

Figure 7-2 is available in the illustration section at the end of the book.

Figure 7-3 is available in the illustration section at the end of the book.

Figure 7-4 is available in the illustration section at the end of the book.

7005. FUNCTIONING OF THE SKELETON AND MUSCLES

a. General. The skeleton and muscles function in close coordination. Muscles move the body, and bones support the muscles.

b. Effect of Exercise on Bones. Continuous exercise, particularly among younger people, usually strengthens the bones, allowing them to withstand greater stress and strain. Bones which are not used lose much of their strength, a factor which should be considered when individuals return to physical training after a prolonged period of inactivity. People in this category should refrain from activities which might result in bone breakage before the bone is exercised back to normal condition. The condition known as "march fracture" is usually a result of this condition.

c. Muscular Strength. When exercised vigorously to improve strength, muscle grows in size. In general, the larger the muscle, the stronger the muscle. Furthermore, trained muscle functions more efficiently than untrained muscle. Trained muscles contract more vigorously and with less effort. Regular and strenuous exercise also toughens muscle, making it firmer and able to stand more strain.

d. **Muscular Endurance.** Muscles naturally become fatigued during continuous, repeated work. Through specific exercises, such as chinups or situps, local muscle groups become exhausted long before an individual fatigues. Training to develop muscular endurance enables people to continue a relatively heavy load of exercise over a long period of time. Lengthened exercise periods improve muscular endurance.

e. Muscular Coordination. Improved speed and strength result in part from improved muscular coordination. An unskilled individual may use irrelevant muscles to perform a particular activity, increasing the amount of work without increasing the mechanical output. This increase in skill is highly desirable, but it should be offset by greater effort in duration or intensity to compensate for the loss in overload due to increased skill.

f. Muscular Fatigue. When the rate of work is increased, the energy required is proportionately much greater than the increase in rate. For example, if an individual doubles running speed, the amount of power demanded to do this is increased eight times.

g. Circulation in Muscles Regular, strenuous exercise causes the creation of new capillaries and the opening of inactive, latent capillaries, increasing blood circulation within the muscle by as much as 400 percent. This increases the supply of food materials and oxygen to the muscle, improving its endurance. This process takes about 8 to 12 weeks of regular conditioning in young adults and longer as age increases.

7006. THE CIRCULATORY AND RESPIRATORY SYSTEMS

a. The Circulatory System. The circulatory system transports blood to all parts of the body, removes waste products for disposal, and delivers protecting and repairing substances where needed. The circulatory system includes the heart and blood vessels.
(1) The Heart. The heart is a pump which forces blood through the blood vessels to the parts of the body. The heart is a little larger than a fist and is located in the left center of the chest between the lungs.
(2) Blood Vessels. The vessels carrying blood away from the heart are the arteries. They eventually divide into capillaries, the very small vessels through which diffusion and osmosis take place. The capillaries gradually increase in size, forming the veins which carry blood back to the heart.

b. The Respiratory System. The respiratory system performs the function of breathing, whereby oxygen is inhaled and carbon dioxide is exhaled. The respiratory system consists of the mouth, nose, trachea, lungs, and diaphragm.
o Trachea. The trachea, or windpipe, is a hollow, tube-like structure that carries air from the mouth to the lungs.
o Lungs. The lungs are elastic bags in the chest where the exchange of oxygen and carbon dioxide takes place.
o Diaphragm. The diaphragm is a thin, sheet-like muscle just below the lungs. During inhalation, the diaphragm flattens out and lowers, allowing the lungs to expand and fill with air. During exhalation, the diaphragm raises into a dome shape, helping to expel air from the lungs.

7007. CARDIOVASCULAR FUNCTIONING

a. General. Cardiovascular functioning is the combined functioning of the circulatory and respiratory systems. The chief organs of these systems, the heart and lungs, function together to provide oxygen to the body.(See fig. 7-5.)

Figure 7-5 is available in the illustration section at the end of the book.

b. Heart Action. The heart, the organ which propels blood through the blood vessels, is the chief organ of cardiovascular endurance. A person tires quickly if the heart lacks the capacity to circulate the blood through the body. The heart is a muscular organ and is developed by exercise just like any other muscle. The heart cannot be exercised alone; any exercise that puts a sufficient load on the heart, such as running or cycling, will also exercise other body parts. A conditioned heart tends to beat more slowly but pump more blood when at rest. This is known as an increase in "stroke volume,"a desirable condition because it enables the heart to pump more blood with a slower contraction rate.

c. Functions of the Lungs. Strenuous and regular exercise can improve the efficiency with which the lungs transmit oxygen to the blood by as much as 25 percent In a poorly conditioned person, some of the alveoli (air sacs) within the lungs are closed or collapsed. The forced breathing created by exercise over a period of weeks causes the air sacs to slowly expand, increasing the ability to absorb oxygen.

d. Relationship of Heart and Lungs. The heart pumps carbon dioxide-laden blood through the pulmonary artery to the lungs. In the lungs the carbon dioxide is exchanged for oxygen and the purified blood is returned to the heart by way of the pulmonary vein. The heart then pumps the blood through the aorta for circulation throughout the body. Blood moving into the muscles exchanges oxygen for carbon dioxide, after which time it is ready for the return trip to the heart.

e. Cardiovascular Functioning in High Altitudes. Marines to be employed in high altitudes should be acclimated in a similar area for 10 to 14 days prior to employment. Persons not accustomed to the rarefied air of higher altitudes tire more quickly and may collapse after rapid physical exertion. Air is much less dense at high altitudes than at sea level. Persons accustomed to sea level inhale only about 80 percent of the oxygen they are used to no matter how hard they breathe. Further, people accustomed to sea level or moderate altitude simply do not have enough red corpuscles in the blood to fulfill their needs at high altitudes. The red blood cells will increase over time.

f. Symptoms of Cardiovascular Malfunctioning. The brain is the first organ to react to a lack of oxygen. Unconsciousness results when the brain is denied sufficient oxygen. "Blacking out" is actually a defense mechanism to keep the body alive.

APPENDIX A
INSTRUCTOR HINTS AND INSTRUCTOR TRAINING

1. General. This appendix informs instructors on administration of exercise periods, commands, the extended rectangular and circle formations, methods of instruction and conduct of exercise activities, explanation of basic positions associated with the various activities, and instructor training.

2. Aim of the Physical Fitness Program. The aim of the physical fitness program is to prepare Marines physically for war. This can be conveniently split into two phases.

a. Preparation Phase

(1) Development of all-around physical fitness.

(2) Development of purposeful physical skills.

(3) Development of mental alertness.

(4) Development of character and leadership.

b. Maintenance Phase. Once the preparation is completed, the program must progress to a maintenance program. This phase revolves around the need to maintain the trained Marine's physical fitness for a particular role. This will include—

(1) Maintenance of a high standard of all-around physical fitness.

(2) Application of purposeful physical skills to combat skills.

(3) Development of leadership and self-confidence.

(4) Development of mental toughness and endurance.

3. Physical and Mental Development

a. Marines need to be developed physically and mentally. The physical training (PT) instructor must ensure that the program accomplishes this.

b. The system of the body functions best in response to progressive exercise. This means that Marines should master the easier exercises before they attempt the more difficult ones.

c. The body and mind are linked inseparably and must be considered as one in physical fitness training. Physical exercises assist in developing the mind.

d. The confidence in oneself and in one's comrades attained by physical achievement helps to promote the indomitable will to win. The Marine must be taught to realize this. It will help during arduous training, which at all times must be imaginative, realistic, and filled with enthusiasm. All physical fitness training must be associated with developing the qualities required of a Marine.

4. Planning and Preparation of Programs The following factors must be considered when compiling a program (see also ch. 2)~

a. Aim or desired skill level.

b. Duration of the complete training program and number of physical training lessons.

c. Purpose of included exercises,drills, or practical applications.

d. Beginning/inventory fitness levels.

e. Medical facilities, medical personnel, and Marine Corps or local command regulations governing their proximity to the training.
f. Number of Marines to be trained.
g. Availability of instructors, assistants, and facilities.
h. Apparatus and equipment available.
i. Alternative training areas for varying weather conditions,
j. Stages of progression and tests of achievement.

5. Five Basic Principles of Program Design
a. Regularity. Regularity of exercise is far more important than the amount of exercise performed on a given day. A Marine should exercise a minimum of three times per week. Human muscle begins to atrophy after 3 days without exercise.
b. Balance. The program must be balanced in terms of total physical fitness. Imbalance is created when, for example, an individual merely runs. While the cardiovascular system is improved, general upper body strength may be lacking. Conversely, a person who just does weight training may develop strength, but not cardiovascular endurance.
c. Overload. If total body fitness is to be developed, then the muscular and cardiovascular systems must be made to perform more work than previously. This is done by increasing the frequency, intensity, and duration of the exercise. The body adapts to the amount of stress placed upon it and becomes stronger and more efficient.
d. Progression. To be successful, any type of training program must be progressive. This is an area where untrained persons conducting PT can cause injury to Marines resulting in failure of the PT program. Close supervision must be made of untrained instructors. The result of progressing too fast is that the class begins to reject PT because it is no longer enjoyable or motivating. The program should start at a slow pace, and the pace should increase as proficiency improves. Average persons will encounter three stages of progressiom the toughening stage, the slow improvement stage, and the sustaining stage. (See ch. 1.)
e. Variety. Some programs fail because the training becomes boring. Variety is essential and perhaps presents the greatest challenge to the commander and staff.

6. Grouping. Providing for different levels of physical fitness is particularly recommended in the early stages of conditioning. Older Marines and those in poorer physical condition should be expected to attain a group level of fitness, but they should be given more time to do it.
a. Homogeneous Grouping One simple method of providing for the difference in levels is to group them according to their condition. A two-group classification would divide Marines into highly conditioned and average groups. A finer classification could be obtained by dividing them into three groups—a highly conditioned, a moderately conditioned, and an unconditioned group.
b. Performance or Age Grouping. The segregation of Marines into different exercise groups should be based on physical fitness test scores or on the level of fitness they demonstrate. They may also be grouped at first according to age. A common classification by ages is under 30, 30 to 34, and 35 and above.

7. Preparatory Commands and Commands of Execution. The prepreparatory command describes and specifies what is required. The command of execution calls into action what has been prescribed. All preparatory commands are given with a rising inflection. The interval between commands is long enough to permit the average Marine to understand the first one before the second one is given.

8. Extended Rectangular Formation. The traditional formation for carrying on many physical training activities is the extended rectangular formation (fig. A-1). This formation is more compact than any other. It is the best type to use for large numbers of Marines because it is simple and easy to assume.

a. For the formation of oneplatoon, the base man paces off five paces from the stand, faces left and moves five paces, halts, and again faces left. With the base man positioned facing the stand, the platoon leader then commands. **FALL OUT AND FALL IN ON THE BASE MAN**. At this command, all Marines run to the designated area and reform. This procedure is preferred to marching the unit into position. If more control is desired, the unit may march at double time to the vicinity of the base man and then be directed to fall out and fall in on him. Much valuable time is wasted in the field by needless maneuvering of Marines at quick time in an effort to position the platoon or unit on the exact spot for the exercises.

b. A company-size unit assumes the extended rectangular formation from a column of threes or fours at normal intervals between squads. This extension can also be executed from a company mass without interval between platoons. In extending either a platoon or company-size unit, take your place at the head of the column and command:

(1) TAKE INTERVAL TO THE LEFT, MARCH. At this command, the Marines in the right flank file stand fast with arms extended sideward. All others turn to the left and run forward at double time. After taking a sufficient number of steps, all Marines face the front with both arms extended sideward. The distance between finger-tips is about 12 inches, and dress is right.

Figure A-1 is available in the illustration section at the end of the book.

(2) ARMS DOWNWARD, MOVE. At this command, the arms are lowered smartly to the sides.

(3) LEFT, FACE

(4) EXTEND TO THE LEFT, MARCH. At this command, those in the right flank file stand fast with arms extended sideward. All others turn to the left and run forward at double time. Spacing is the same as in (1) above, and dress is right.

(5) ARMS DOWNWARD, MOVE. Same as in (2) above.

(6) RIGHT, FACE.

(7) FROM FRONT TO REAR, COUNT OFF. At this command, the leading Marine in each column turns the head to the right rear, calls of **ONE** and faces the front. Successive Marines in each column call off in turn **TWO, THREE, FOUR, FIVE** in the same manner.

(8) EVEN NUMBERS TO THE LEFT, UNCOVER. At this command, each even-numbered Marine stride jumps sideward to the left, squarely in the center of the interval. In doing this, each one swings the left leg sideward, jumps from the right foot to the left foot, and smartly brings the right into position against the left.

c. To assemble the unit, you commandASSEMBLE TO THE RIGHT, MARCHAt this command, all return to their original position in the column at double time and reform on the base man.

d.It is recommended that the area for grounding equipment and arms be at the edge of, or well away from, the area to be used for exercising. To conserve time and ensure proper position of the unit, the base man or, if the unit is composed of several platoon-sizegroups, the various base men may precede the unit and establish their positions in relation to the instructor's stand.

9. Circle Formation.The circle formation is effective for conducting various exercise activities (fig. A-2). This formation has an advantage in that supervision of all Marines is facilitated and a moving formation is available which provides control. Guerrilla exercises, grass drills, and some forms of running are examples of activities which are more easily conducted in the circle formation than in the extended rectangular formation.

a. It is not advisable to have more than 60 Marines in a circle. When more must be accommodated, separate circles should be used.Concentric circles may be employed to reduce the size of the circle or to accommodate more persons. If concentric circles are formed, the different circles are made by designating squads for each circle.Each additional circle requires more Marines than the one inside it.For example, one squad of a platoon may form the inner circle and theremaimng three squads the outer circle.

b. When a platoon is to form a circle, the commands ai€IRCLEFORMATION, MARCH, FOLLOW MEUpon this command, the right flank squad of the column moves forward at double time with the leader of the platoon group gradually forming a circle in a counterclockwise direction. Each succeeding file falls in behind that on the right. After the rough outline of the circle is formed, the leader commandsPICK UP A 5-YARD INTERVAL.

Figure A-2 is available in the illustration section at the end of the book.

c.The group may be halted and faced toward the center, or, if instruction is not necessary, the exercise activity may be executed without stopping the platoon.

lO.Leadership Techniques

a. Unless you experience all the exercises, you cannot appreciate how arduous they are, what movements are most strenuous and difficult, where the errors in performance are likely to occur, and what the proper cadence should be.

b. You must give everyone careful supervision and participate in the exercises to show that you can do them. When you participate, your assistant instructors should supervise because it is difficult for you to supervise and exercise at the same time.

c. Marines should never be kept too long in one position, especially a constrained one. They should never have to perform an exercise more times than they can do it without losing the proper form. You should insist upon proper form in the execution of all exercise activities.Even slight deviations from the proper form reduce the value of the exercise.

d. Avoid long explanations.As a rule, it should be necessary to give a full explanation of new exercises only. Explain the most essential features of an exercise first; add details later. Too many details at one time are more likely to confuse than to assist them.Minor corrections should be made to the entire class while the exercise is in progress (for exampHEADS **UP,** KNEES STRAIGHT). If necessary, follow this direction by the name of the person who is particularly at fault. If a Marine requires special attention, givethat person separate instruction by an assistant instructor to avoidwasting the time of the group.

e. The heavy demand on your voice can be lightened by training assistant instructors to assume some of the instruction. Using mass cadence is also an effective method of lessening the demand on your voice.

“Illustrations”

DAY	TIME	ACTIVITY
MONDAY	60 MIN	RUNNING ACTIVITY
TUESDAY	60 MIN	COMBATIVES
WEDNESDAY	60 MIN	CONDITIONING & LOG DRILLS
THURSDAY	60 MIN	CONDITIONING & RELAYS
FRIDAY	60 MIN	RUNNING & CONDITIONING
SATURDAY	60 MIN	RUNNING ACTIVITY

Figure 2-1. Sample A- -Single Activity.

DAY	TIME	ACTIVITY	REMARKS
MONDAY	20 MIN 20 MIN 20 MIN	RIFLE DRILL COMBATIVES RUNNING	PROGRESSIVE ACTIVITY SYSTEM
TUESDAY	30 MIN	GUERRILLA EXERCISES TABLE I	SINGLE ACTIVITY
WEDNESDAY	30 MIN	RIFLE DRILL	SINGLE ACTIVITY
THURSDAY	30 MIN	CIRCUIT INTERVAL TRAINING	SINGLE ACTIVITY
FRIDAY	20 MIN 20 MIN 20 MIN	STRENGTH CIRCUIT CONDITIONING DRILL CROSS-COUNTRY RUN	ROTATION ACTIVITY
SATURDAY	60 MIN	WARM UP RUN AND GRASS DRILL	SINGLE ACTIVITY

Figure 2-2. Sample B- -Various Time Periods, Activities, and Systems.

DAY	TIME	ACTIVITY
MONDAY	45 MIN	RUNNING – CROSS-COUNTRY
TUESDAY	75 MIN	TEAM ATHLETICS – TOUCH FOOTBALL
WEDNESDAY	45 MIN	RUNNING – CROSS-COUNTRY
THURSDAY	20 MIN 40 MIN	LOG DRILL DUAL COMBATIVES
FRIDAY	90 MIN	SPEED MARCH
SATURDAY	90 MIN	TEAM CONTESTS

Figure 2-3. Sample C--Sustaining Activities.

Route	Kilometers from SP	Recommended rate of march (kmph)	Remarks
SP: RJ 8th Div Rd—Superhighway	......	4	Bridge; hard surface; two guides.
RJ Jamestown—Hersey Rd	2,25	4	Traffic heavy; two guides; Jamestown Rd bears to right.
RJ Jamestown—Yankee Rd	7.10	4	Two guides
RJ Jamestown—Lightning Rd	10.50	4	Two guides
RJ Jamestown—Sunshine Rd	12.40	4	Light traffic; one guide Sunshine Rd; gravel, poor traction when wet.
RP; RJ Sunshine—Sedan Rd	14.80	4	Two guides

Figure 3-1.Example of Route Reconnaissance Report.

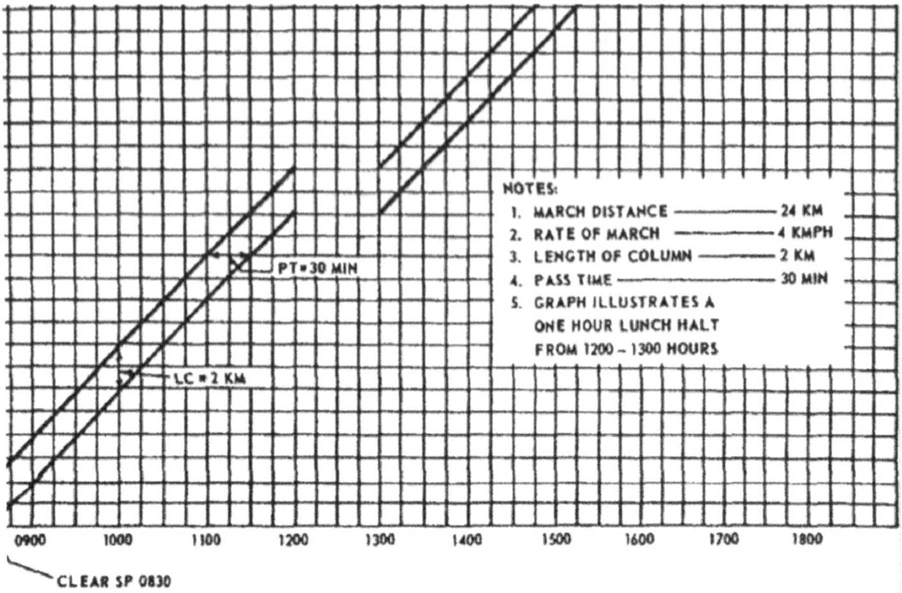

Figure 3-2.Road Movement Graph.

Figure 3-3.The Pace-Setter.

	SINGLE FILE	COLUMN OF TWO
2 m/MAN APART	24	12
5 m/MAN APART	54	27

GAP DISTANCES			
SERIAL:	(3-1) × 100 m	=	200 m
MARCH UNIT:	(12-1-2-) × 50 m	=	450 m
	TOTAL COLUMN GAP		650 m

Figure 3-4.Conditioning Drill One.

Figure 3-5.Conditioning Drill Two.

Figure 3-6.Conditioning Drill Three.

Figure 3-7. Rifle Drill.

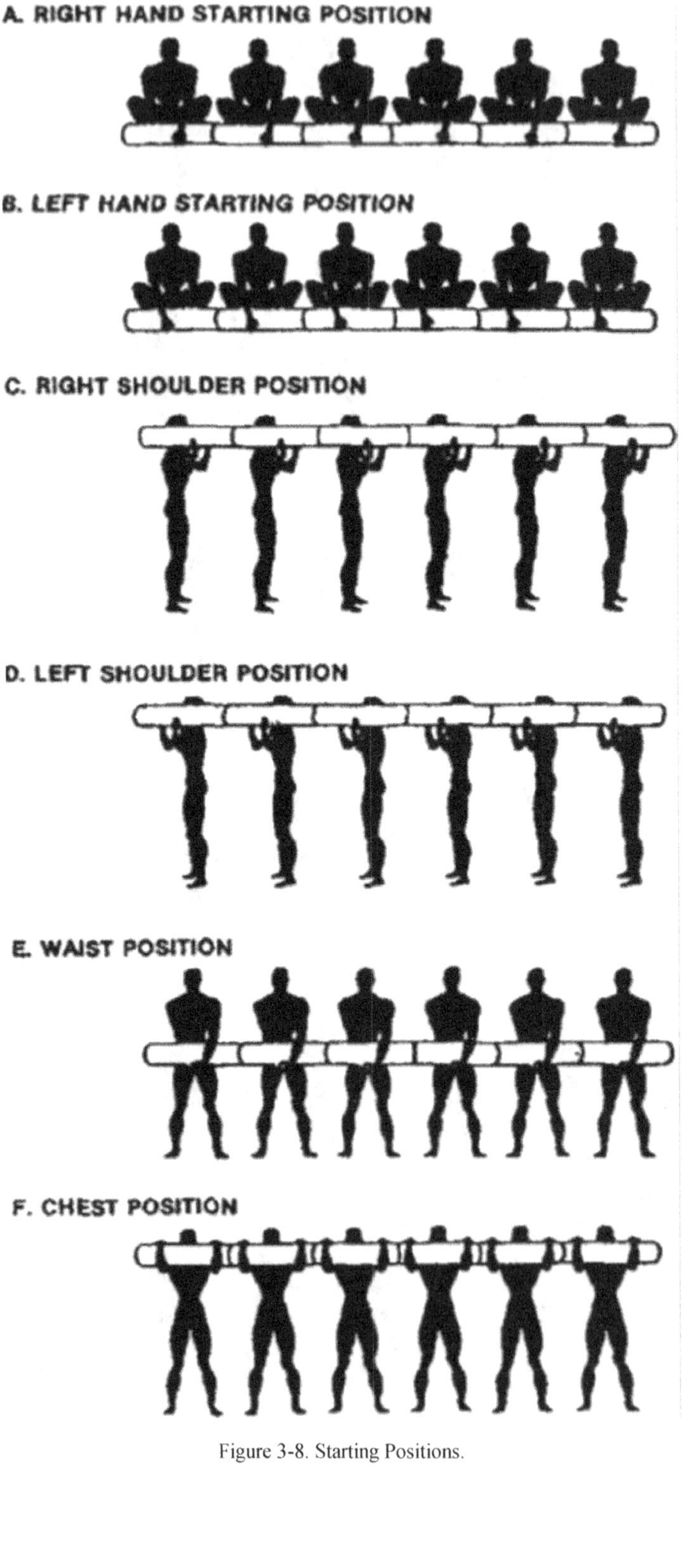

Figure 3-8. Starting Positions.

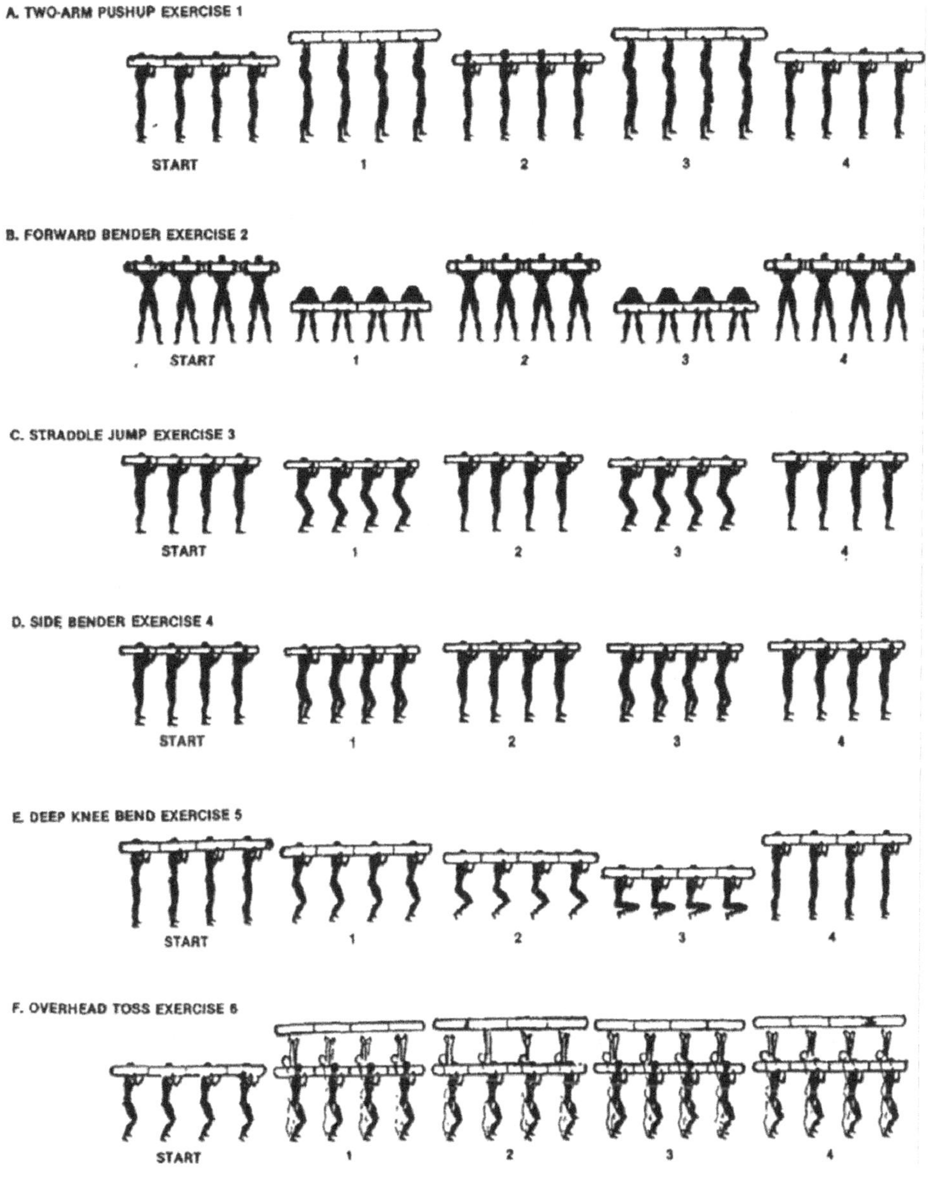

Figure 3-9. Log Drill

Figure 3-10.Basic Positions for Grass Drill.

Figure 3-11.Grass Drill One.

Figure 3-12.Grass Drill Two.

Figure 3-13.Guerrilla Table I.

Figure 3-14.Guerrilla Table II.

Figure 3-15.Proper Running Form.

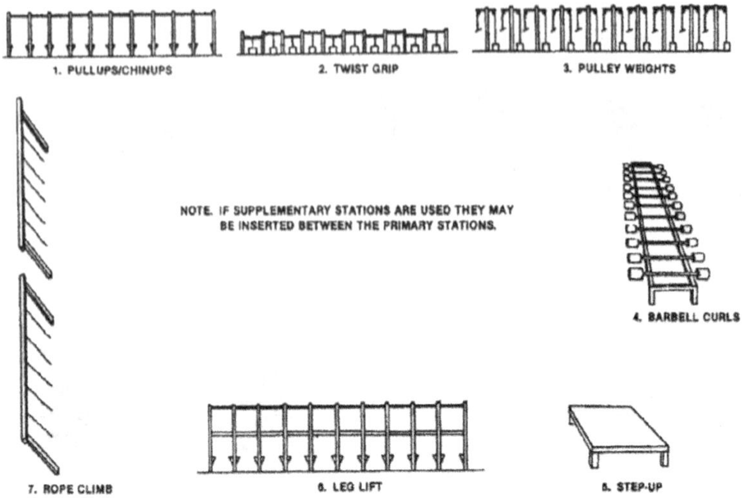

Figure 3-16. Fixed Strength Circuit.

STATION	ITEM	NO.	SPECIFICATION
1	BARBELL		1¼-INCH PIPE 5 FEET LONG WITH CONCRETE FILLED NO. 10 CANS
2	JUMP ROPE		¼- OR ⅜-INCH ROPE, 10 FEET LONG
3	TWIST GRIP		HANDLE 12 INCHES LONG, ROPE 4 FEET LONG, NO. 10 CAN CONCRETE FILLED.
4	INCLINE PLANE		¾-INCH PLYWOOD PLATFORM 2 FEET WIDE AND 6 FEET, 6 INCHES LONG, ELEVATED 10 INCHES AT ONE END. STRAP TO HOLD FEET DOWN.
5	WAR CLUB		HEAD IS 6 BY 12 INCHES, HANDLE IS 14 INCHES LONG BY 1¼ INCHES IN DIAMETER, ABOUT 20 POUNDS.
6	BICYCLE RIDE		PLYWOOD BOARD OR PLATFORM 2 BY 3 FEET WITH 2 BY 2 RUNNERS.
7	STEP – UP		A BOX OR STURDY PLATFORM 18 INCHES HIGH, 18 INCHES WIDE, 24 INCHES LONG.
8	ISOMETRIC PULL		TWO HANDLES 12 INCHES LONG WITH 4 FEET (BETWEEN HANDLES) OR LIGHT WIRE CABLE OR ¼-INCH ROPE

Figure 3-17.Movable Strength Circuit.

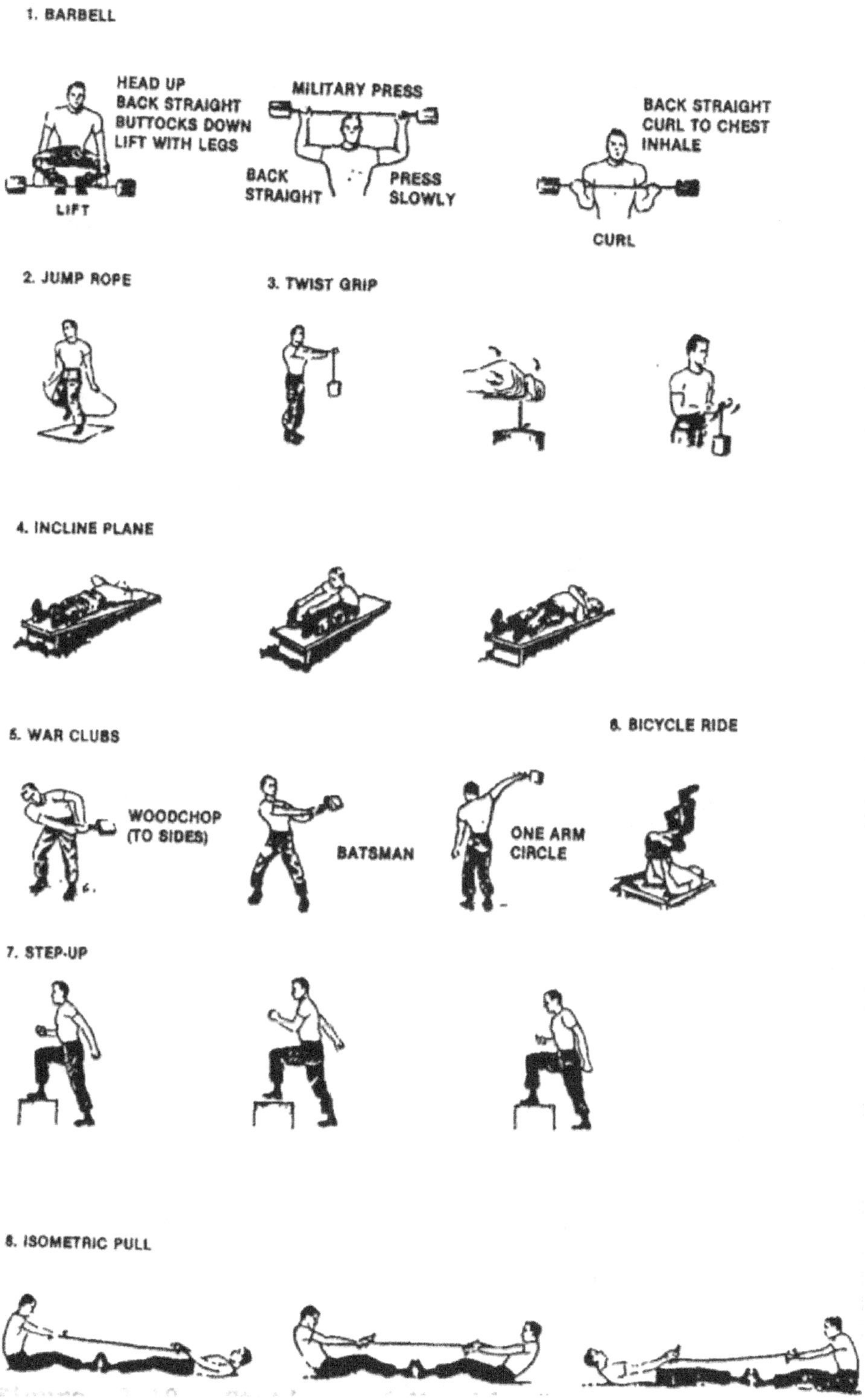

Figure. 3-18.Stations of Movable Strength Circuit.

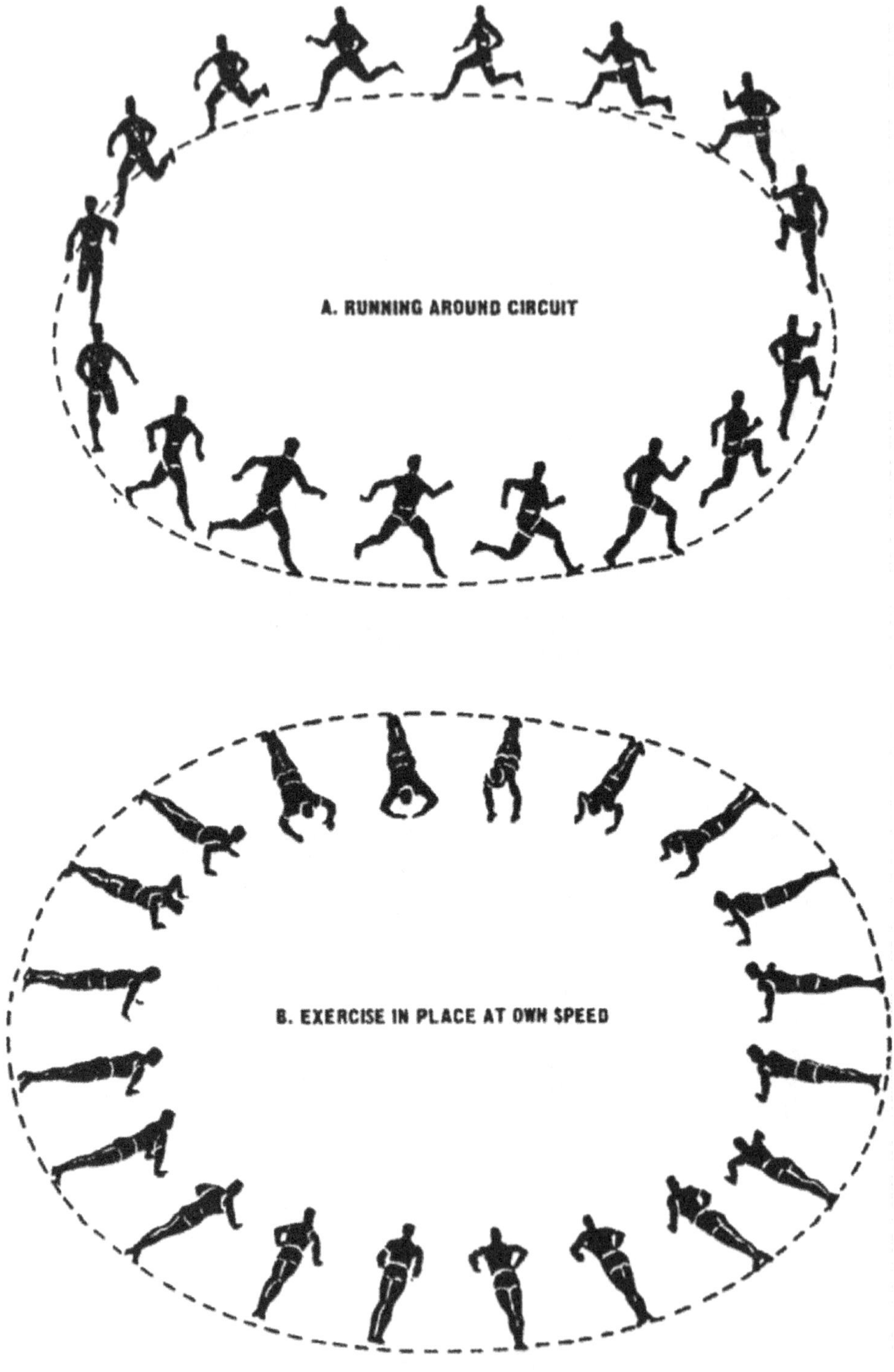

Figure 3-19. Circuit-IntervalTable.

Figure 3-20. Jumping.

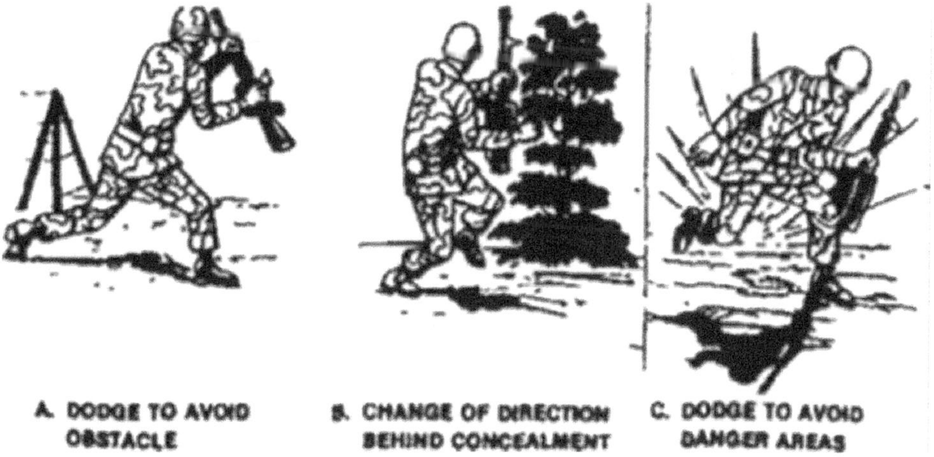

Figure 3-21.Dodging.

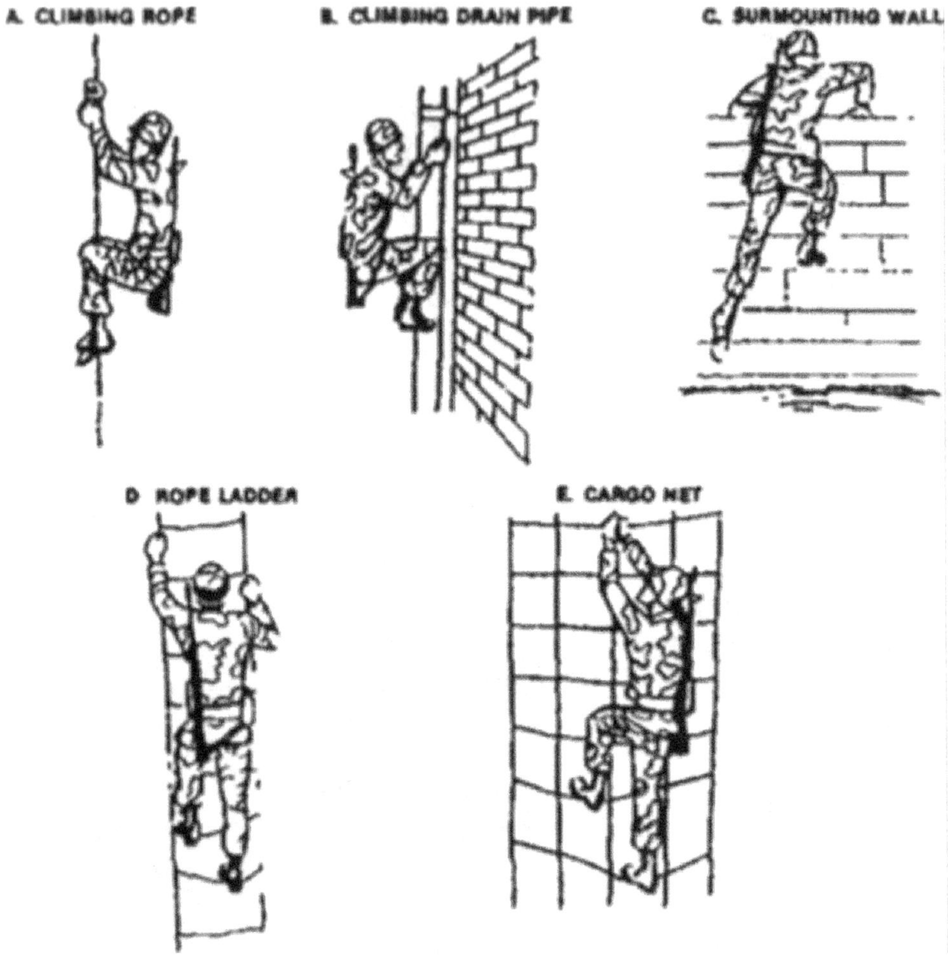

Figure 3-22.Climbing and Surmounting Vertical Objects.

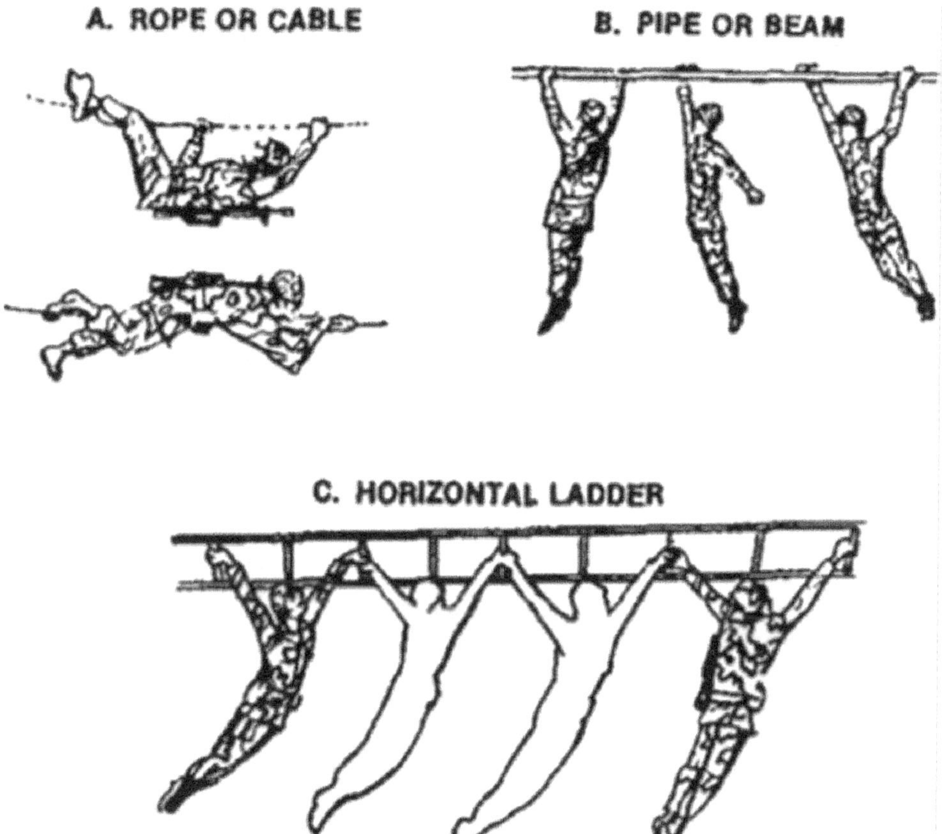

Figure 3-23. Traversing Horizontal Objects.

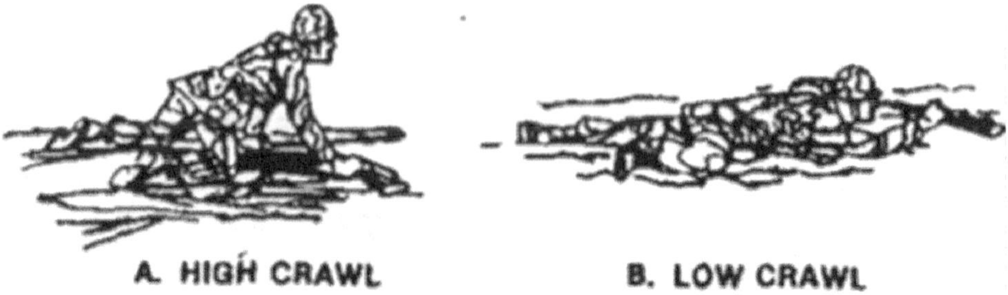

Figure 3-24.Crawling.

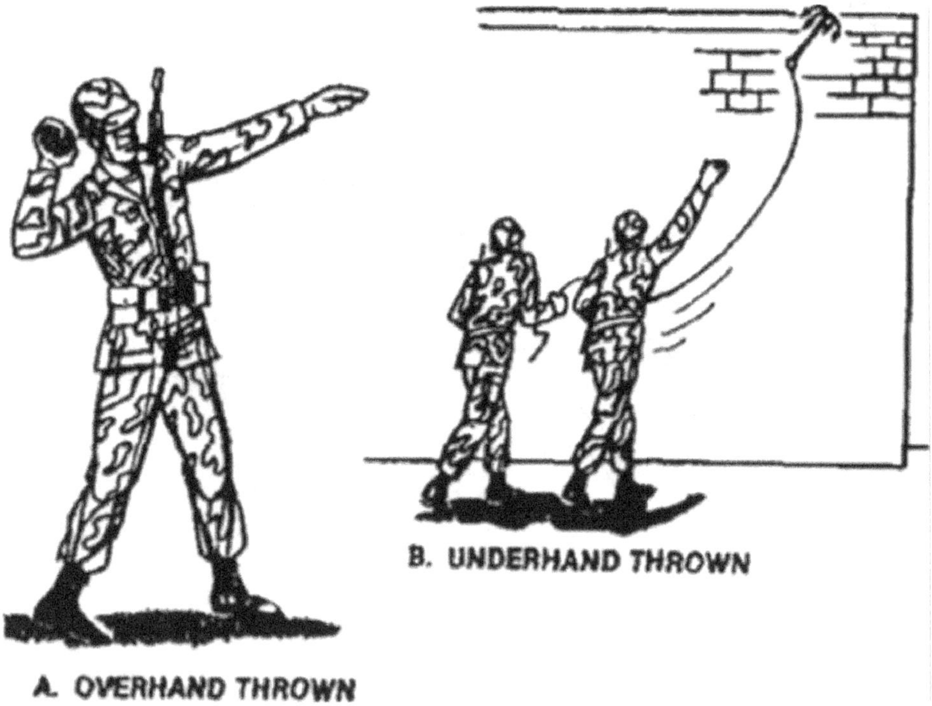

Figure 3-25.Throwing.

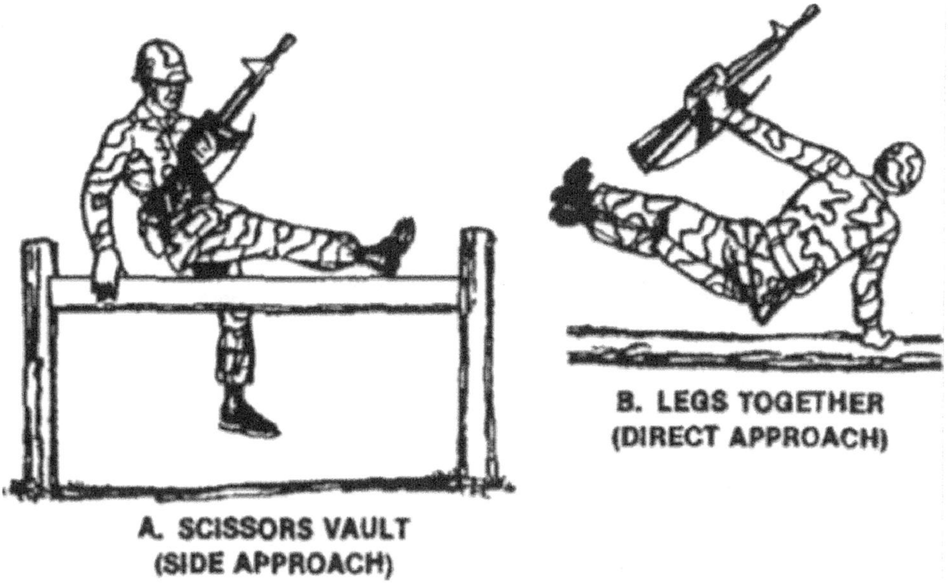

Figure 3-26. Vaulting

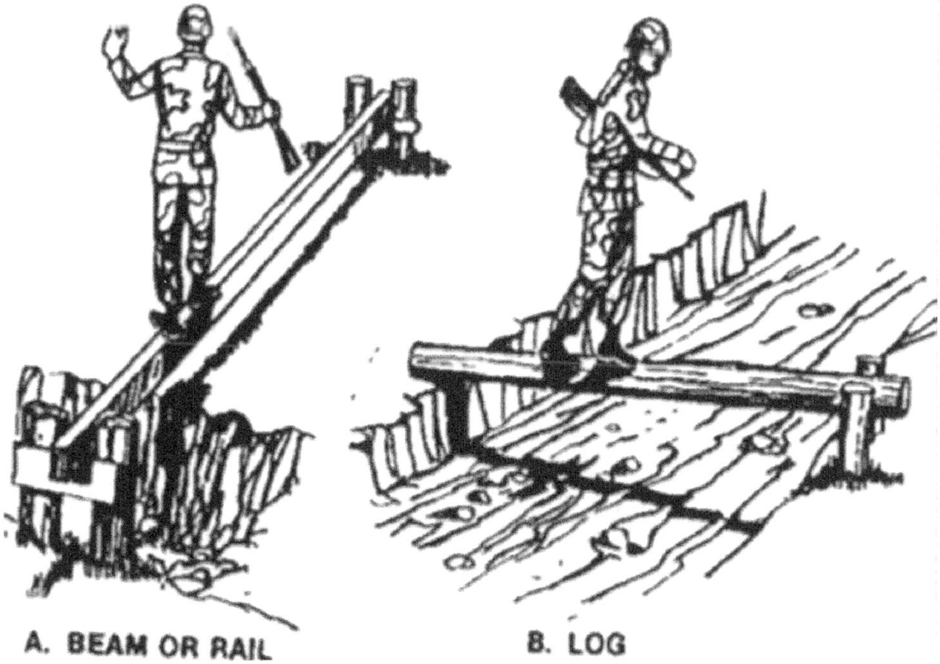

Figure 3-27.Balancing.

Figure 3-28.Falling.

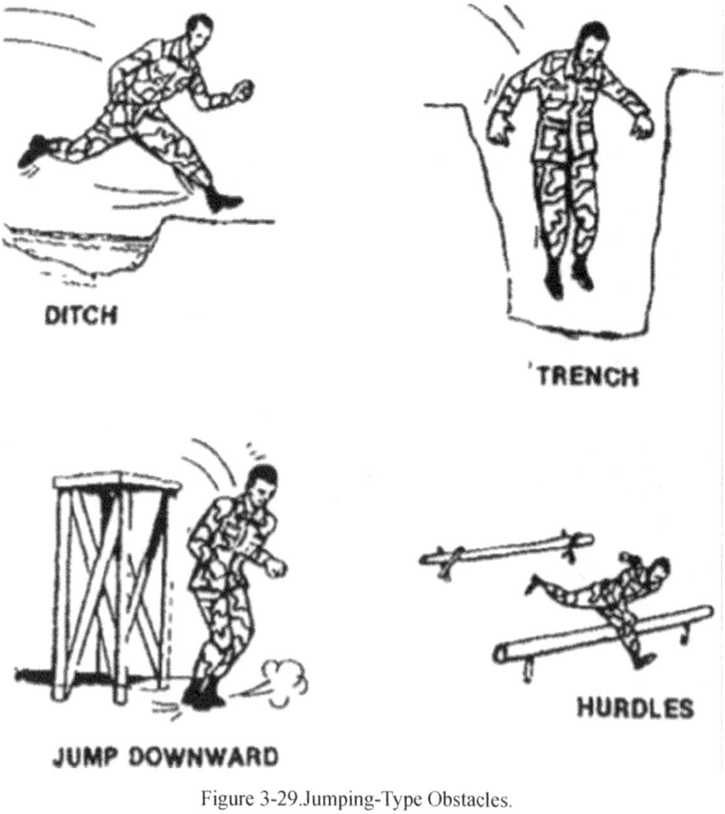

Figure 3-29.Jumping-Type Obstacles.

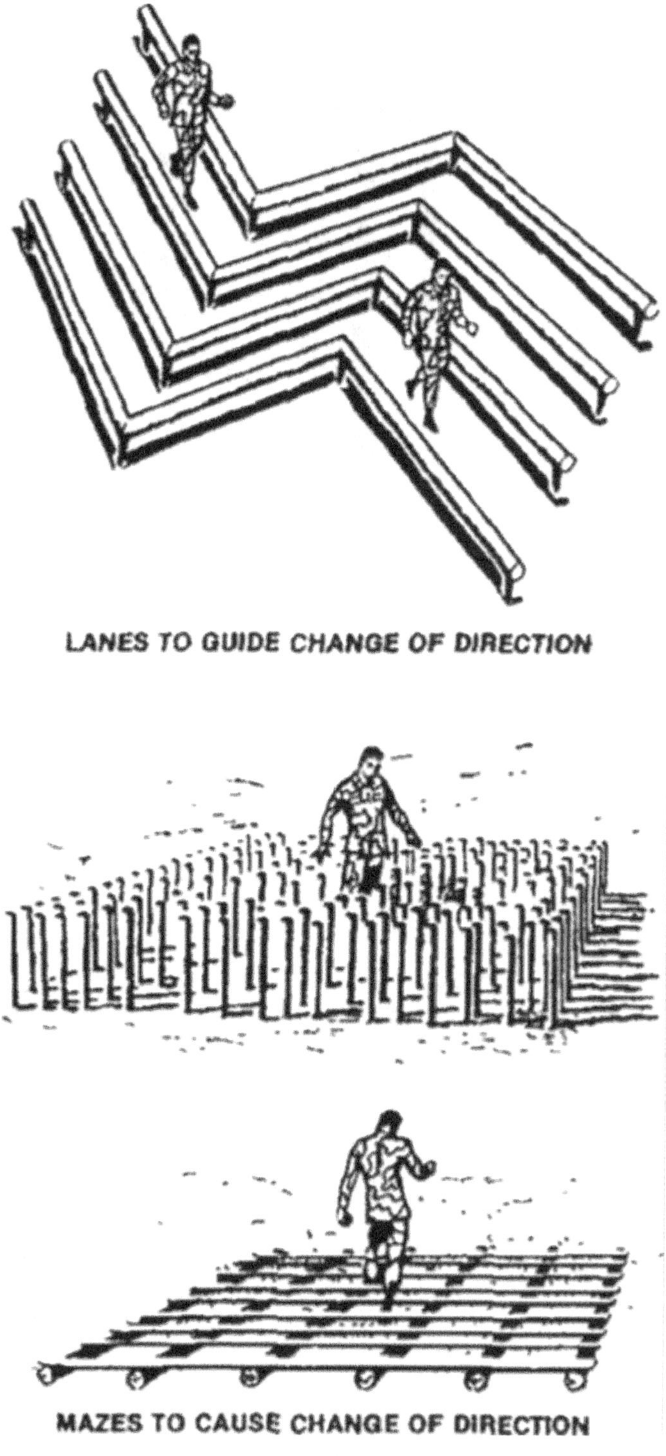

Figure 3-30.Dodging-Type Obstacles.

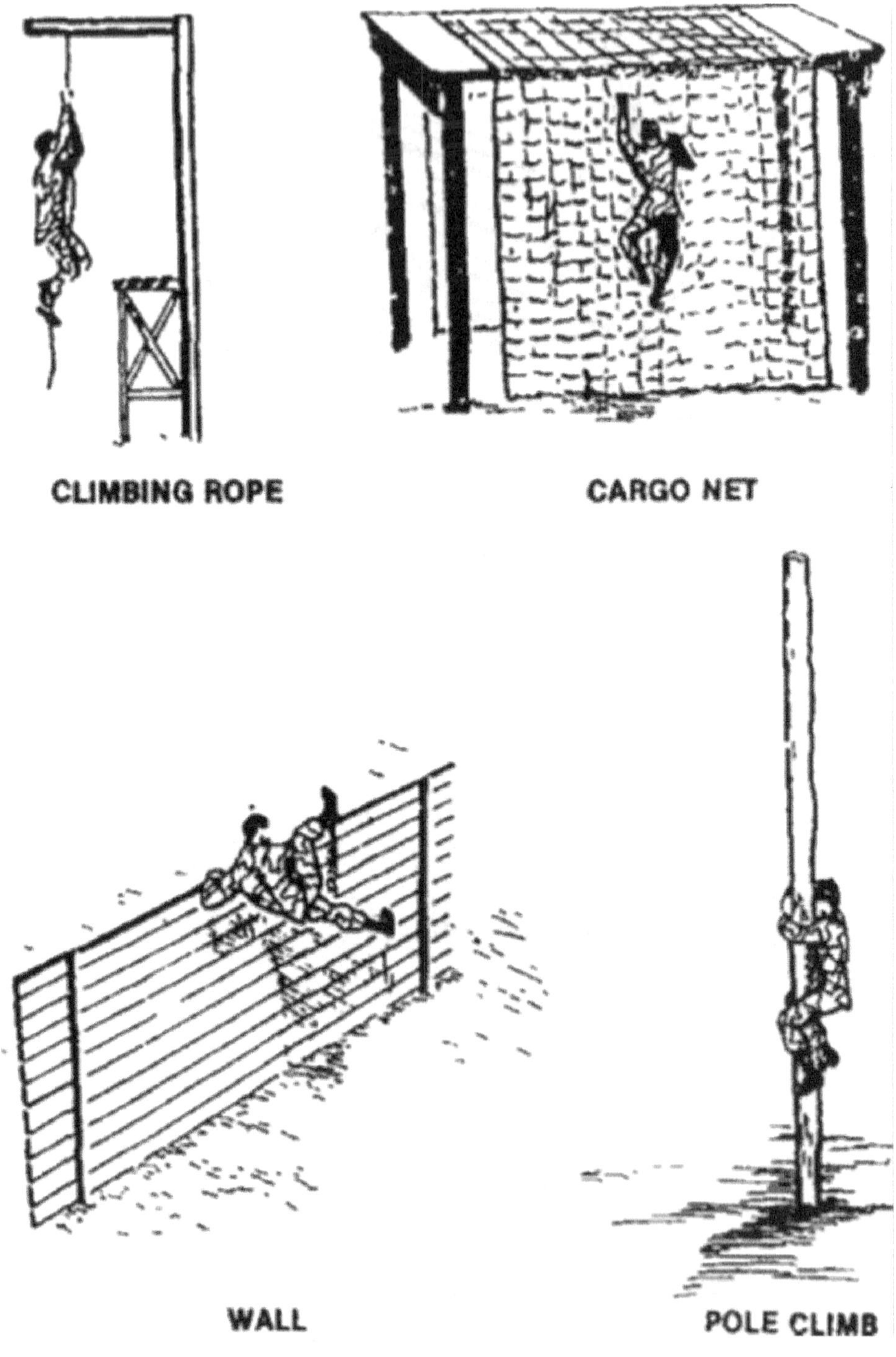

Figure 3-31. Vertical Climbing Obstacles.

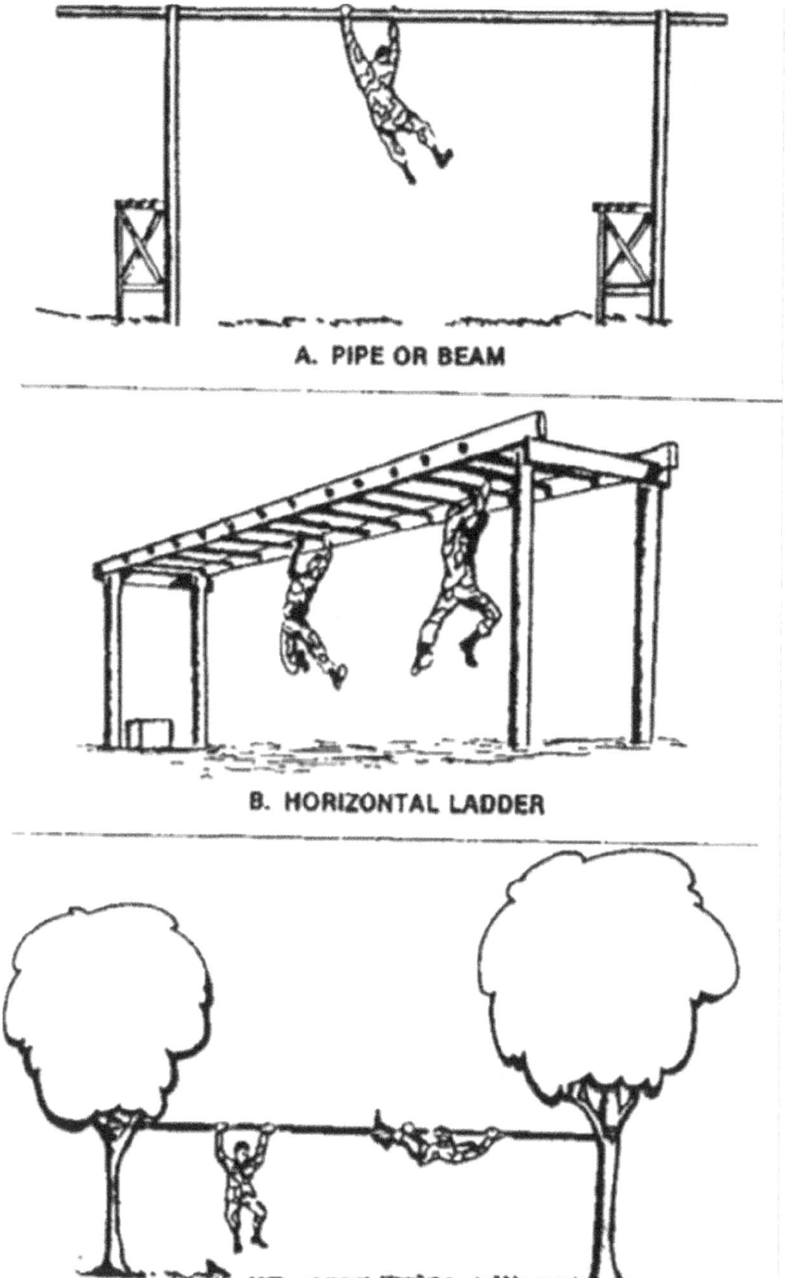

Figure 3-32.Horizontal Type Obstacles.

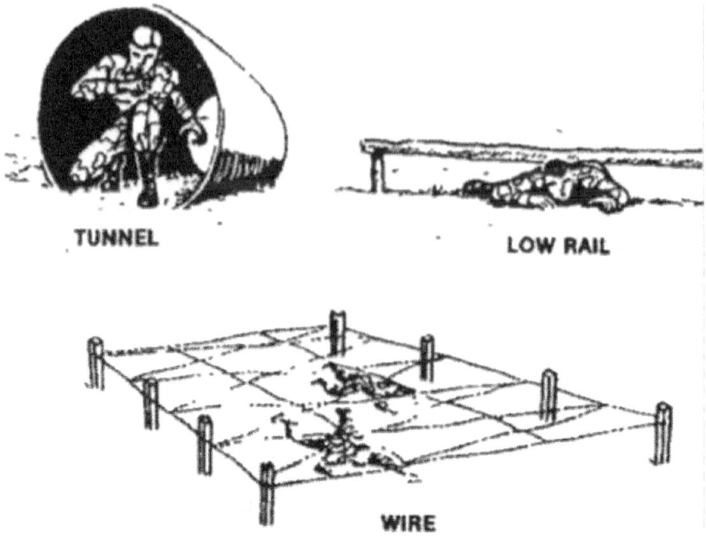

Figure 3-33.Crawling-Type Obstacles.

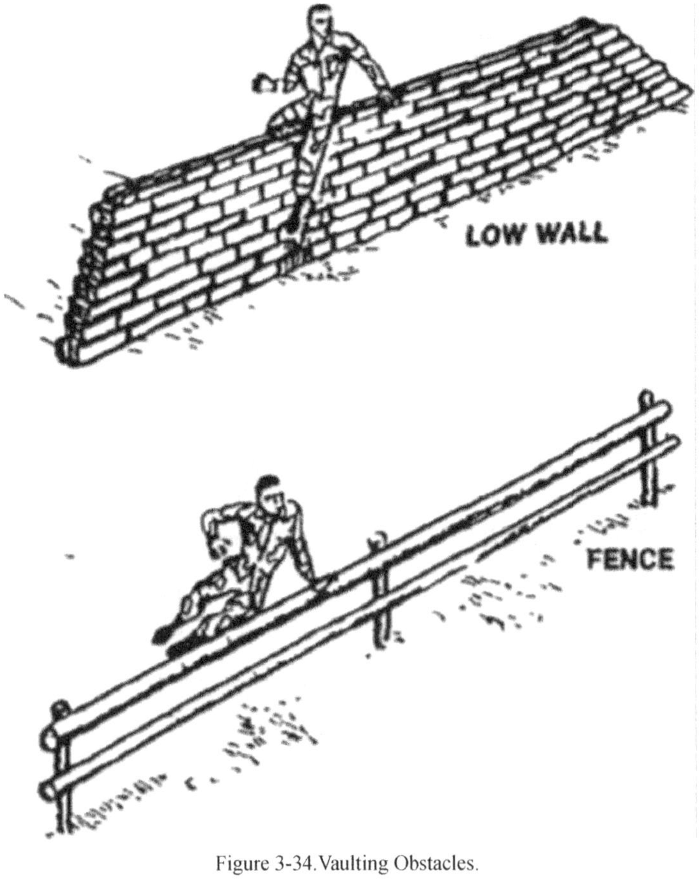

Figure 3-34.Vaulting Obstacles.

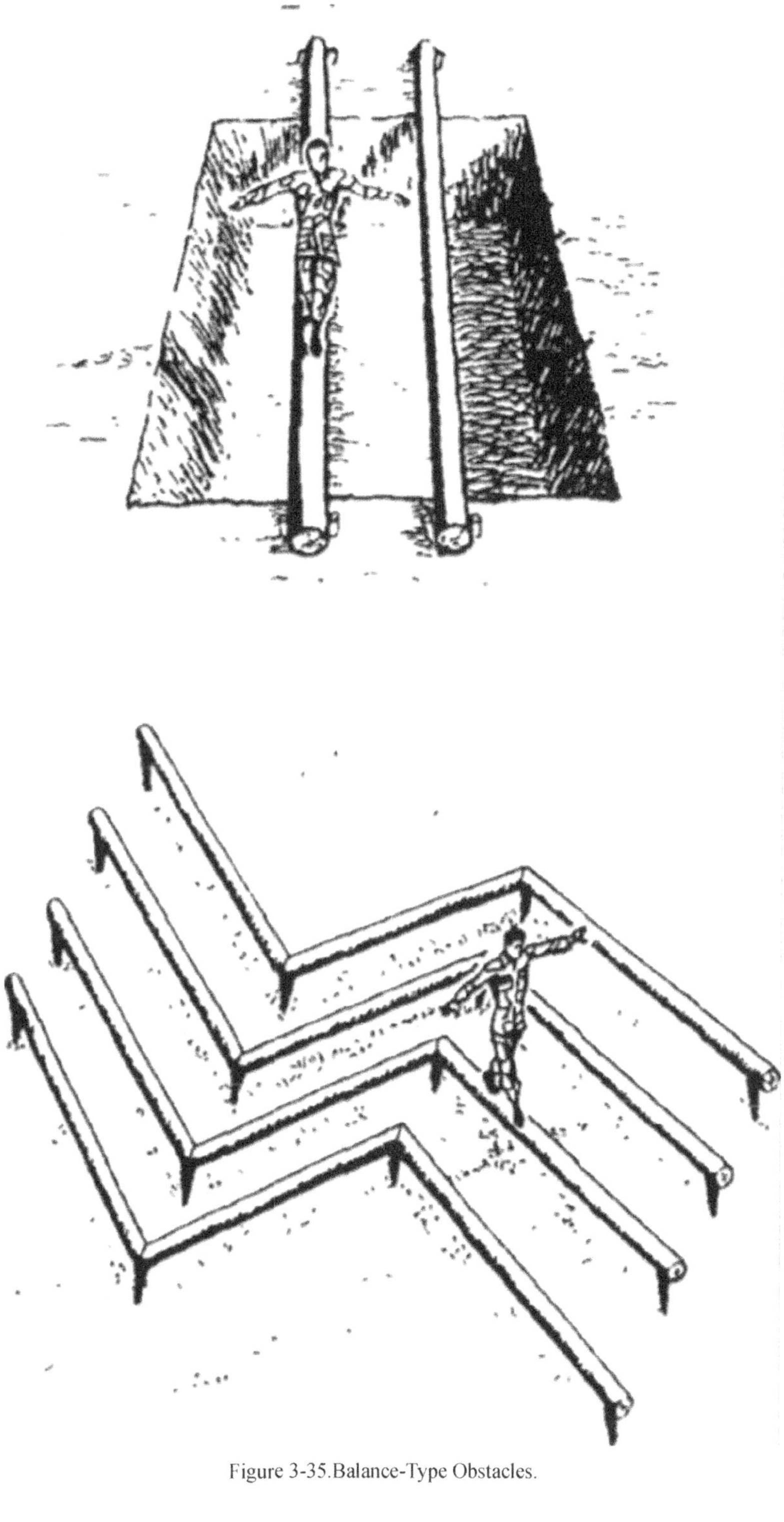

Figure 3-35.Balance-Type Obstacles.

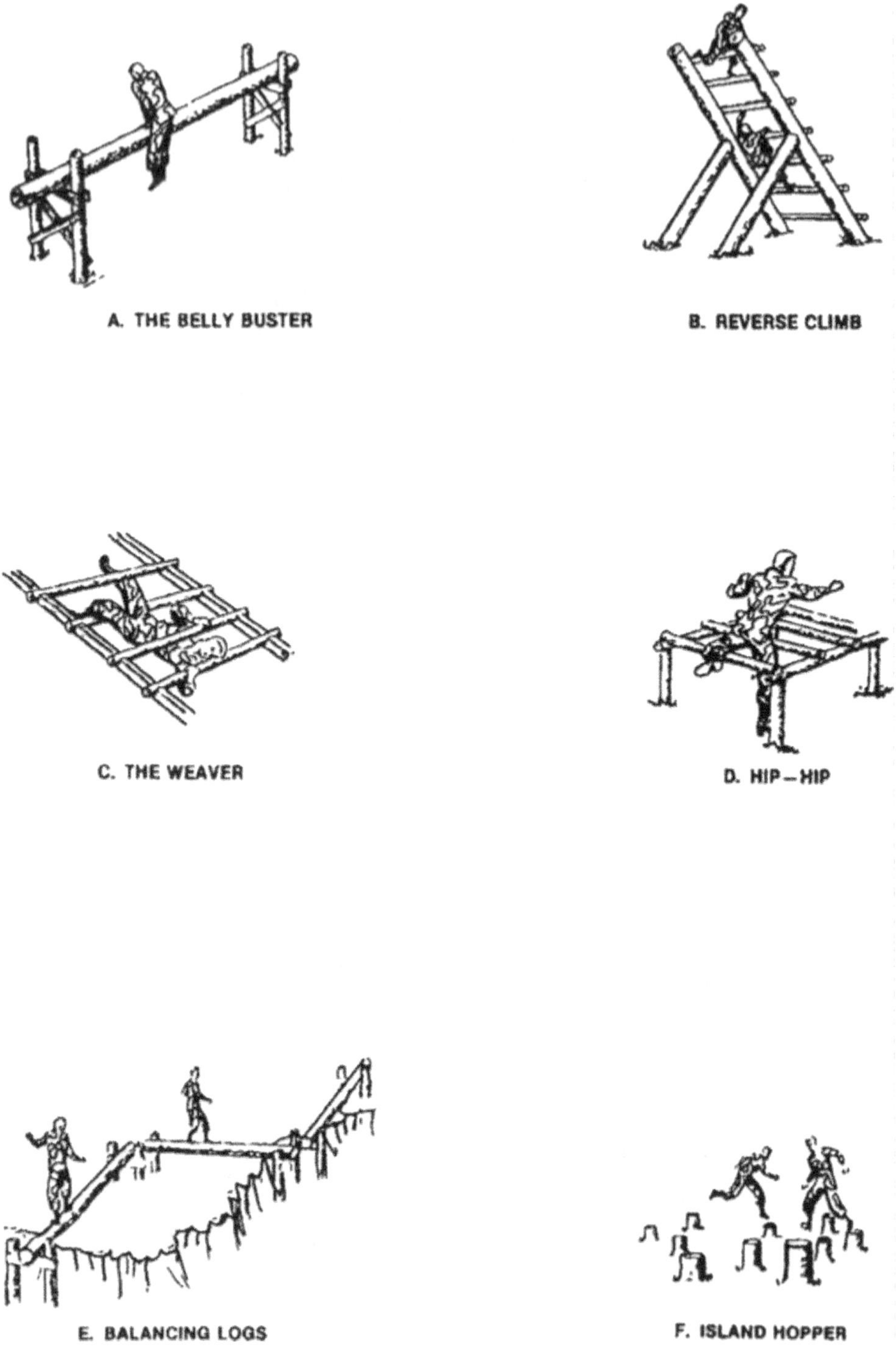

Figure 3-36.Red Group.

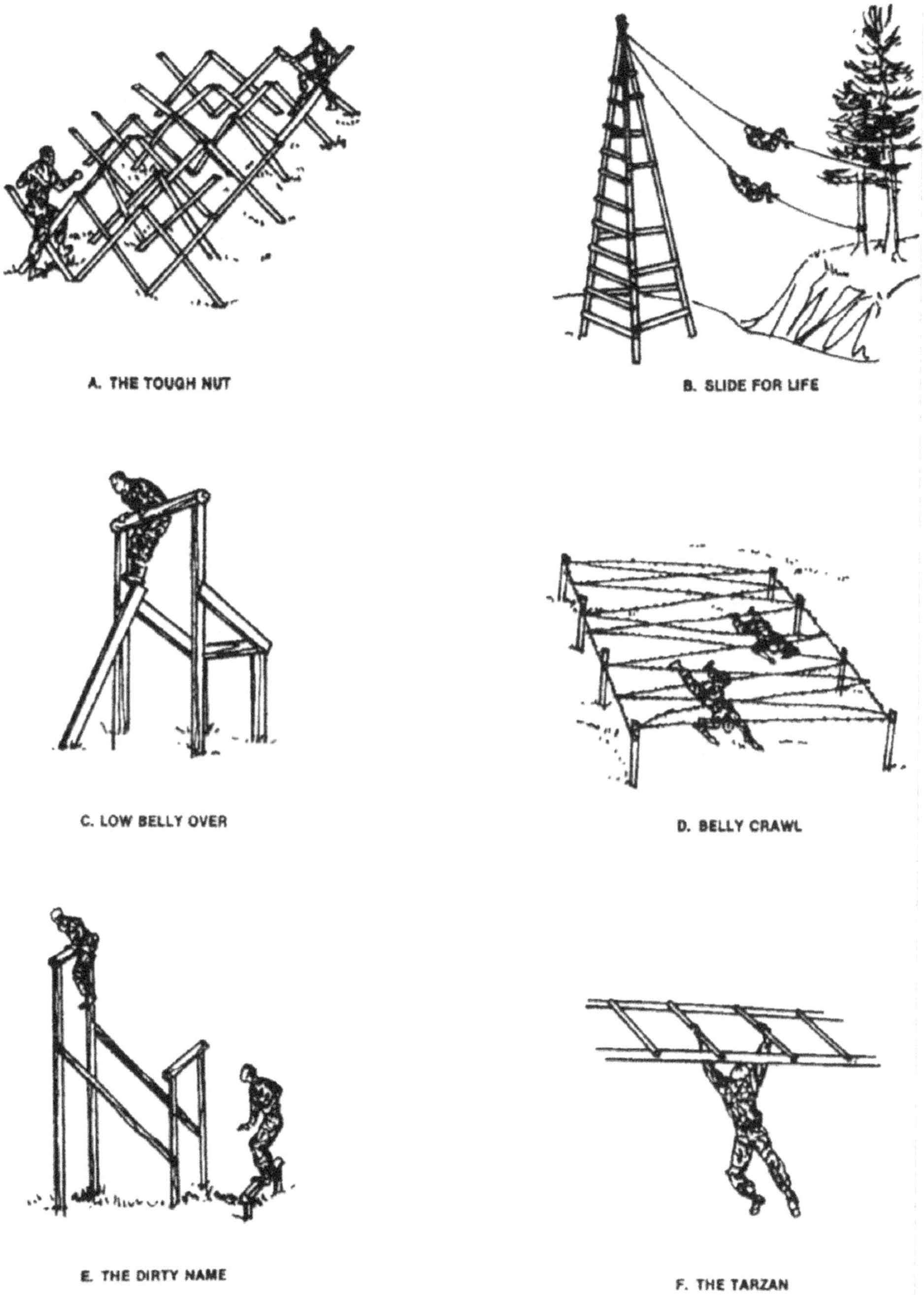

Figure 3-37.White Group.

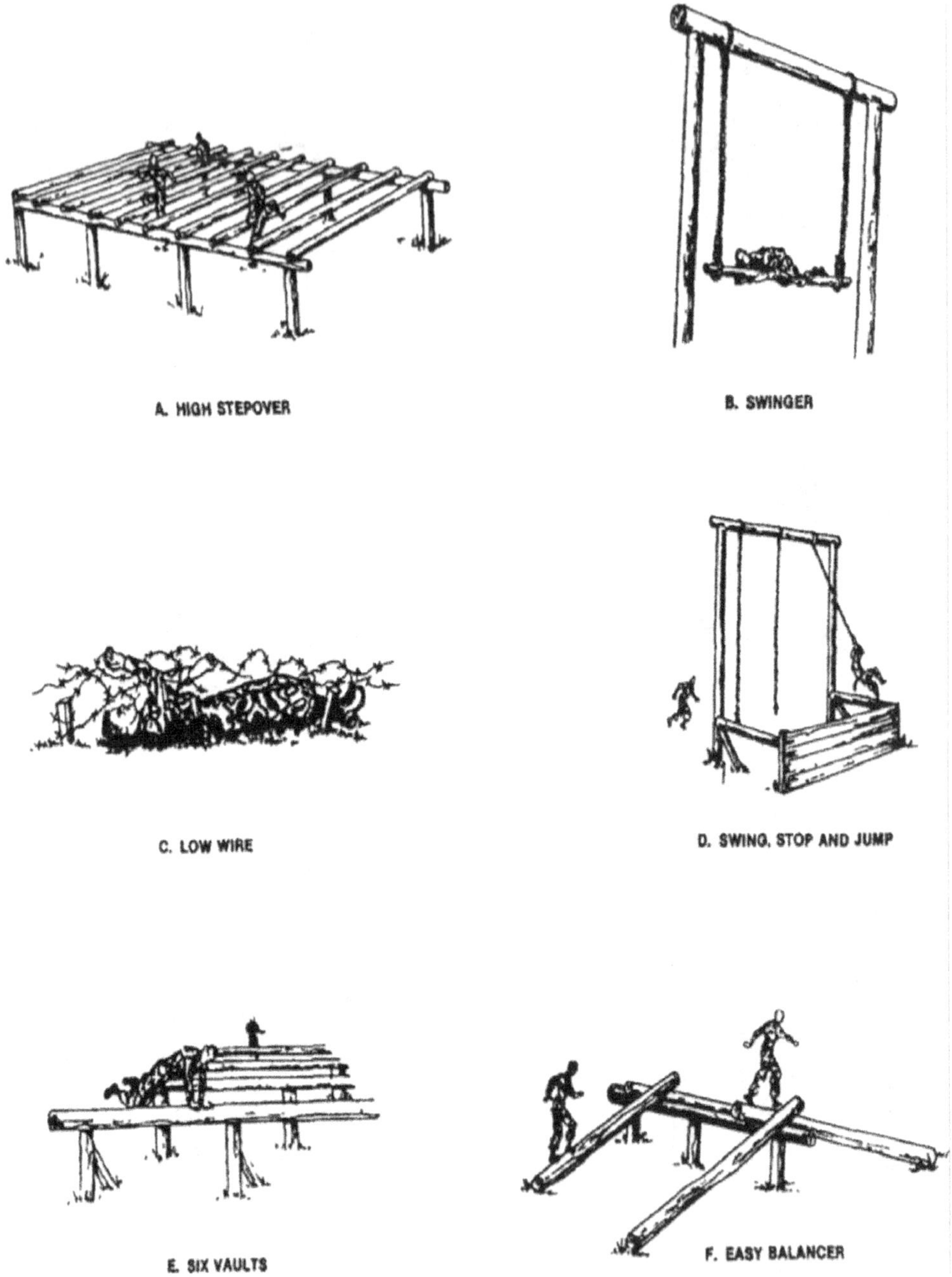

Figure 3-38.Blue Group.

A. INCLINE WALL

B. SKYSCRAPER

C. JUMP AND LAND

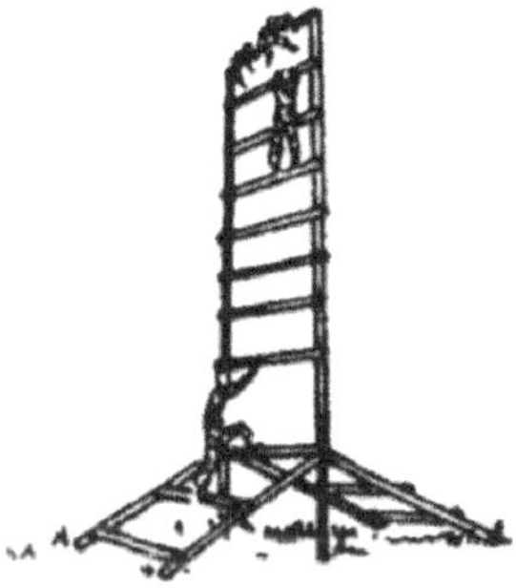

D. CONFIDENCE CLIMB

E. BELLY ROBBER

F. THE TOUGH ONE

Figure 3-39.Black Group.

Figure 3-40. Bench Conditioning, Table I.

Figure 3-41. Bench Conditioning, Table II.

PROGRESSION GUIDE

AGE GROUP	LEVEL	EXERCISES 1	2	3	4	5	6
17	A	15	18	14	15	15	250
to	B	13	16	13	13	13	235
29	C	11	14	12	11	11	[illegible]
30	A	13	14	12	13	13	200
to	B	11	13	11	11	11	185
39	C	9	12	10	9	9	165
40	A	11	11	10	11	11	150
to	B	9	10	9	9	9	135
44	C	7	9	8	7	7	120
45	A	9	8	8	9	9	100
to	B	7	7	7	7	7	90
49	C	5	6	6	5	5	80
50	A	7	6	6	7	7	75
to	B	5	5	5	5	5	70
59	C	3	4	4	3	3	60
60	A	4	5	4	4	4	50
and	B	3	4	3	3	3	40
over	C	2	3	2	2	2	30
Minutes for each exercise		2	1	1	1	2	5

Table I. Progression Guide.

Table I. Exercises.

PROGRESSION GUIDE

AGE GROUP	LEVEL	EXERCISES 1	2	3	4	5	6
17	A	17	17	17	9	19	300
to	B	15	15	15	8	17	270
29	C	13	[illegible]	13	7	15	245
30	A	15	15	15	8	17	235
to	B	13	13	13	7	15	210
39	C	11	11	11	6	13	190
40	A	13	13	13	7	15	175
to	B	11	11	11	6	13	155
44	C	9	10	9	5	11	135
45	A	11	11	11	6	13	125
to	B	9	9	9	5	11	110
49	C	7	7	7	4	9	100
50	A	9	9	9	5	11	95
to	B	7	7	7	4	9	85
59	C	5	5	5	3	7	75
60	A	6	7	7	4	9	70
and	B	5	5	5	3	7	60
over	C	4	4	4	2	5	50
Minutes for each exercise		1	1	1	1 1/2	1 1/2	6

Table II. Progression Guide.

Table II. Exercises.

PROGRESSION GUIDE

AGE GROUP	LEVEL	EXERCISES 1	2	3	4	5	6
17	A	10	19	19	16	10	350
to	B	9	17	17	15	9	315
29	C	8	15	15	14	8	280
30	A	9	17	17	14	9	270
to	B	8	15	15	13	8	240
39	C	7	13	13	12	7	210
40	A	8	15	15	12	8	200
to	B	7	13	13	11	7	180
44	C	6	11	11	10	6	160
45	A	7	13	13	10	7	150
to	B	6	11	11	9	6	135
49	C	5	9	9	8	5	120
50	A	6	11	11	8	6	115
to	B	5	9	9	7	5	105
59	C	4	7	7	6	4	95
60	A	5	9	9	7	5	90
and	B	4	7	7	6	4	80
over	C	3	5	5	4	3	70
Minutes for each exercise		1½	1	1	1½	1	6

Table III. Progression Guide.

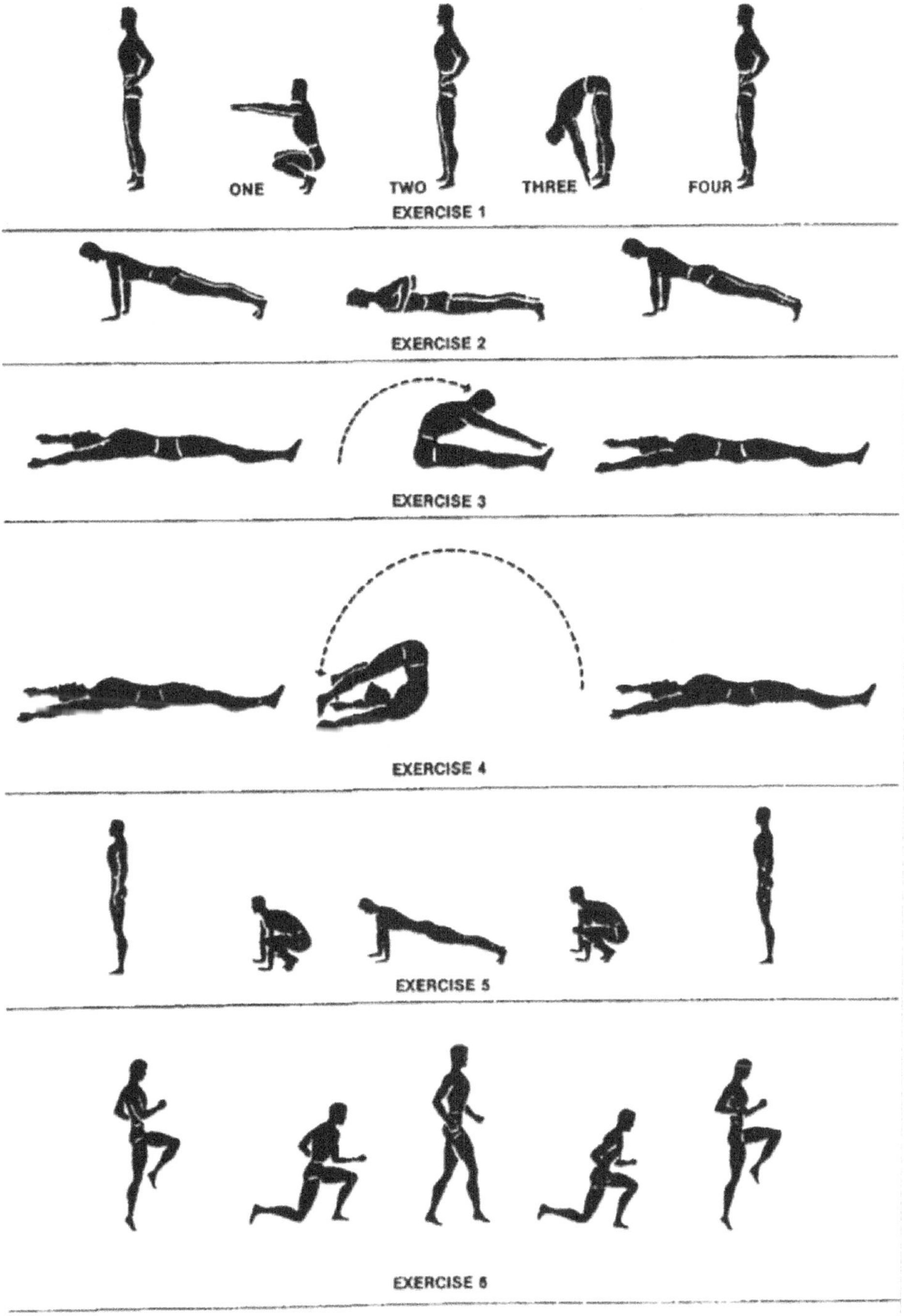

Table III. Exercises.

PROGRESSION GUIDE

AGE GROUP	LEVEL	EXERCISES 1	2	3	4	5	6
17	A	12	9	12	24	25	400
to	B	11	8	11	22	23	380
29	C	10	7	10	21	21	360
30	A	11	8	11	23	23	305
to	B	10	7	10	21	21	290
39	C	9	6	9	20	20	275
40	A	10	7	10	20	21	225
to	B	9	6	9	18	18	215
44	C	8	5	8	16	14	205
45	A	8	6	8	16	16	175
to	B	7	5	7	14	14	165
49	C	6	4	6	12	12	155
50	A	6	5	6	13	13	135
to	B	5	4	5	11	11	130
59	C	4	3	4	10	10	120
60	A	5	4	5	10	10	100
and	B	4	3	4	9	9	95
over	C	3	2	3	8	8	90
Minutes for each exercise		1	2	1	1	1	6

Table IV. Progression Guide.

Table IV. Exercises.

PROGRESSION GUIDE

AGE GROUP	LEVEL	EXERCISES 1	2	3	4	5	6
17	A	14	13	28	14	30	450
to	B	13	12	27	13	28	430
29	C	12	11	26	12	26	410
30	A	12	12	25	12	26	350
to	B	11	11	24	11	24	330
39	C	10	10	23	10	22	310
40	A	11	11	23	11	23	250
to	B	10	10	21	10	21	240
44	C	9	9	19	9	19	230
45	A	9	9	20	9	20	200
to	B	8	8	18	8	18	190
49	C	7	7	16	7	16	180
50	A	7	7	16	7	16	170
to	B	6	6	14	6	14	155
59	C	5	5	12	5	12	140
60	A	6	6	12	6	12	115
and	B	5	5	11	5	10	110
over	C	4	4	9	4	9	105
Minutes for each exercise		2	1	1	2	1	5

Table V. Progression Guide.

EXERCISE 1

EXERCISE 2

EXERCISE 3

EXERCISE 4

EXERCISE 5

EXERCISE 6

Table V. Exercises.

PROGRESSION GUIDE

AGE GROUP	LEVEL	EXERCISES 1	2	3	4	5	6
17	A	17	15	32	32	35	500
to	B	16	14	30	30	33	480
29	C	15	13	28	28	31	460
30	A	15	13	30	30	31	400
to	B	14	12	28	28	29	380
39	C	13	11	26	26	27	360
40	A	13	10	27	27	27	310
to	B	12	9	25	25	25	285
44	C	11	8	23	23	23	265
45	A	11	9	23	23	23	250
to	B	10	8	21	21	21	230
49	C	9	7	19	19	19	210
50	A	9	8	19	19	19	200
to	B	8	7	17	17	17	190
59	C	7	6	15	15	15	175
60	A	8	7	15	15	17	140
and	B	7	6	13	13	15	130
over	C	5	5	10	10	12	120
Minutes for each exercise		2	1	1	1	1	6

Table VI. Progression Guide.

Table VI. Exercises.

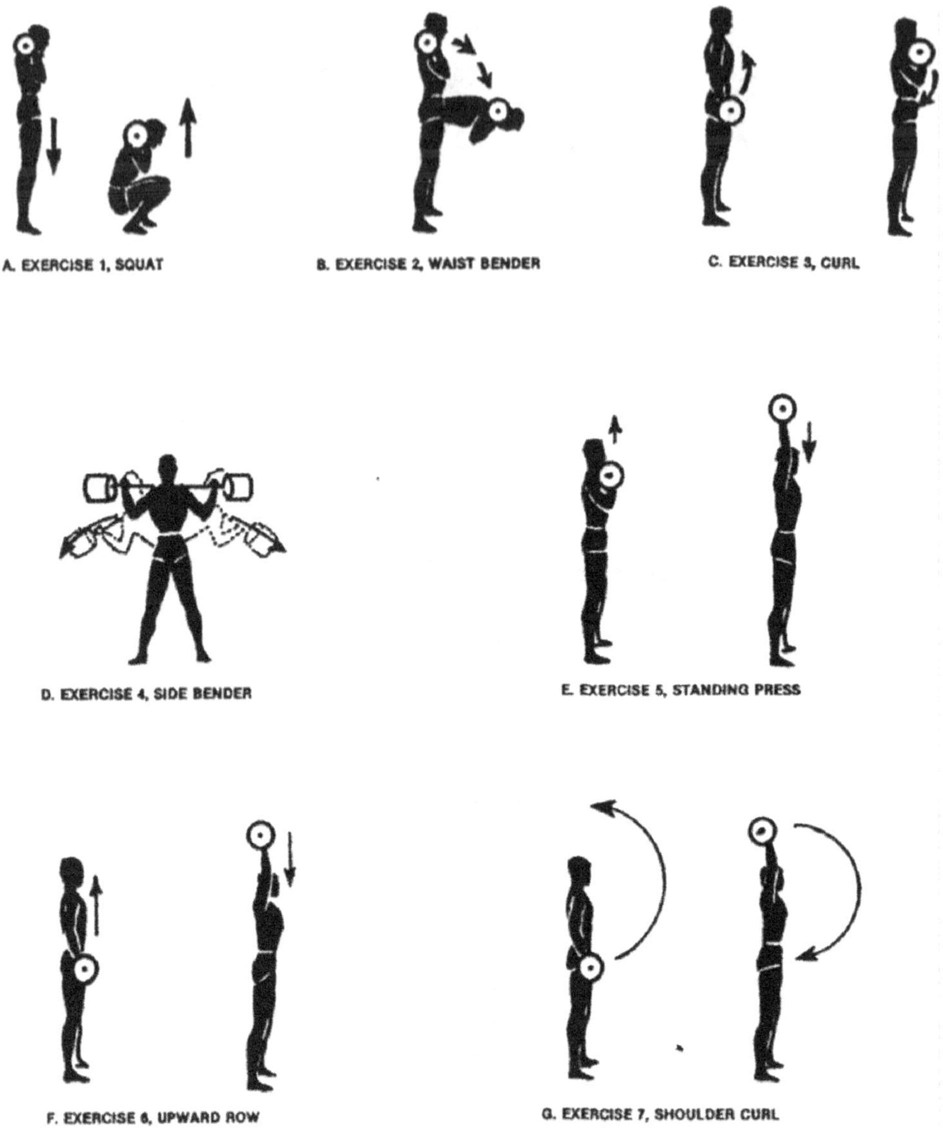

Figure 3-42. Weight Training.

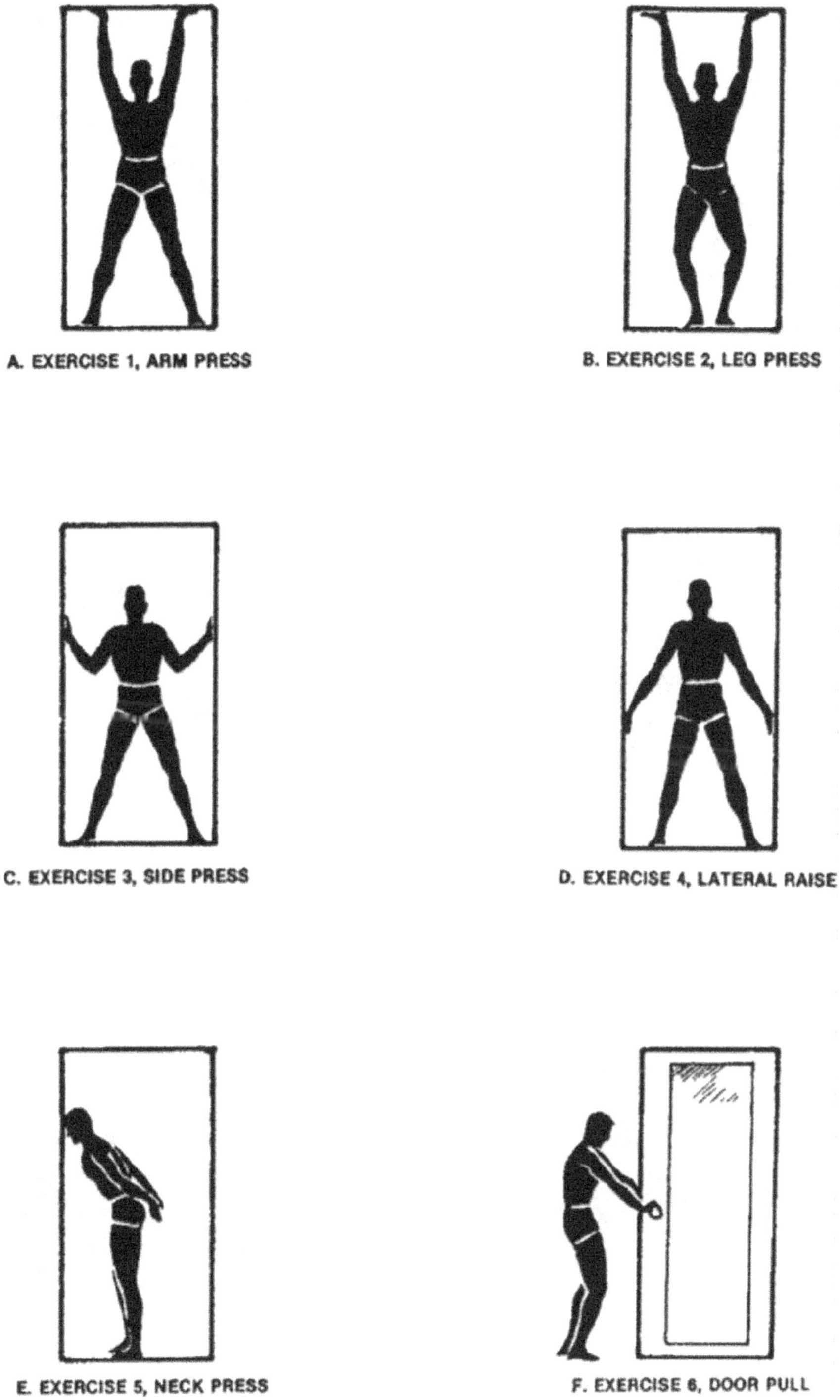

Figure 3-43. Isometric Exercises.

Figure 5-1.Combatives Table I.

Figure 5-2.Combatives Table II.

Figure. 5-3.Combatives Table III.

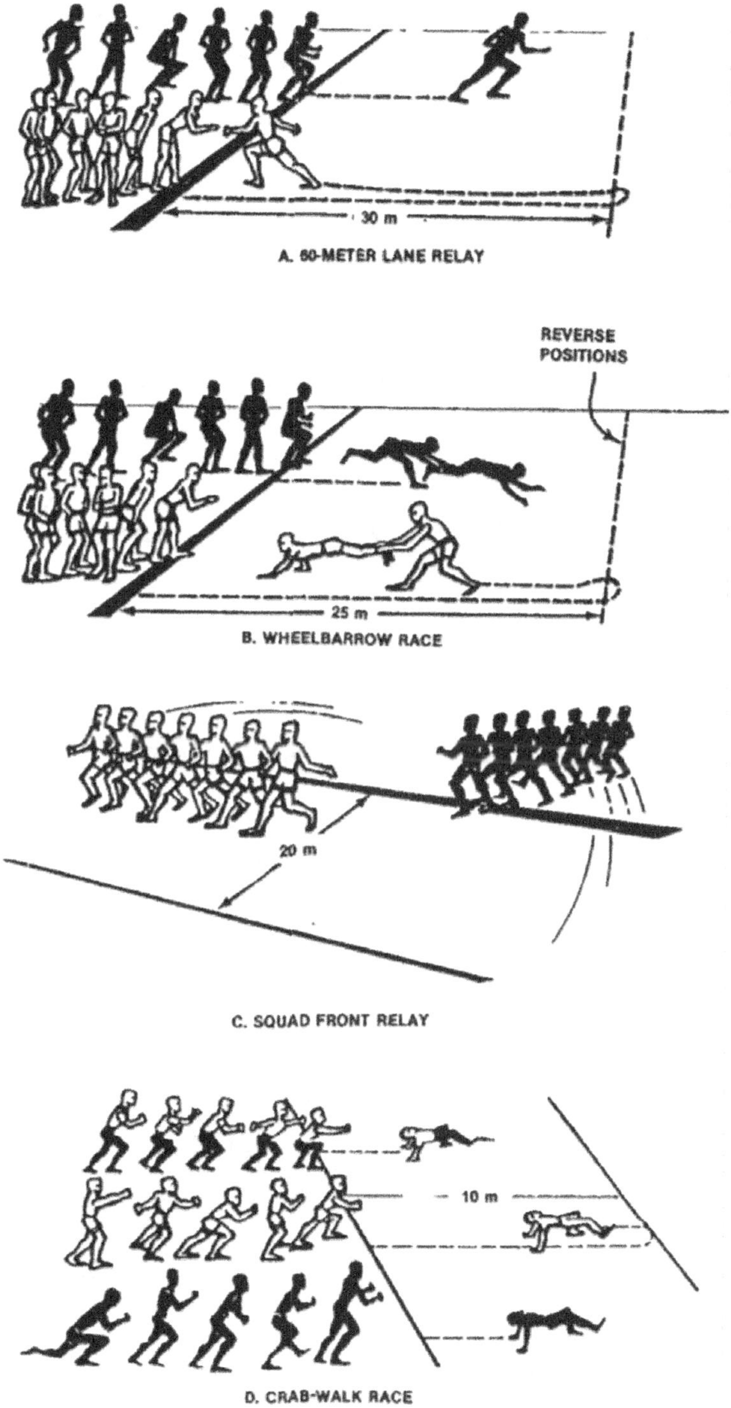

Figure 5-4.Relay Table I.

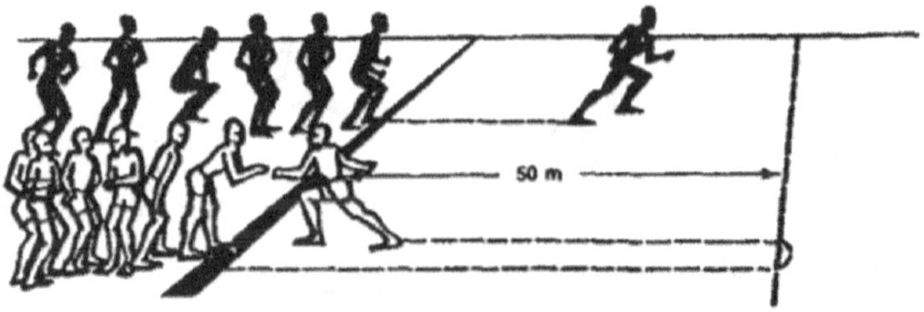

A. 100-METER LANE RELAY

B. FROG-JUMP RELAY

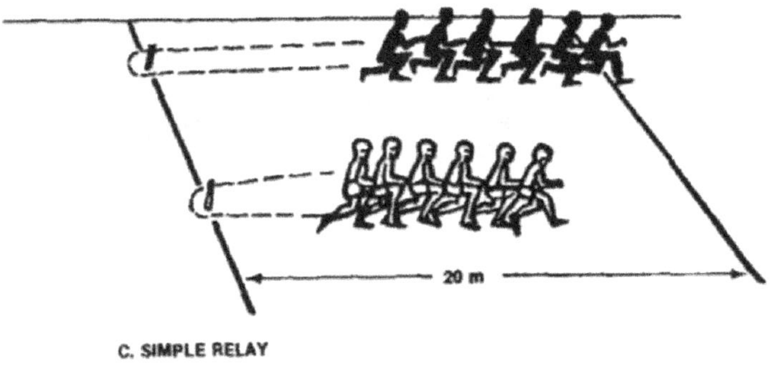

C. SIMPLE RELAY

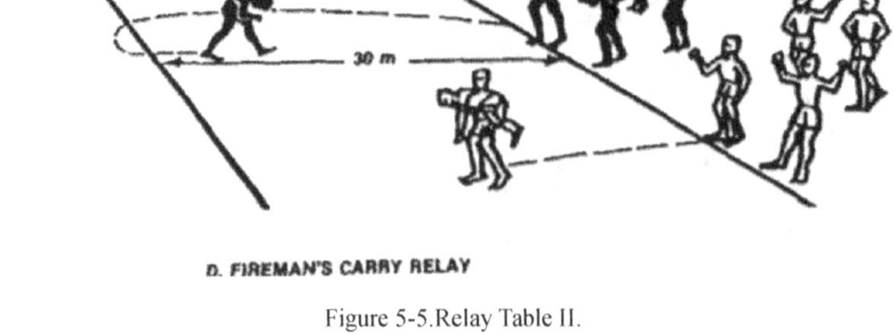

D. FIREMAN'S CARRY RELAY

Figure 5-5.Relay Table II.

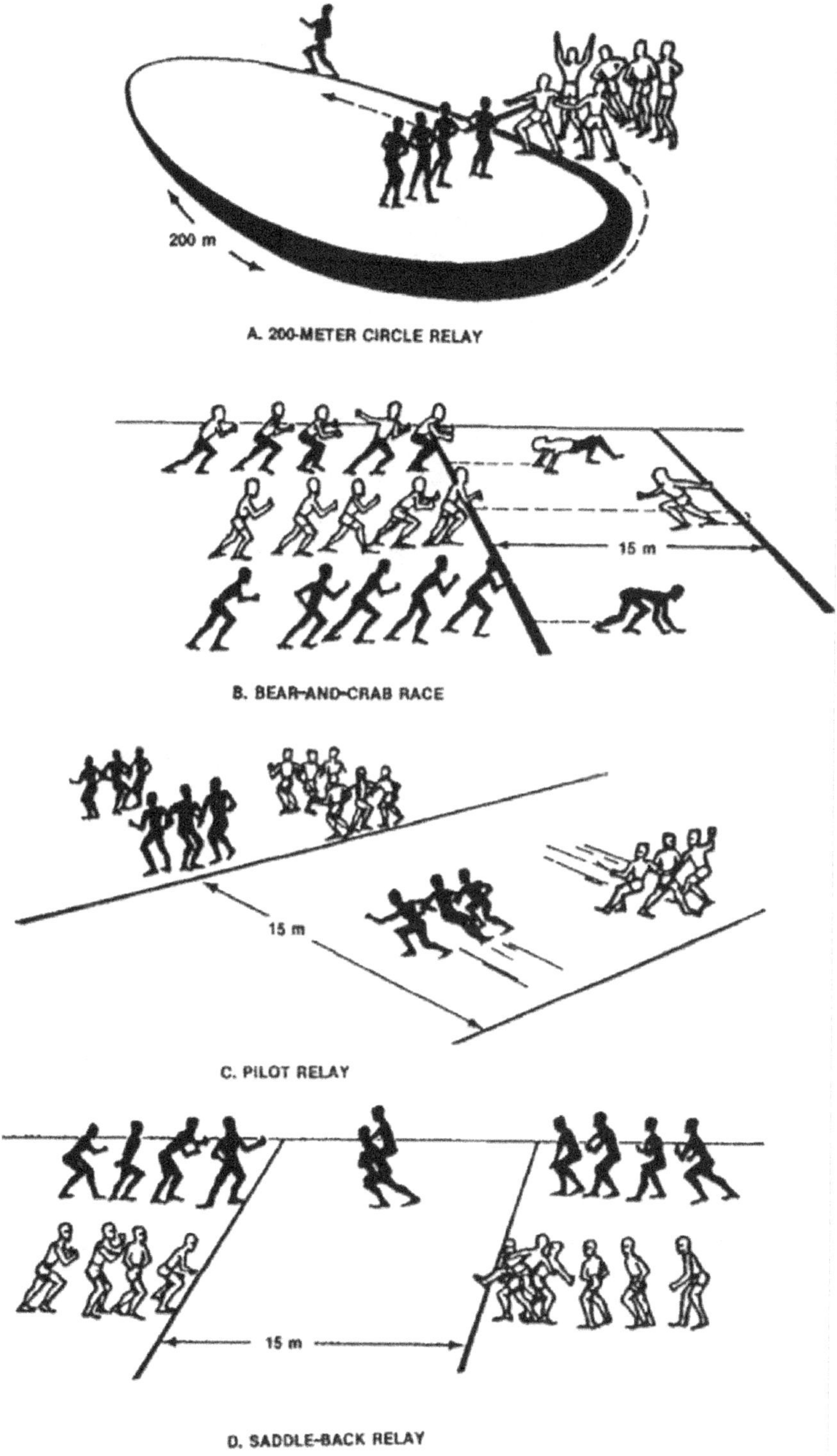

Figure 5-6.Relay Table III.

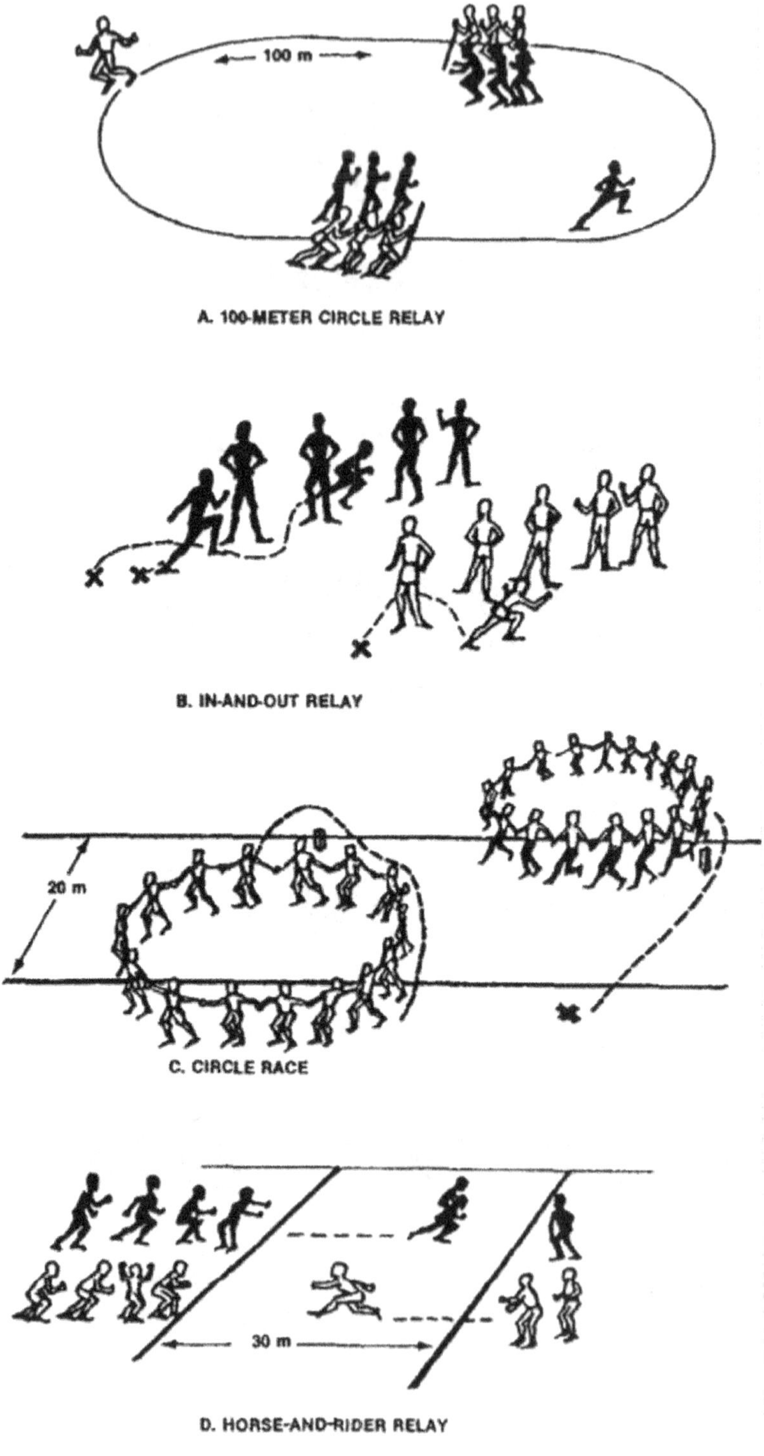

Figure 5-7.Relay Table IV.

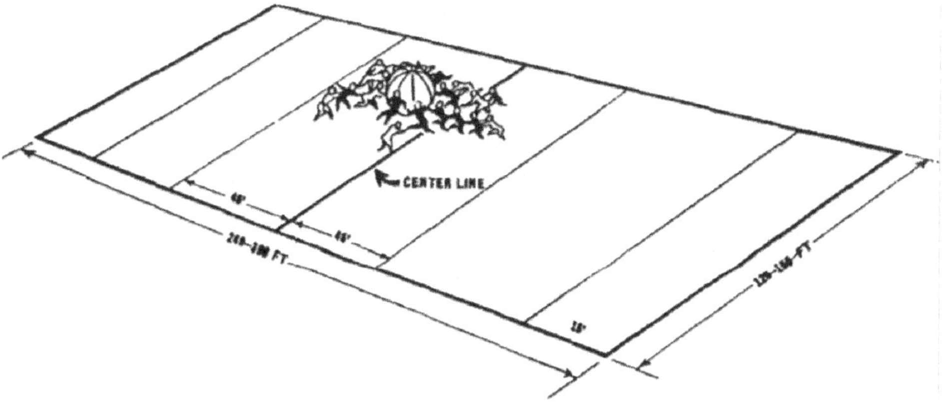

Figure 5-8.Pushball.

Figure 5-9.Line Rush.

Figure. 5-10. Human Tug-of-War.

Figure 5-11. Master of the Ring.

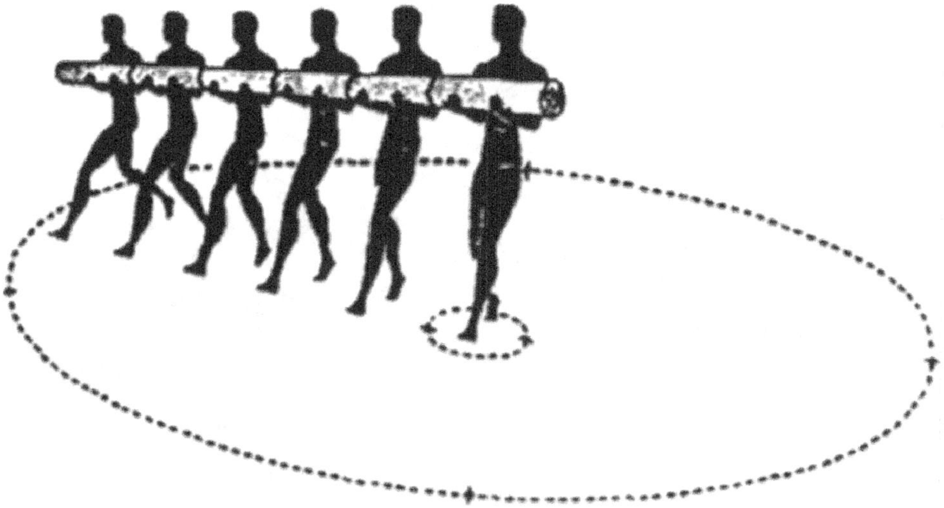

Figure 5-12.Log Pivot Circle.

Figure 5-13.Log-Rolling Race.

Figure 5-14. Prone Pushing Contest.

Figure 5-15. Shuttle Relay Race.

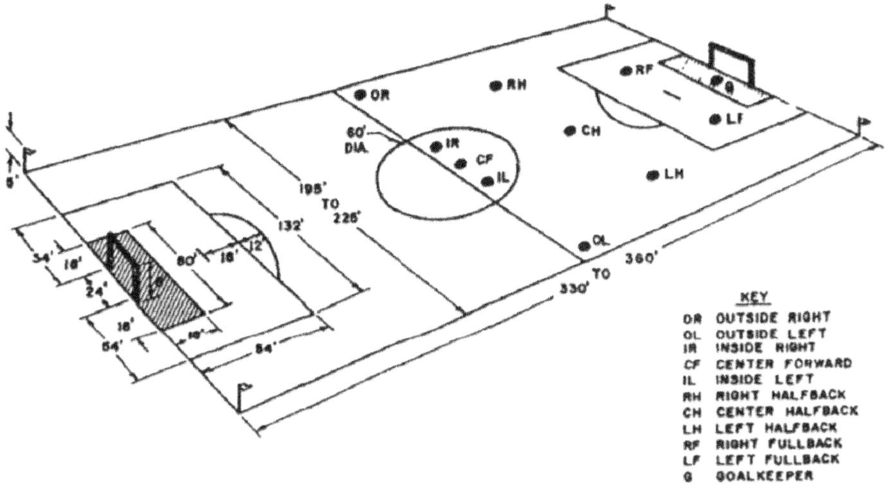

Figure 5-16. Soccer.

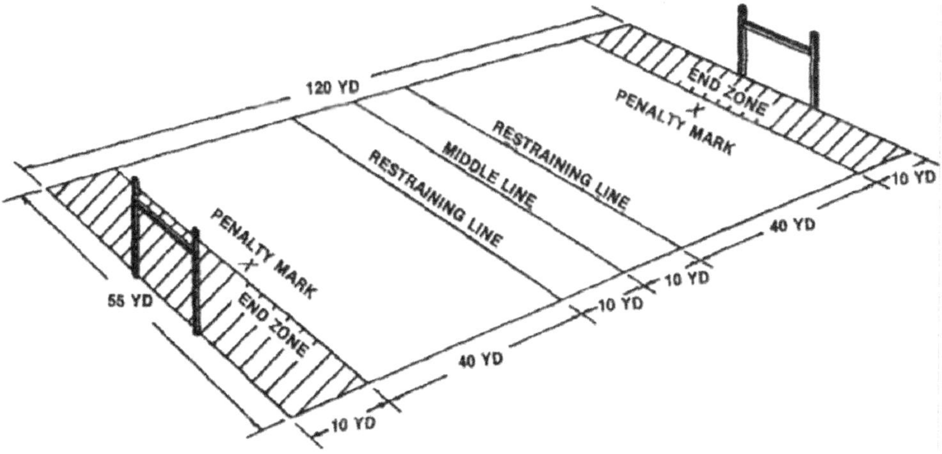

Figure 5-17.Speedball Field.

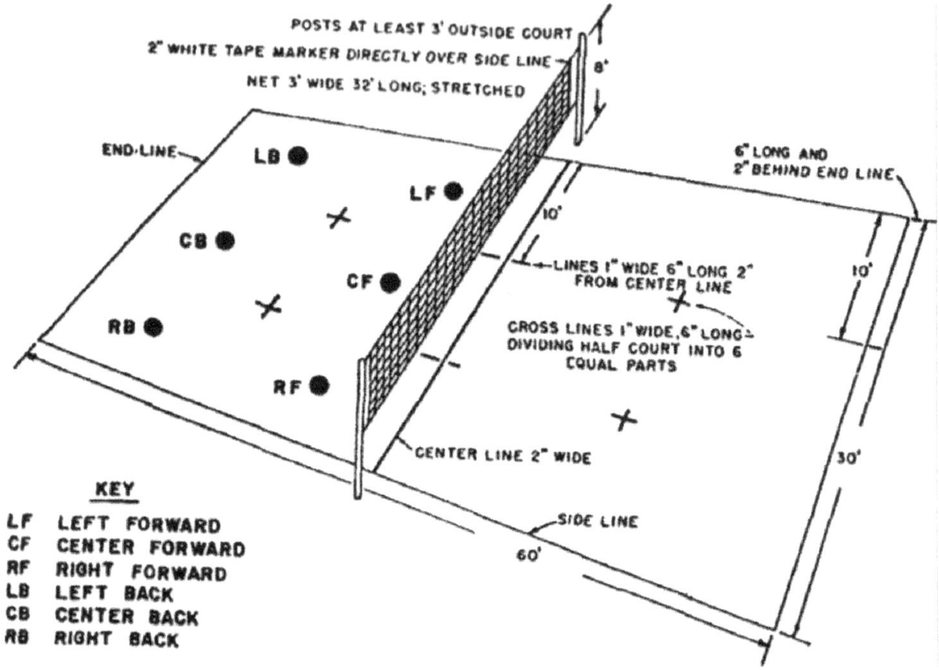

Figure 5-18. Volleyball.

PTS	UPHILL CLIMB	ROPE CLIMB	EVACUATION	FIRE AND MANEUVER	FORCED MARCH
100	1:20-BELOW	0:10-BELOW	0:25-BELOW	1:20-BELOW	26:00-BELOW
99	1:21-1:22	0:11		1 21-1:22	26:01-26:12
98	1:23-1:24	0:12	0:26	1:23-1:24	26:13-26:24
97	1:25-1:26	0:13		1:25-1:26	26:25-26:36
96	1:27-1:28	0:14	0:27	1:27-1:28	26:37-26:48
95	1:29-1:30	0:15		1:29-1:30	26:49-27:00
94	1:31-1:32	0:16	0:28	1:31-1:32	27:01-27:12
93	1:33-1:34	0:17		1 33-1.34	27:13-27:24
92	1:35-1:36	0:18	0:29	1:35-1.36	27:25-27:36
91	1:37-1:38	0:19		1:37-1:38	27:37-27:48
90	1:39-1:40	0:20	0:30	1:39-1:40	27:49-28:00
89	1:41-1:42	0:21	0:31	1:41-1:42	28:01-28:12
83	1:43-1:44	0:22	0:32	1:43-1:44	28:13-28:24
87	1:45-1:46	0:23	0:33	1:45-1:46	28:25-28:36
86	1:47-1:48	0:24	0:34	1.47-1:48	28:37-28:48
85	1:49-1:50	0:25	0:35	1.49-1:50	28:49-29:00
84	1:51-1:52	0:26	0:36	1:51-1:52	29:01-29:12
83	1:53-1:54	0:27	0.37	1:53-1:54	29:13-29:24
82	1:55-1:56	0:28	0:38	1:55-1:56	29:25-29:36
81	1:57-1:58	0:29	0:39	1:58-1:58	29:37-29:48
80	1:59-2.00	0:30	0:40	1:59-2:00	29:49-30:00
79	2.01-2:02	0:31	0:41	2:01-2:02	30:01-30:12
78	2:03-2:04	0:32	0:42	2:03-2.04	30:13-30:24
77	2:05-2:06	0.33	0:43	2:05-2:06	30:25-30:36
76	2:07-2:08	0.34	0.44	2:07-2:08	30:37-30:48
75	2:09-2:10	0:35	0:45	2:09-2:10	30:49-31:00
74	2:11-2:12	0:36	0:46	2:11-2 12	31.01-31:12
73	2:13-2:14	0:37	0.47	2.13-2:14	31:13-31:24
72	2:15-2:16	0:38	0:48	2.15-2:16	31:25-31:36
71	2:17-2:18	0:39	0:49	2:17-2:18	31:37-31:48
70	2:19-2:20	0:40	0:50	2:19-2.20	31:49-32.00
69	2.21-2:22	0:41	0:51	2:21-2:22	32:01-32:12
68	2:23-2:24	0:42	0:52	2:23-2:24	32:13-32:24
67	2:25-2.26	0:43	0:53	2:25-2:26	32:25-32:36
66	2:27-2:28	0:44	0:54	2.27-2:28	32.37-32:48
65	2:29-2:30	0.45	0.55	2.29-2 30	32.49-33:00

Figure 6-1.Physical Readiness Test Scoring Table for Male.

	EXAMPLE 1	EXAMPLE 2	EXAMPLE 3
Event #1	100 pts	100 pts	99 pts
Event #2	100 pts	99 pts	86 pts
Event #3	85 pts	75 pts (F)	80 pts
Event #4	90 pts	79 pts (F)	100 pts
Event #5	79 pts	85 pts	90 pts
Pts	454 pts	438 pts	455 pts
Final Score	79.0% (F)	69.0% (F)	91.0%
(F) = Failure			

Figure 6-2.Scoring Examples.

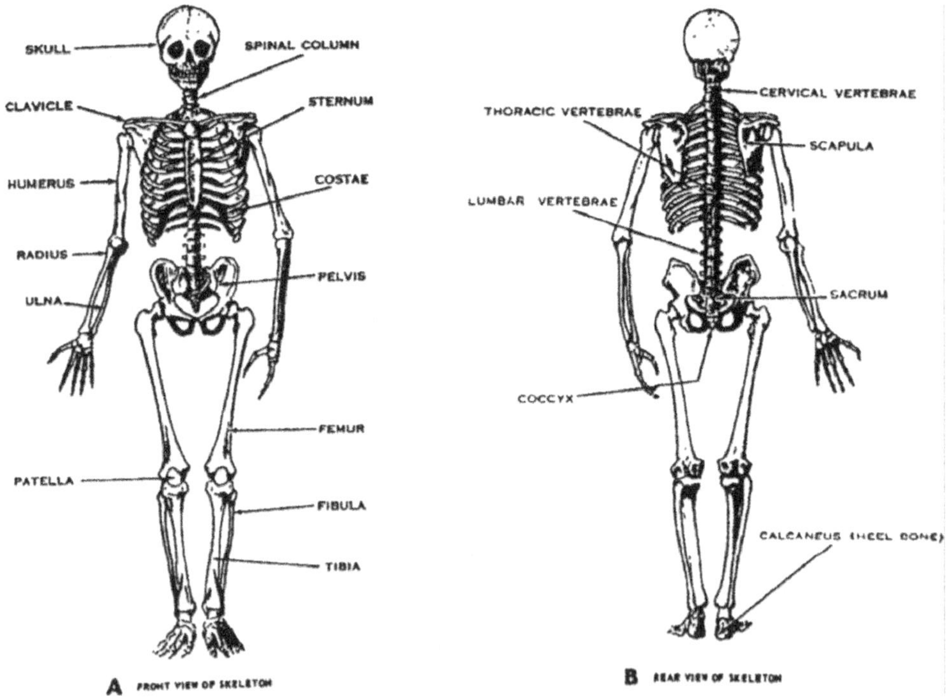

Figure 7-1.Front and Rear Views of Skeleton.

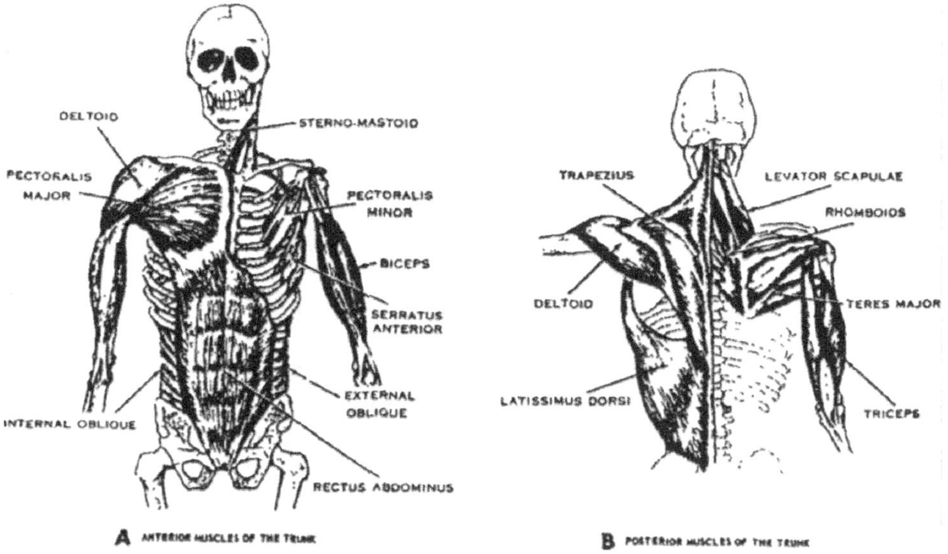

Figure 7-2.Muscles of the Trunk.

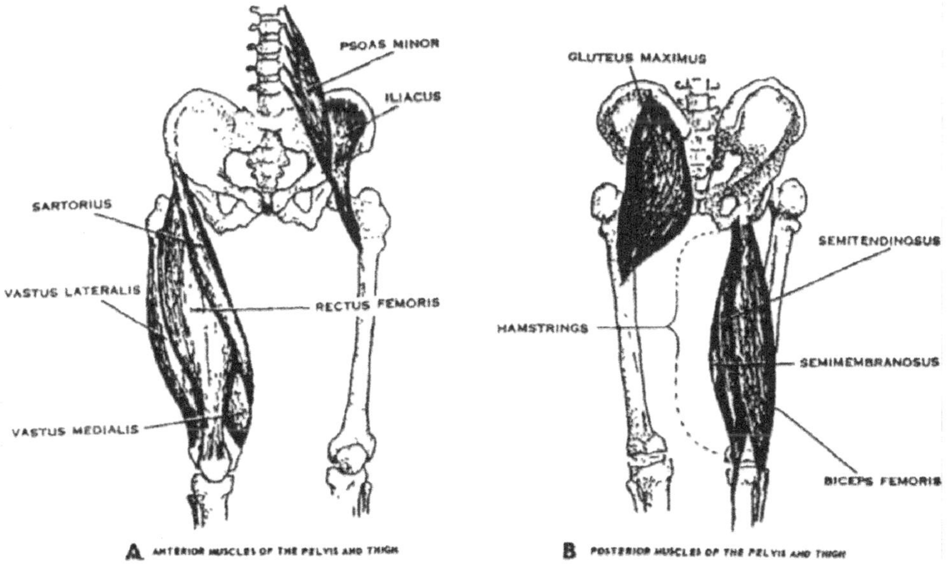

Figure 7-3.Anterior and Posterior Muscles of the Pelvisand Thigh.

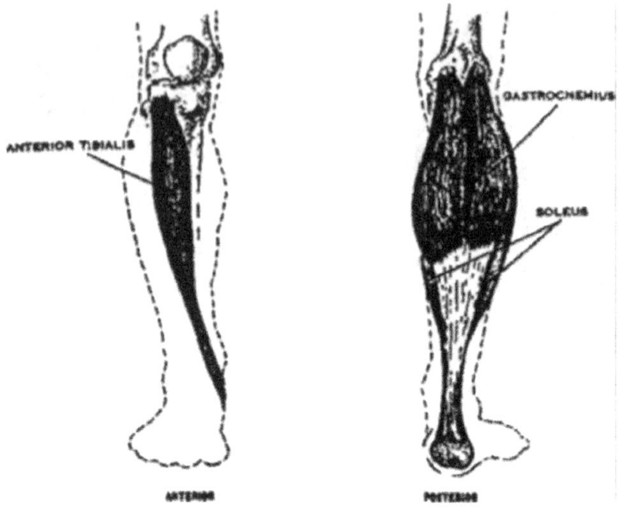

Figure 7-4. Muscles of Lower Leg.

Figure 7-5.Circulation of Blood from Heart to the Body.

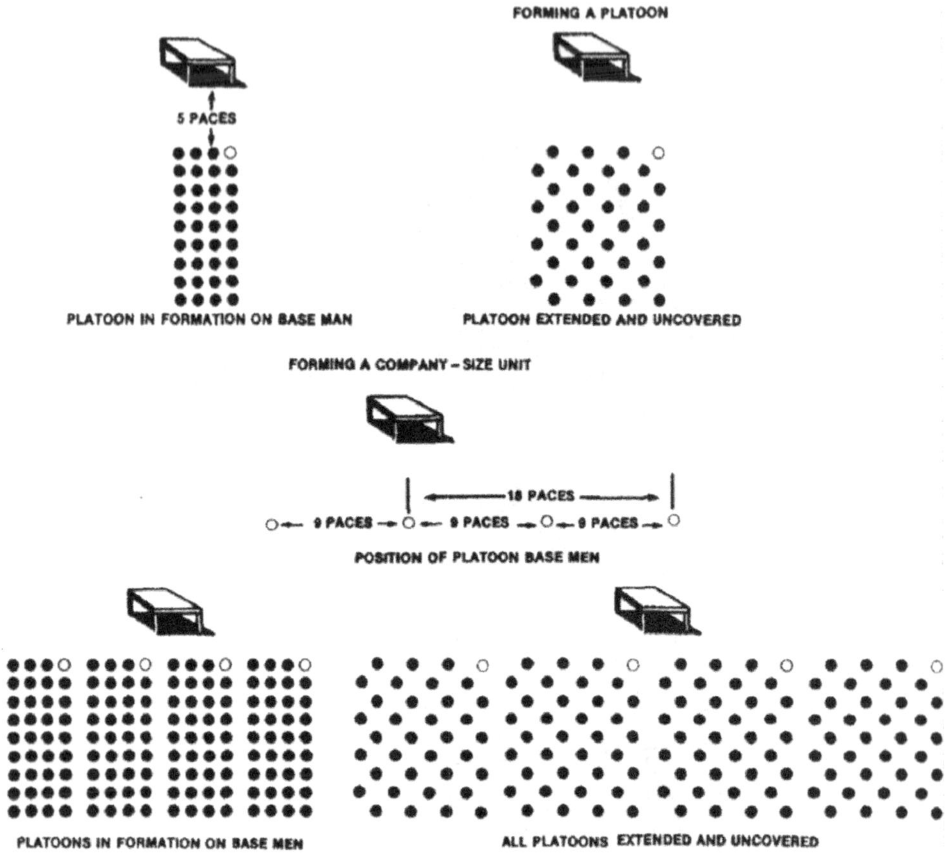

Figure A-l. Forming the Extended Rectangular Formation.

Figure A-2.Circle Formation.

Table of Contents

Chapter 3. Weapons Handling

Chapter 4. Fundamentals of Marksmanship

Chapter 5. Rifle Firing Positions

Chapter 6. Use of Cover and Concealment

Chapter 7. Rifle Presentation

Chapter 8. Effects of Weather

Chapter 9. Zeroing

Chapter 10. Engagement Techniques

Appendices

CHAPTER 1. INTRODUCTION TO RIFLE MARKSMANSHIP

All Marines share a common warfighting belief: "Every Marine a rifleman." This simple credo reinforces the belief that all Marines are forged from a common experience, share a common set of values, and are trained as members of an expeditionary force in readiness. As such, there are no "rear area" Marines, and no one is very far from the fighting during expeditionary operations. The Marine rifleman of the next conflict will be as in past conflicts: among the first to confront the enemy and the last to hang his weapon in the rack after the conflict is won.

1001. ROLE OF THE MARINE RIFLEMAN

Marine Corps forces are employed across the entire range of military operations. At one end is war, which is characterized by large-scale, sustained combat operations. At the other end of the scale are those actions referred to as military operations other than war (MOOTW). MOOTW focuses on deterring aggression, resolving conflict, promoting peace, and supporting civil authorities. These operations can occur before, during, and after combat operations. Training and preparation for MOOTW should not detract from the Corps' primary mission of training Marines to fight and win in combat. MOOTW normally does not involve combat. However, Marines always need to be prepared to protect themselves and respond to changing threats and unexpected situations. Whenever the situation warrants the application of deadly force, the Marine rifleman must be able to deliver well aimed shots to eliminate the threat. Sometimes the need for a well aimed shot may even be heightened by the presence of noncombatants in close proximity to the target. The proficient rifleman handles this challenge without unnecessarily escalating the level of violence or causing unnecessary collateral damage. The Marine rifleman must have the versatility, flexibility, and skills to deal with a situation at any level of intensity across the entire range of military operations.

To be combat ready, the Marine must be skilled in the techniques and procedures of rifle marksmanship and take proper care of his rifle. Even when equipped with the best rifle in the world, a unit with poorly trained riflemen cannot be depended upon to accomplish their mission. Usually, poorly trained riflemen either fail to fire their weapon or they waste ammunition by firing ineffectively. To send Marines into harm's way without thorough training in the use of their individual weapons carries undue risks for every Marine in the unit. On the other hand, well trained riflemen can deliver accurate fire against the enemy under the most adverse conditions. A well trained rifleman is not only confident that he can help his unit accomplish it's mission, he is confident that he can protect his fellow Marines and himself.

1002. CONDITIONS AFFECTING MARKSMANSHIP IN COMBAT

Many factors affect the application of marksmanship in combat; among them are—

- Most targets are linear in nature and will consist of a number of men or objects irregularly spaced along covered or concealed areas.
- Most targets can be detected by smoke, flash, dust, noise or movement, but will only be visible for a brief moment before taking cover.
- The nature of the target, irregularities of terrain and vegetation will often require a rifleman to use a position other than prone in order to fire effectively on the target.
- The time in which a target can be engaged is often fleeting.

1003. COMBAT MINDSET

In a combat environment, the Marine must be constantly prepared for possible target engagement. When a target presents itself in combat, there may be very little time to take action. A Marine must be able to engage the target quickly and accurately. The unique demands of combat (i.e., stress, uncertainty) dictate that the Marine be both physically and mentally prepared to engage enemy targets. It will not be enough to simply know marksmanship techniques and procedures. The Marine must develop the mental discipline to prepare for enemy contact. In the confusion, noise, and stress of the combat environment, the Marine must have the ability to eliminate any hesitation, fear or uncertainty of action and to focus on the actions required to fire well-aimed shots. This is accomplished through establishment of a combat mindset. The key factors in the development of a combat mindset include both physical and mental preparation.

Physical Preparation

In combat, targets can present themselves without warning. Therefore, it is essential for the Marine to maintain proper balance and control of his weapon at all times so he can quickly assume a firing position, present the weapon, and accurately engage the target. However, speed alone does not equate to effective target engagement. The Marine should fire only as fast as he can fire accurately, never exceeding his physical ability to apply the fundamentals of marksmanship. To be effective in combat, the Marine must train to perfect the physical skills of shooting so those skills become second nature. Mastery of physical skills allow the Marine to concentrate on the mental aspects of target engagement; e.g., scanning for targets, detection of targets, selection and use of cover. The more physical skills that a Marine can perform automatically, the more concentration he can give to the mental side of target engagement.

Mental Preparation

While combat is unpredictable and constantly changing, the Marine can prepare himself mentally for confrontation with the enemy. The stress of battle, coupled with the often limited time available to engage targets, requires concentration on the mental aspects of target engagement; e.g., scanning for targets, detection of targets, and the selection and use of cover.

Knowledge of the Combat Environment

The Marine must be constantly aware of the surroundings to include the terrain, available cover, possible areas of enemy contact, backdrop of the target, etc. This awareness will enable the Marine to select and assume a firing position and to quickly and accurately engage targets.

Plan of Action

In combat, the situation will dictate the action to be taken. The Marine must understand the situation, identify and evaluate possible courses of action, and then develop a plan for target engagement that accomplishes the mission.

Confidence

The Marine must believe in his ability to engage targets accurately in any combat situation. A Marine's level of confidence is rooted in the belief that future challenges will be overcome—particularly the challenge of firing well aimed shots in a combat environment where the enemy may be returning fire. A key factor in a Marine's level of confidence is the degree to which he has mastered the techniques and procedures of the rifle marksmanship. Mastery of rifle marksmanship can only be obtained by quality instruction. Quality instruction is the foundation for practical application of the marksmanship fundamentals during range and field firing.

Chapter 2. Introduction to the M16A2 Service Rifle

Note

+The procedures in this manual are written for right-handed Marines; left-handed Marines should reverse instructions as necessary.

2001. DESCRIPTION

The M16A2 service rifle is a lightweight, 5.56 millimeter (mm), magazine-fed, gas-operated, air-cooled, shoulder-fired rifle. (Fig. 2-1 shows a right-side view and fig. 2-2, on page 2-2, shows the left-side view.) The rifle fires in either semiautomatic (single-shot) mode or a three-round burst through the use of a selector lever. The M16A2 service rifle has a maximum effective range of 550 meters for individual or point targets. The bore and chamber are chrome-plated to reduce wear and fouling. The handguards are aluminum-lined and are vented to permit air to circulate around the barrel for cooling purposes and to protect the gas tube. An aluminum receiver helps reduce the overall weight of the rifle. The trigger guard is equipped with a spring-loaded retaining pin that, when depressed, allows the trigger guard to be rotated out of the way for access to the trigger while wearing heavy gloves. An ejection port cover prevents dirt and sand from getting into the rifle through the ejection port. The ejection port cover should be closed when the rifle is not being fired. It is automatically opened by the action of the bolt carrier. The muzzle compensator serves as a flash suppressor and assists in reducing muzzle climb.

2002. OPERATIONAL CONTROLS

Selector Lever

The selector lever has three settings—safe, semi, and burst. The setting selected depends on the firing situation. See figure 2-3 on page 2-2.

Safe

The selector lever in the safe position prevents the rifle from firing.

Semi

The selector lever in the semi position allows one shot to be fired with each pull of the trigger.

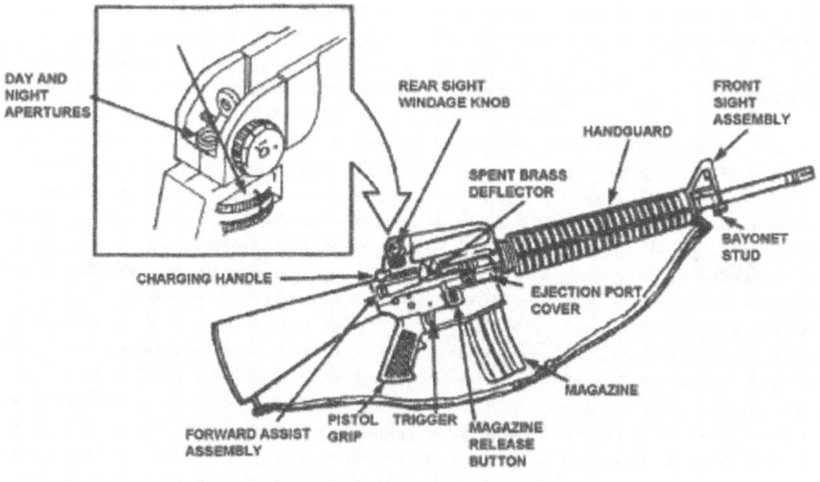

Figure 2-1. M16A2 Service Rifle (Right Side View).

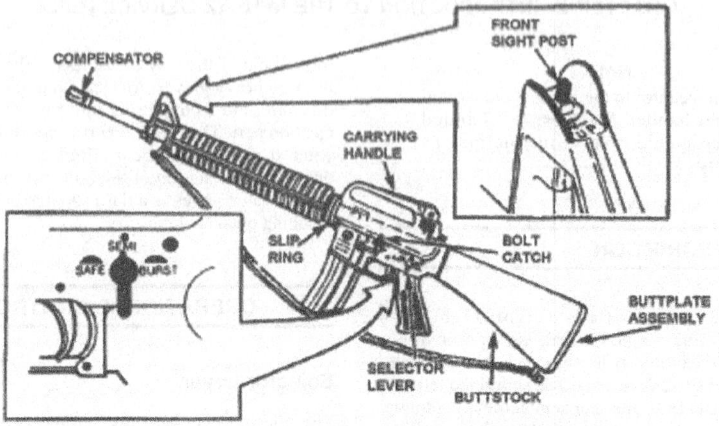

Figure 2-2. M16A2 Service Rifle (Left Side View).

Burst

The selector lever in the burst position allows the rifle to continue its cycle of operation until interrupted by the burst cam. With each pull of the trigger, the burst cam limits the maximum number of rounds fired to three. The burst cam is not "self-indexing." If burst is selected, the burst cam does not automatically reset to the first shot position of the three-round burst. One, two or three shots may be fired on the first pull of the trigger. Each subsequent pull of the trigger results in a complete three-round burst unless the trigger is released before the cycle is complete. If the trigger is released during the burst and the three-round cycle is interrupted, the next pull of the trigger fires the rounds remaining in the interrupted three-round cycle.

Magazine Release Button

The magazine release button releases the magazine from the magazine well. See figure 2-4.

Charging Handle

When the charging handle is pulled to the rear, the bolt unlocks from the barrel extension locking lugs and the bolt carrier group moves to the rear of the receiver. See figure 2-5.

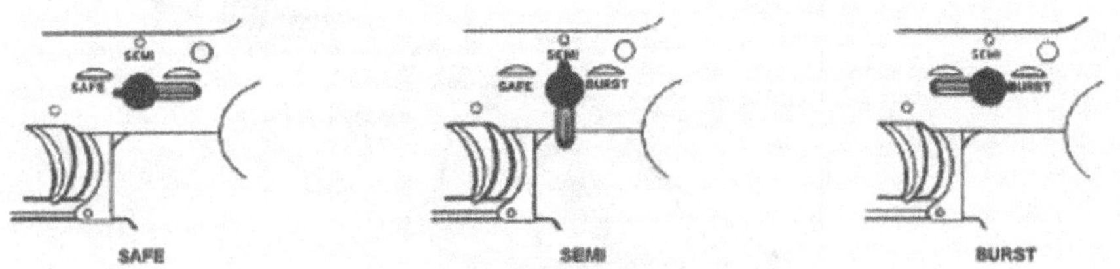

Figure 2-3. Selector Lever.

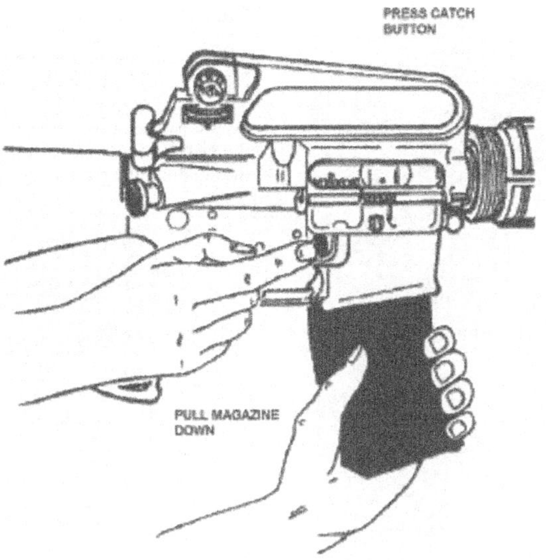

Figure 2-4. Magazine Release Button.

Bolt Catch

If the charging handle is pulled to the rear when the lower portion of the bolt catch is depressed, the bolt carrier group will lock to the rear. When the bolt carrier group is locked to the rear and the upper portion of the bolt catch is depressed, the bolt carrier group will slide forward, driven by the buffer assembly and action spring, into the firing position.

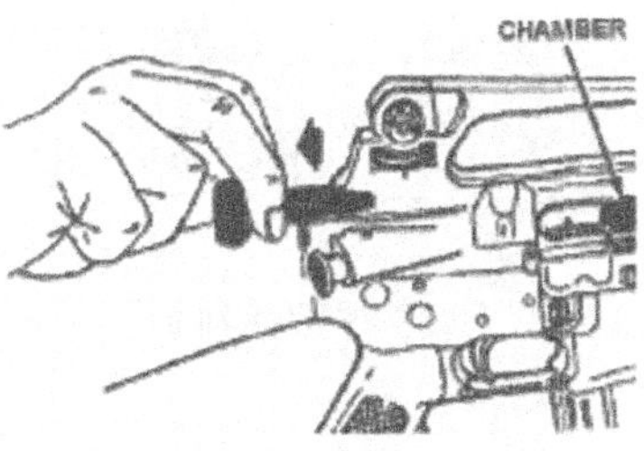

Figure 2-5. Charging Handle.

2003. CYCLE OF OPERATION

Firing

The hammer releases and strikes the head of the firing pin, driving the firing pin into the round's primer. The primer ignites the powder in the cartridge. Gas generated by the rapid burning of powder propels the projectile through the barrel. After the projectile passes the gas port, a portion of the expanding gas enters the gas port and gas tube. The gas tube directs the gas rearward into the bolt carrier key and causes the bolt carrier to move rearward. See figure 2-6 on page 2-4.

Unlocking

Figure 2-7 on page 2-4 illustrates unlocking of the bolt. As the bolt carrier moves to the rear, the bolt cam pin follows the path of the cam track located in the bolt carrier. This causes the bolt assembly to rotate until the bolt-locking lugs are no longer aligned behind the barrel extension locking lugs.

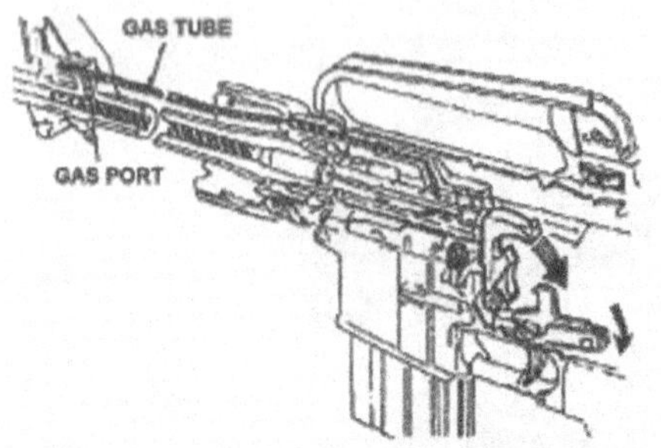

Figure 2-6. Firing.

Figure 2-7. Unlocking.

Extracting

As the bolt carrier group continues to move to the rear, the extractor claw withdraws the cartridge case from the chamber. See figure 2-8.

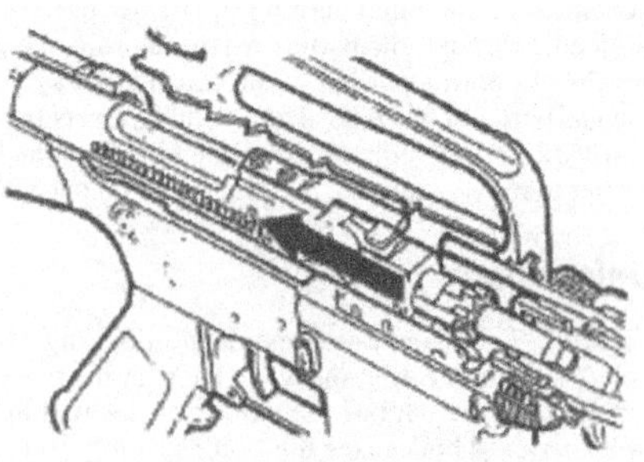

Figure 2-8. Extracting.

Ejecting

The ejector, located in the bolt face, is compressed into the bolt body by the base of the cartridge case. The rearward movement of the bolt carrier group allows the nose of the cartridge case to clear the front of the ejection port. The cartridge case is thrown out by the action of the ejector and spring. See figure 2-9.

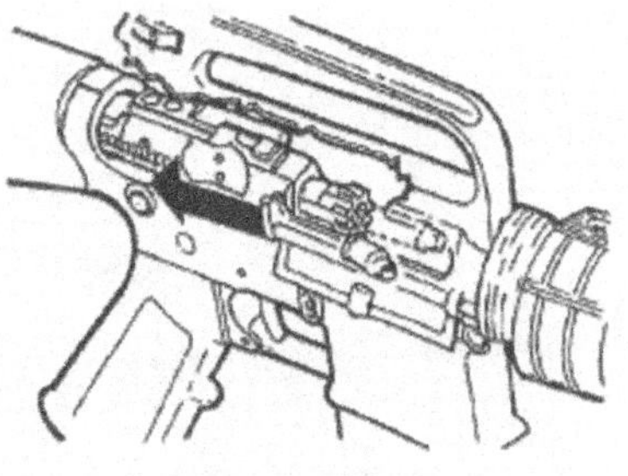

Figure 2-9. Ejecting.

Cocking

Continuing its rearward travel, the bolt carrier overrides the hammer, forces it down into the receiver, compresses the hammer spring, and causes the disconnector to engage the lower hammer hook. See figure 2-10.

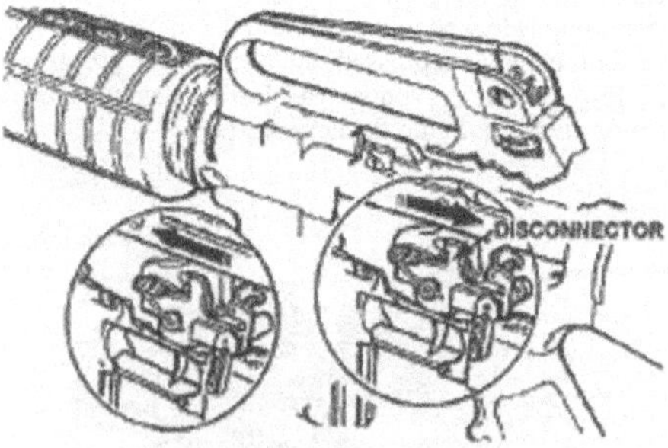

Figure 2-10. Cocking.

Feeding

Once rearward motion causes the bolt carrier group to clear the top of the magazine, the expansion of the magazine spring forces a round into the path of the bolt. After the action spring overcomes and absorbs the rearward motion of the bolt carrier group, it expands and sends the buffer assembly and bolt carrier group forward with enough force to strip a round from the magazine. See figure 2-11.

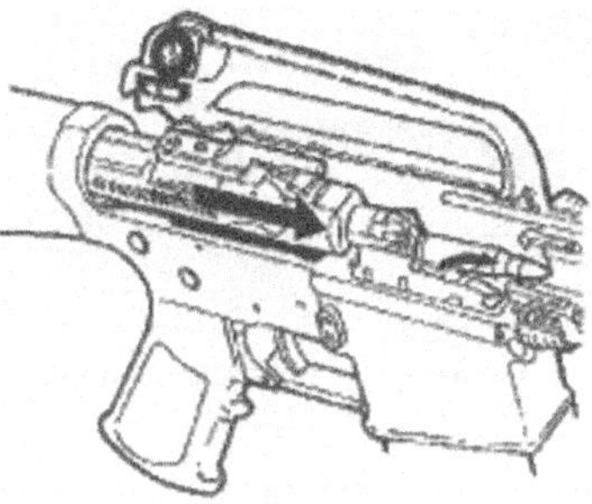

Figure 2-11. Feeding.

Chambering

As the bolt carrier group continues to move forward, pushing a fresh round in front of it, the face of the bolt thrusts the new round into the chamber. The extractor claw grips the rim of the cartridge case. The ejector is forced into its hole, compressing the ejector spring. See figure 2-12.

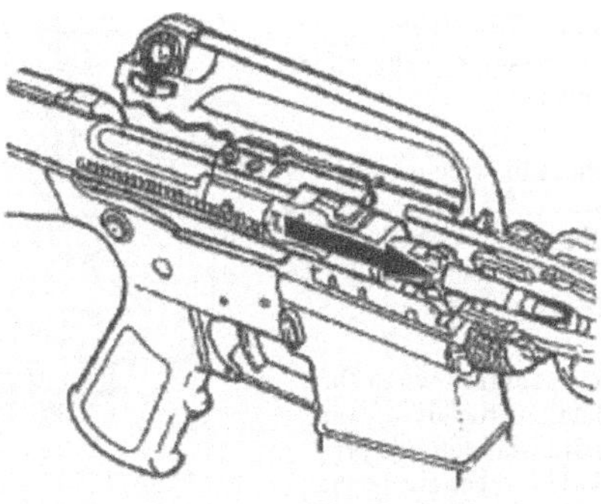

Figure 2-12. Chambering.

Locking

As the bolt carrier group continues to move forward, the bolt-locking lugs are forced against the barrel extension and the bolt cam pin is forced along the cam track. The bolt rotates and aligns the bolt locking lugs behind the barrel extension locking lugs. The weapon is ready to fire. See figure 2-13.

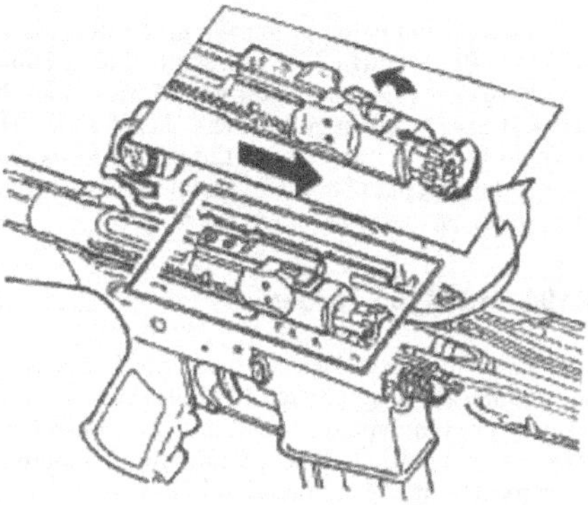

Figure 2-13. Locking.

2004. Ammunition

Four types of ammunition are authorized for use with the M16A2 service rifle: ball (M193 and M855), tracer (M196 and M856), dummy (M199), and blank (M200) (see fig. 2-14).

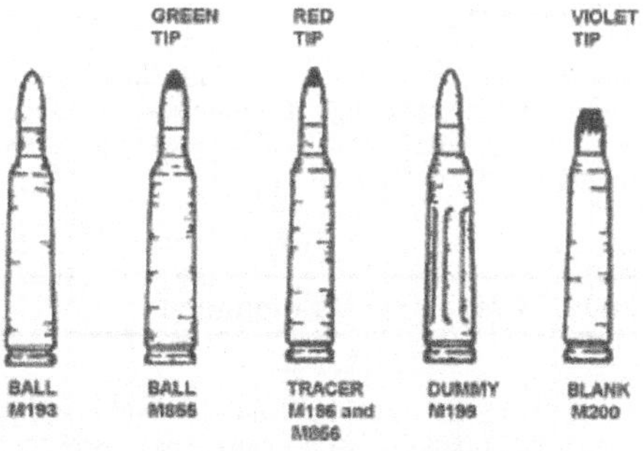

Figure 2-14. Authorized Ammunition.

M193 Ball

This ammunition is a 5.56mm centerfire cartridge with a 55-grain gilded-metal jacket, lead alloy core bullet. The primer and case are waterproofed. The M193 ball ammunition has no identifying marks.

M855 Ball

This ammunition is the primary ammunition for the M16A2 rifle. Identified by a green tip, its 5.56mm centerfire cartridge has better penetration than the M193. It has a 62-grain gilded-metal jacket bullet. The rear two-thirds of the core of the projectile is lead alloy and the front one-third is a solid steel penetrator. The primer and case are waterproofed.

M196 and M856 Tracer

This ammunition has the same basic characteristics as ball ammunition. Identified by a bright red tip, its primary uses include observation firing, incendiary effect, and signaling. Tracer ammunition should be intermixed with ball ammunition in a ratio no greater than 1:1. The preferred ratio is one tracer to four balls (1:4) to prevent metal fouling in the bore.

M199 Dummy

This ammunition has six grooves along the side of the case. It contains no propellants or primer. The primer well is open to prevent damage to the firing pin. The dummy cartridge is used during dry fire and other training purposes.

M200 Blank

This ammunition has the case mouth closed with a seven-petal rosette crimp. It contains no projectile. Blank ammunition, identified by its violet tip, is used for training purposes.

2005. Preventive Maintenance

Normal care and cleaning of the rifle will result in proper functioning of its all parts. Only issue-type cleaning materials maybe used. Improper maintenance can cause stoppages, reducing combat readiness and effectiveness.

Main Group Disassembly

The M16A2 service rifle is disassembled into three main groups (see fig. 2-15). Before disassembling the M16A2 service rifle—

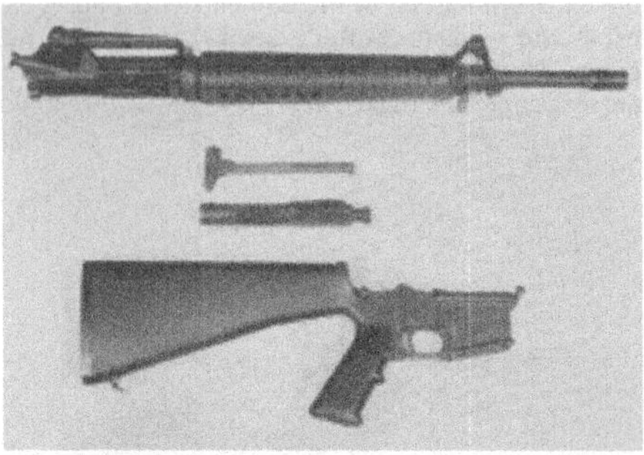

Figure 2-15. Three Main Groups.

- Ensure the rifle is in Condition 4 (see para. 3002).
- Remove the sling.

Note

To facilitate control and ease of disassembly, the handguards may be removed using the "buddy system" as described in Technical Manual (TM) 05538C-10/1A, *Operator's Manual w/Components List Rifle, 5.56-mm, M16A2E W/E (1005-01-128-9936).*

Upper Receiver

To disassemble the upper receiver—

- Place the rifle on the buttstock. Press down on the slip ring with both hands. Pull the handguards free (see fig. 2-16).

Use caution when the handguards are off the rifle. Handguards provide protection for the gas tube. Damage to the gas tube adversely affects the functioning of the rifle.

Figure 2-16. Removing the Handguards.

- Move the take down pin from the left to the right as far as it will go to allow the lower receiver to pivot down from the upper receiver.
- Move the receiver pivot pin from left to right as far as it will go and separate the upper and lower receivers.
- Pull back the charging handle and bolt carrier about 3 inches and remove the bolt carrier group.
- Remove the charging handle by sliding it back and down, out of the upper receiver.

No further disassembly is conducted on the upper receiver group.

Bolt Carrier

To disassemble the bolt carrier—

- Remove the firing pin retaining pin.
- Push the bolt back into the bolt carrier to the locked position.
- Tap the base of the bolt carrier against the palm of your hand so the firing pin will drop out.
- Rotate the bolt cam pin one-quarter turn and lift the bolt cam pin out.
- Withdraw the bolt assembly from the carrier.
- Press on the extractor's rear and use the firing pin to push out the extractor-retaining pin. Remove the extractor and spring (the spring is permanently attached to the extractor). (See fig. 2-17.)

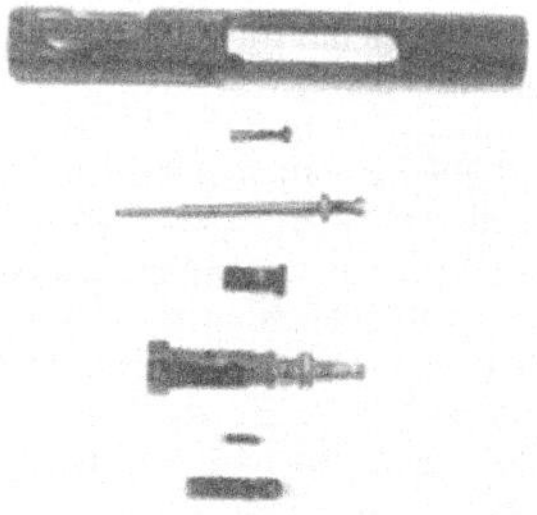

Figure 2-17. Bolt Carrier Disassembled.

CAUTION

Be careful not to damage the tip of the firing pin while pushing out the extractor-retaining pin.

Note

The extractor assembly has a rubber insert within the spring. Do not attempt to remove it. If the spring comes loose, put the large end of the spring in the extractor and seat it. Push in the extractor pin.

Lower Receiver

To disassemble the lower receiver—

- Press in the buffer and depress the buffer retainer.

Note

It may be necessary to use the edge of the charging handle to depress the buffer retainer.

- Press the hammer downward and ease the buffer and action spring forward and out of the receiver.
- Separate the parts. See figure 2-18.

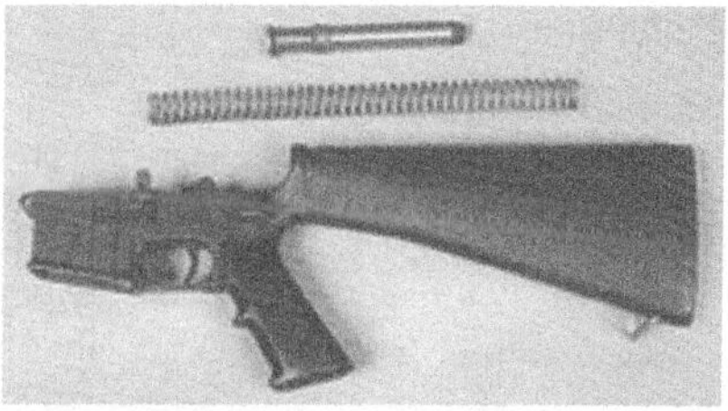

Figure 2-18. Lower Receiver Disassembled.

No further disassembly of the lower receiver is performed.

Note

In combat situations, the rifle may be partially disassembled in any sequence. However, combat situations are the exception, not the rule. Under normal operating circumstances, disassemble the rifle in the sequence just performed. Any further disassembly of the rifle is to be performed by a qualified armorer.

Magazine Disassembly

The magazine should be disassembled regularly for cleaning to avoid the possibility of malfunction or stoppage of the rifle caused by dirty or damaged magazines. To disassemble the magazine—

- Pry up and push base plate out from the magazine.
- Jiggle the spring and follower to remove. Do not remove the follower from the spring. See figure 2-19.

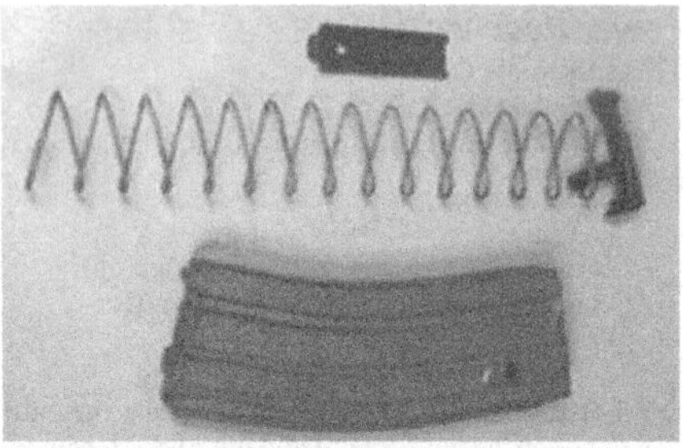

Figure 2-19. Magazine Disassembled.

Cleaning

Cleaning Materials

The following cleaning materials are used in preventive maintenance (see fig. 2-20):

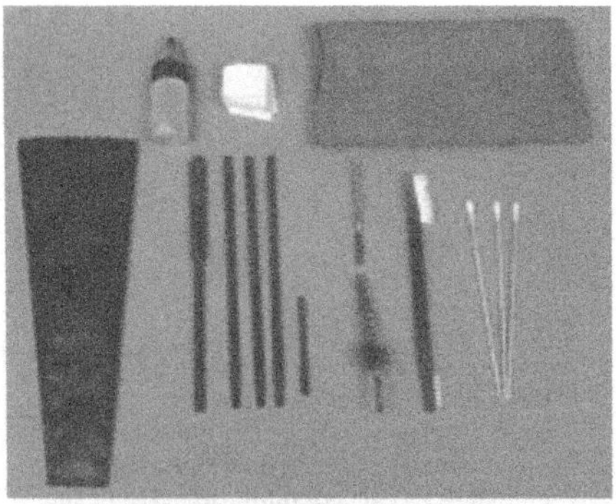

Figure 2-20. Cleaning Materials.

- Cleaner, lubricant, and preservative (CLP). Always shake the bottle well before use.
- Rod in three sections and a handle assembly.
- Patch holder section, swabs, patches, pipe cleaners, and clean rags.
- Brushes: bore, chamber, and general purpose.

Cleaning the Upper Receiver

Basic cleaning of the upper receiver group should include the following:

- Attach the three rod sections together but leave each one about two turns short of being tight.
- Attach the patch holder onto the rod.
- Point the muzzle down and insert the non-patch end of the rod into the chamber. Attach the handle to the cleaning rod section and pull a CLP-moistened 5.56mm patch through the bore.
- Attach the bore brush to the rod but leave it two turns short of being tight. Put a few drops of CLP on the bore brush. Insert the rod into the barrel from the chamber end, attach the handle, and pull the brush through the bore. Repeat 3 times. Remove bore brush and attach the patch holder to the rod with a CLP moistened patch insert the rod into the barrel from the chamber end, attach the handle, and pull the patch through the bore.
- Inspect the bore for cleanliness by holding the muzzle to your eye and looking into the bore.
- Repeat the above steps until the patches come out of the bore clean.
- Attach the chamber brush and one section of the cleaning rod to the handle. Moisten it well with CLP and insert it into the chamber.
- Scrub the chamber and bolt lugs using a combination of a plunging and clockwise rotating action.

Note

Do not reverse direction of the brush while it is in the chamber.

- Clean the interior portion of the upper receiver with the general-purpose brush and CLP.
- Dry the bore, chamber, and the interior of the receiver with rifle patches, swabs, and clean rags until they come out clean. Then moisten all interior surfaces with CLP.
- Wipe the barrel, gas tube, and handguards clean with a rag.

Cleaning the Bolt Carrier Group

- Clean the outer and inner surfaces of the bolt carrier with a general-purpose brush.
- Clean the bolt carrier key with a pipe cleaner.
- Clean the locking lugs, gas rings, and exterior of the bolt with the general-purpose brush.
- Insert a swab into the rear of the bolt and swab out the firing pin recess and gas ports.

- Clean the extractor with the general-purpose brush, ensuring all the carbon is removed from underneath the extractor lip.
- Clean extractor pin, firing pin, and firing pin retaining pin using the general-purpose brush and CLP.
- Clean charging handle assembly with the general-purpose brush and patches.

Cleaning the Lower Receiver

- Wipe dirt from the firing mechanism using a general-purpose brush, clean patch, pipe cleaners, and swabs.
- Clean the outside of the receiver with the general-purpose brush and CLP. Clean the buttplate and rear sling swivel, ensuring drain hole is clear of dirt.
- Wipe the inside of the buffer tube, buffer, and action spring.
- Wipe the inside of the magazine well with a rag.
- Wipe out the inside of the pistol grip and ensure that it is clean.

Cleaning the Magazine

- Clean the inside of the magazine with the general-purpose brush and CLP.
- Wipe dry.
- Keep the spring lightly oiled.

Inspection

While cleaning the rifle, and during each succeeding step in the preventive maintenance process, inspect each part for cracks and chips and ensure parts are not bent or badly worn. Report any damaged part to the armorer. Inspection is a critical step to ensure the combat readiness of your rifle. It is performed normally during rifle cleaning (prior to lubrication), however, it can be performed throughout the preventive maintenance process.

Lubrication

Lubrication is performed as part of the detailed procedure for preventive maintenance. Lubrication procedures are also performed in preparation for firing.

Lubricant

In all but the coldest arctic conditions, CLP is the lubricant for the rifle. Remember to remove excess CLP from the bore and chamber before firing.

- Lightly lube means that a film of CLP barely visible to the eye should be applied.
- Generously lube means the CLP should be applied heavily enough that it can be spread with the finger.

Upper Receiver

- Lightly lube the inside of the upper receiver, bore, chamber, outer surfaces of the barrel, and surfaces under the handguard.
- Depress the front sight detent and apply two or three drops of CLP to the front sight detent. Depress several times to work lubrication into the spring.
- Lubricate the moving parts and elevation screw shaft of the rear sight.

Bolt Carrier Group

- Generously lube the outside of the cam pin area, the bolt rings, and outside the bolt body.
- Lightly lube the charging handle and the inner and outer surfaces of the bolt carrier.

Lower Receiver

- Lightly lube the inside of the lower receiver extension.
- Generously lube the moving parts inside the lower receiver and their pins.

Reassembly

Reassembling the Rifle

- Return all cleaning gear into the buttstock of the rifle and close the buttplate.
- Connect the buffer and action spring and insert them into the buffer tube/stock.
- Place the extractor and spring back on the bolt. Depress the extractor to align the holes and reinsert the extractor pin.
- Insert the bolt into the carrier. Do not switch bolts between rifles.
- Hold the bolt carrier with the bolt carrier key at 12 o'clock. Insert the bolt into the bolt carrier with the extractor at 12 o'clock.

- Rotate the bolt counterclockwise until the cam pinhole aligns to the cam pin slot in the bolt carrier.

WARNING WARNING WARNING WARNING

WARNING: Ensure the cam pin is installed in the bolt group or the rifle may explode while firing.

- Insert the bolt cam pin through the bolt carrier and into the bolt. Rotate the cam pin 1/4 turn right or left. Pull the bolt forward until it stops.
- Drop in the firing pin from the rear of the bolt carrier and seat it.
- Replace the firing pin retaining pin. Ensure the head of the firing pin retaining pin is recessed inside the bolt carrier. The firing pin should not fall out when the bolt carrier group is turned upside down.
- Place the charging handle in the upper receiver by lining it up with the grooves in the receiver. Push the charging handle partially in.
- With the bolt in the unlocked position, place bolt carrier key into the groove of the charging handle.
- Push the charging handle and bolt carrier group into the upper receiver until the charging handle locks.
- Join the upper and lower receivers and engage the receiver pivot pin.
- Ensure the selector lever is on safe before closing the upper receiver. Close the upper and lower receiver groups. Push in the takedown pin.
- Install the handguards.
- Attach the sling.

Reassembling the Magazine

To reassemble the magazine—

- Insert the follower and jiggle the spring to install.
- Slide the base under all four tabs until the base catches. Make sure the printing is on the outside.

2006. Function Check

A function check is performed to ensure the rifle operates properly. To perform a function check:

- Place the weapon in Condition 4 (see para. 3002).
- Pull the charging handle to the rear and release. Ensure the selector lever is on safe and pull the trigger. The hammer should not fall.
- Place the selector lever on semi. Pull the trigger and hold it to the rear. The hammer should fall. Pull the charging handle to the rear and release. Release the trigger and pull again. The hammer should fall.
- Pull the charging handle to the rear and release. Place the selector lever on burst. Pull the trigger and hold it to the rear. The hammer should fall. Pull the charging handle to the rear three times and release. Release the trigger and pull again. The hammer should fall.
- Pull the charging handle to the rear and release. Place the selector lever on safe.

2007. User Serviceability Inspection

Individual Marines must perform user serviceability inspections on their weapons before firing them. This inspection ensures the weapon is in an acceptable operating condition.

- Place rifle in Condition 4 (see para. 3002).
- Conduct a function check.
- Check the rifle to ensure the following:
- Compensator is tight.
- Barrel is tight.
- Front sight post is straight.
- Front sight post is adjustable.
- Handguards are serviceable.
- Rear sight elevation and windage knobs are adjustable and have distinct clicks.
- Stock is tight on the lower receiver.
- Weapon is properly lubricated for operational conditions.
- Barrel is clear of obstructions.
- Gas rings are serviceable.
- Ensure magazines are serviceable.
- Load the rifle with an empty magazine. Ensure that the magazine can be seated.
- Without depressing the bolt catch, pull the charging handle to the rear. Ensure that the bolt locks to the rear.
- Depress the upper portion of the bolt catch and observe bolt moving forward on an empty chamber.

Ensure the bolt moves completely forward and locks in the chamber.

- Repeat this procedure with all magazines.

2008. Field Maintenance

Preventive maintenance in the field is performed when detailed disassembly and cleaning is not practical due to operational tempo or the level of threat. To perform limited field preventive maintenance—

- Place the rifle in Condition 4 (see paragraph 3002).
- Break the rifle down by removing the rear take down pin and rotating the upper receiver and barrel forward.
- Remove the bolt carrier group.
- Do not disassemble the bolt carrier group further.
- Clean the bolt carrier group.
- Clean the upper and lower receiver groups (without further disassembly).
- Clean the bore and chamber.
- Lubricate the rifle.
- Reassemble the rifle and perform a user serviceability inspection.

2009. Cleaning the Rifle in Various Conditions

The climatic conditions in various locations require special knowledge about cleaning and maintaining the rifle. The conditions that will affect the rifle the most are: hot, wet tropical; hot, dry desert; arctic or low temperature; and heavy rain and fording.

Hot, Wet Tropical

- Perform normal maintenance.
- Clean and lubricate your rifle more often. Inspect hidden surfaces for corrosion. Pay particular attention to spring-loaded detentes.
- Use lubricant more liberally.
- Unload and check the inside of the magazine more frequently. Wipe dry and check for corrosion.
- When practical, keep the rifle covered.

Hot, Dry Desert

Hot dry climates are usually areas that contain blowing sand and fine dust. Dust and sand will get into the rifle and magazines, causing stoppages. It is imperative to pay particular attention to the cleaning and lubrication of the rifle in this type of climate.

Corrosion is less likely to form in these environments, and lubrication will attract more dirt. For this reason, use lubrication more sparingly.

Whenever practical, keep the rifle covered.

Arctic or Low Temperature

- Clean and lubricate the rifle in a warm room, with the rifle at room temperature, if possible. Lubricating oil, arctic weapons (LAW) can be used below a temperature of 0 degrees Fahrenheit and must be used below -35 degrees Fahrenheit.
- Keep the rifle covered when moving from a warm to a cold environment to allow gradual cooling of the rifle. This prevents the condensation of moisture and freezing. Condensation will form on the rifle when it is moved from outdoors to indoors. If possible, leave the rifle in a protected but cold area outdoors. When bringing the rifle inside to a warm place, it should be disassembled and wiped down several times as it warms.
- Always try to keep the rifle dry.
- Unload and perform a function check every 30 minutes, if possible, to help prevent freezing of functional parts.
- Do not lay a warm rifle in snow or ice.
- Keep the inside of the magazine and ammunition wiped dry. Moisture will freeze and cause stoppages.

Heavy Rain and Fording

- Keep the rifle dry and covered when practical.
- Keep water out of the barrel if possible. If water does get in, drain and (if possible) dry with a patch. If water is in the barrel, point the muzzle down and break the seal by doing a chamber check so the water will drain. If water is in the stock of the weapon, ensure the drain hole in the stock is clear so the water can run out.
- Perform normal maintenance.

CHAPTER 3. WEAPONS HANDLING

Weapons handling procedures provide a consistent and standardized way for a Marine to handle, operate, and employ the rifle safely and effectively. Proper weapons handling procedures ensure the safety of Marines by eliminating negligent discharges and reinforcing positive identification of targets before engagement. Weapons handling procedures apply at all levels of training and during combat operations.

Note

+The procedures in this manual are written for right-handed Marines; left-handed Marines should reverse instructions as necessary.

3001. Safety Rules

Safe handling of the rifle is critical. If proper weapons handling procedures are not used, a Marine risks his safety and the safety of his fellow Marines. During combat, a Marine must react quickly, safely, and be mentally prepared to engage targets. To ensure that only the intended target is engaged, a Marine must apply the following safety rules at all times:

Rule 1—Treat every weapon as if it were loaded. When a Marine takes charge of a rifle in any situation, he must treat the weapon as if it were loaded, determine its condition (see para. 3003), and continue applying the other safety rules.

Rule 2—Never point a weapon at anything you do not intend to shoot. A Marine must maintain muzzle awareness at all times.

Rule 3—Keep your finger straight and off the trigger until you are ready to fire. A target must be identified before taking the weapon off safe and moving the finger to the trigger.

Rule 4—Keep the weapon on safe until you intend to fire. A target must be identified before taking the weapon off safe. This rule is intended to eliminate the chance of the weapon discharging by accident (e.g., brush snagging the trigger).

3002. Weapons Conditions

A weapon's readiness is described by one of four conditions. The steps in the loading and unloading process take the rifle through four specific conditions of readiness for live fire.

Condition 1. Safety on, magazine inserted, round in chamber, bolt forward, ejection port cover closed.

Condition 2. Not applicable to the M16A2 rifle.

Condition 3. Safety on, magazine inserted, chamber empty, bolt forward, ejection port cover closed.

Condition 4. Safety on, magazine removed, chamber empty, bolt forward, ejection port cover closed.

3003. Determining a Weapon's Condition (Chamber Check)

A Marine must know the condition of his weapon at all times. When a Marine takes charge of a weapon in any situation, he must determine its condition. Situations include coming across an unmanned rifle in combat, taking charge of any weapon after it has been unmanned (e.g., out of a rifle rack, stored in a vehicle), or taking charge of another Marine's weapon. To determine the condition of the weapon in any of these situations, the Marine should:

- Determine if a magazine is present.
- Ensure the rifle is on safe.

- Conduct a chamber check.
- Bring the left hand back against the magazine well.
- Extend the fingers of the left hand and cover the ejection port (see fig. 3-1).

Figure 3-1. Position of Hand.

- Grasp the charging handle with the index and middle fingers of the right hand.
- Pull the charging handle slightly to the rear and visually and physically inspect the chamber (see fig. 3-2). *Right-handed Marines*, insert one finger of the left hand into the ejection port and feel whether a round is present. *Left-handed Marines*, insert the thumb of the right hand into the ejection port and feel whether a round is present.

Figure 3-2. Chamber Check.

Note

The same procedure is used in daylight as during low visibility. A chamber check may be conducted at any time.

CAUTION

Pulling the charging handle too far to the rear while inspecting the chamber may cause double feed or ejection of one round of ammunition.

- Release the charging handle and observe the bolt going forward.
- Tap the forward assist.
- Close the ejection port cover (if time and the situation permit).
- Remove the magazine (if present) and observe if ammunition is present. If time permits, count the rounds. Reinsert the magazine into magazine well.

3004. Weapons Commands

Weapons commands dictate the specific steps required to load and unload the rifle. Six commands are used in weapons handling:

Load. This command is used to take the weapon from Condition 4 to Condition 3.

Make Ready. This command is used to take the weapon from Condition 3 to Condition 1.

Fire. This command is used to specify when a Marine may engage targets.

Cease-Fire. This command is used to specify when a Marine must stop target engagement.

Unload. This command is used to take the weapon from any condition to Condition 4.

Unload and Show Clear. This command is used when an observer must check the weapon to verify that no ammunition is present before the rifle is placed in Condition 4.

Loading the Rifle

Perform the following steps to load the rifle (take the rifle to Condition 3):

- Ensure the rifle is on safe.
- Withdraw the magazine from the magazine pouch.
- Observe the magazine to ensure it is filled.
- Fully insert the magazine in the magazine well. Without releasing the magazine, tug downward on the magazine to ensure it is seated.
- Close the ejection port cover.
- Fasten the magazine pouch.

Making the Rifle Ready

Perform the following steps to make the rifle ready for firing (take the rifle to Condition 1):

- Pull the charging handle to the rear and release. There are two methods of doing this:
- Grip the pistol grip firmly with the right hand and pull the charging handle with the left hand to its rearmost position and release (see fig. 3-3).

  Or grip the handguards firmly with the left hand and pull the charging handle with the right hand to its rearmost position and release (see fig. 3-4).

Figure 3-3. Pulling the Charging Handle with the Left Hand.

Figure 3-4. Pulling the Charging Handle with the Right Hand.

- To ensure ammunition has been chambered, conduct a chamber check (see para. 3003) to ensure a round has been chambered.
- Check the sights (to ensure proper battlesight zero [BZO] setting, correct rear sight aperture, etc.).
- Close ejection port cover.

Fire

On the command "Fire," aim the rifle, take the rifle off safe, and pull the trigger.

Cease-Fire

On the command "Cease Fire," perform the following steps:

- Place your trigger finger straight along the receiver.
- Place the weapon on safe.

Unloading the Rifle

Perform the following steps to unload the rifle (take the rifle to Condition 4):

- Ensure the weapon is on safe.
- Remove the magazine from the rifle and retain it on your person.

- Cup the left hand under the ejection port, rotate the weapon until the ejection port faces down.
- Pull the charging handle to the rear and catch the round in the left hand (see fig. 3-5).

Figure 3-5. Catching the Round.

- Lock the bolt to the rear.
- Put the weapon on safe if the selector lever would not move to safe earlier.
- Ensure the chamber is empty and that no ammunition is present.
- Depress the bolt catch and observe the bolt moving forward on an empty chamber (see fig. 3-6).

Figure 3-6. Observing the Chamber.

- Close the ejection port cover.
- Check the sights (for proper BZO setting, correct rear sight aperture, etc.).
- Place any ejected round into the magazine and return the magazine to the magazine pouch and close the magazine pouch.

Unloading and Showing the Rifle Clear

Perform the following steps to unload the rifle and show it clear to an observer (take the rifle to Condition 4). The Marine—

- Ensures the weapon is on safe.
- Removes the magazine from the rifle and retains it.
- Cups the left hand under the ejection port, rotates the weapon until the ejection port faces down.
- Pulls the charging handle to the rear and catches the round in the left hand.
- Locks the bolt to the rear and ensures the chamber is empty and that no ammunition is present.
- Has another Marine inspect the weapon to ensure no ammunition is present (see fig. 3-7).

Figure 3-7. Observer Inspection.

The observer—

- Visually inspects the chamber to ensure it is empty, no ammunition is present, and the magazine is removed.
- Ensures the weapon is on safe.
- Acknowledges the rifle is clear.

The Marine, after receiving acknowledgment that the rifle is clear—

- Depresses the bolt catch and observes the bolt moving forward on an empty chamber.
- Closes the ejection port cover.
- Checks the sights (for proper BZO setting, correct rear sight aperture, etc.).
- Places any ejected round into the magazine and returns the magazine to the magazine pouch and closes the magazine pouch.

3005. Filling, Stowing, and Withdrawing Magazines

Filling the Magazine with Loose Rounds

Perform the following steps to fill the magazine:

- Remove a magazine from the magazine pouch.
- Place a round on top of the follower.
- Press down until the round is held between the follower and magazine feed lips (see fig. 3-8).

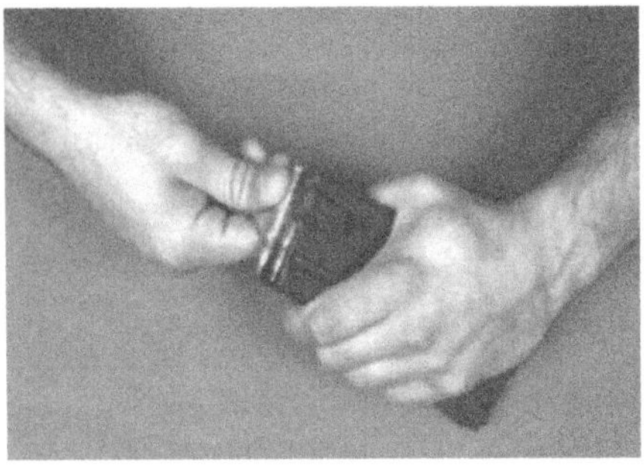

Figure 3-8. Filling the Magazine.

- Repeat until the desired number of rounds is inserted. The recommended number of rounds per magazine is 28 or 29. Thirty rounds in the magazine may prohibit the magazine from seating properly on a closed bolt.
- Tap the back of the magazine to ensure the rounds are seated against the back of the magazine.

Filling the Magazine Using a 10-round Stripper Clip and Magazine Filler

The magazine can also be filled quickly using a 10-round stripper clip and the magazine filler (see fig. 3-9).

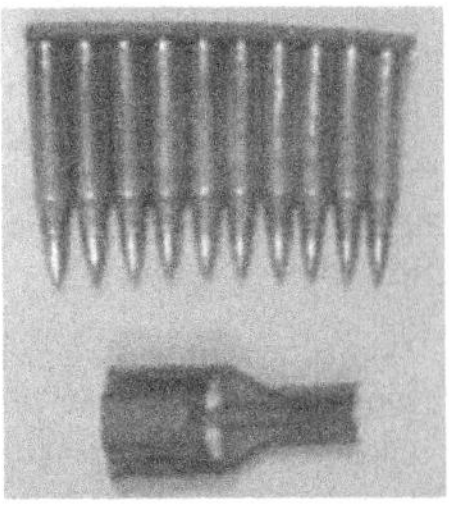

Figure 3-9. Magazine Filler and 10-round

Perform the following steps to fill the magazine with the 10-round stripper clip (see fig. 3-10):

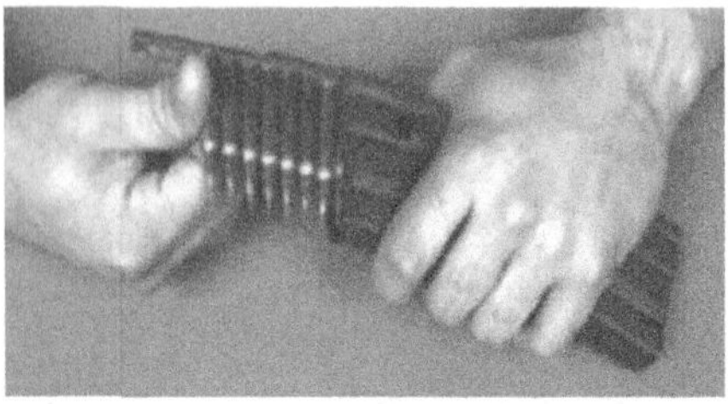

Figure 3-10. Filling the Magazine with a Stripper Clip and Magazine Filler.

- Remove a magazine from the magazine pouch.
- Slide the magazine filler into place.
- Place a 10-round stripper clip into the narrow portion of the magazine filler.
- Using thumb pressure on the rear of the top cartridge, press down firmly until all ten rounds are below the feed lips of the magazine.
- Remove the empty stripper clip while holding the magazine filler in place.
- Repeat until the desired number of rounds is inserted. The recommended number of rounds per magazine is 28 or 29. Thirty rounds in the magazine may prohibit the magazine from seating properly on a closed bolt.
- Remove magazine filler and retain it for future use.
- Tap the back of the magazine to ensure the rounds are seated against the back of the magazine.

Stowing Magazines

Magazine Pouch

In a magazine pouch, filled magazines are stored with rounds down and projectiles pointing away from the body.

Load-bearing Vest

In a load-bearing vest, filled magazines are stored with rounds down and projectiles pointing outboard.

Empty or Partially Filled Magazines

Empty or partially filled magazines are stored with the follower up to allow the selection of filled magazines by touch (i.e., at night).

Withdrawing Magazines

Magazine Pouch

With the right hand, withdraw magazines from the magazine pouch on the right side of the body. With the left hand, withdraw magazines from the magazine pouch on the left side of the body. To withdraw magazines from a magazine pouch—

- Use the thumb and index finger, pinch the magazine pouch release to open the magazine pouch.
- Slide the thumb over the magazines, feeling for a base plate indicating a filled magazine. Continue sliding the thumb until it rests on the back of the magazine.
- Grasp the magazine with the thumb, index finger, and middle finger and lift the magazine directly out of the pouch.
- Once the magazine is clear of the pouch, curl the ring finger and little finger underneath the magazine and rotate it up to observe rounds in the magazine.

Load-bearing Vest

With the right hand, withdraw magazines from the left side of the vest (see fig 3-11). With the left hand, withdraw magazines from the right side of the vest (see fig. 3-12).

To withdraw magazines from a load-bearing vest:

- With the thumb and index finger, unfasten the snap on the vest pouch.
- Slide the thumb over the magazine, feeling for a base plate indicating a filled magazine.
- Rotate the hand over the magazine while sliding the thumb to the back of the magazine.
- Grasp the magazine with the thumb, index finger, and middle finger, while curling the ring finger and little finger on top of the base plate.
- Lift the magazine directly out of the pouch and rotate it up to observe the rounds in the magazine.

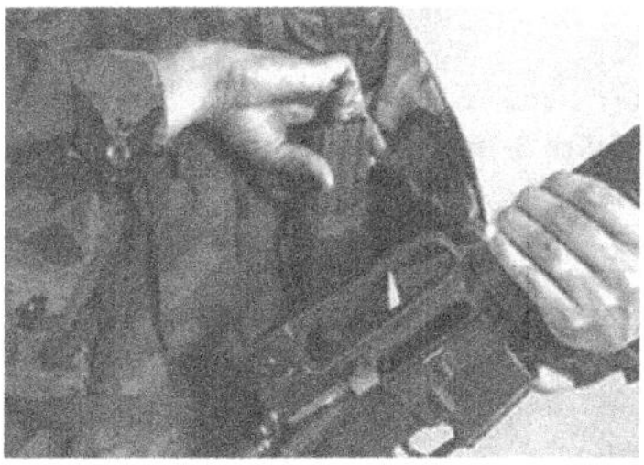

Figure 3-11. Withdrawing a Magazine with the Right Hand.

3006. Reloading the Rifle

Principles of Reloading

The first priority when performing a reload is to get the rifle reloaded and back into action. The second priority when performing a reload is to retain the magazine so when you move, the magazine moves with you. When time permits, retain magazines securely on your person (e.g., in magazine pouch, flak jacket, and cargo pocket). The combat situation may dictate dropping the magazine to the deck when performing a reload. This is acceptable as long as it is picked up before moving to another location.

Take cover before reloading. Always reload before leaving cover to take advantage of the protection provided by cover.

When moving, your focus should be on moving, therefore every effort should be made to not reload while on the move.

When reloading, your focus is on the magazine change.

When reloading, draw the weapon in close to your body so you can see what you are doing and retain positive control of the magazine.

When the new magazine is inserted, tug on it to ensure it is seated. Do not slam the magazine into the weapon hard enough to cause a round to partially pop out of the magazine. This action will cause a double feed and require remedial action.

Figure 3-12. Withdrawing a Magazine

Retain your empty magazines. When there is a lull in the action, refill those magazines so they will be available for future use.

During a lull in the action, replace your magazine when you know you are low on ammunition. This ensures a full magazine of ammunition in the rifle should action resume. Do not wait until the magazine is completely empty to replace it.

Condition 1 Reload

A condition 1 reload is performed when the weapon is in condition 1 by replacing the magazine before it runs out of ammunition. To perform a condition 1 reload, perform the following steps:

- Withdraw a filled magazine from the magazine pouch. With the same hand, press the magazine button and remove the partially filled magazine so it can be retained in the remaining fingers.
- Fully insert the filled magazine into the magazine well and tug downward on the magazine to ensure it is properly seated.
- Store the partially filled magazine in the magazine pouch with rounds up and projectiles pointing away from the body.
- Fasten the magazine pouch.

Dry Reload

A dry reload is required when the magazine in the weapon has been emptied and the bolt has locked to the rear. To perform a dry reload—

- Press the magazine release button.
- Remove the empty magazine and retain it on your person when time permits.
- Fully insert a filled magazine into the magazine well and tug downward on the magazine to ensure it is properly seated.
- Depress the bolt catch to allow the bolt carrier to move forward and observe the round being chambered. This places the rifle in Condition 1.

3007. Remedial Action

If the rifle fails to fire, a Marine performs remedial action. Remedial action is the process of investigating the cause of the stoppage, clearing the stoppage, and returning the weapon to operation.

Observe for Indicators

Once the rifle ceases firing, the Marine must visually or physically observe the ejection port to identify the problem before he can clear it. The steps taken to clear the weapon are based on observation of one of the following three indicators:

Indicator: The bolt is forward or the ejection port cover is closed. See figures 3-13 and 3-14.

Figure 3-13. Bolt Forward.

Figure 3-14. Ejection Port Cover Closed.

To return the weapon to operation—

- Seek cover if the tactical situation permits.
- Tap—Tap the bottom of the magazine.
- Rack—Pull the charging handle to the rear and release it.
- Bang—Sight in and attempt to fire.

Indicator: Brass is obstructing chamber area (usually indicating a double feed or failure to eject). See figure 3-15.

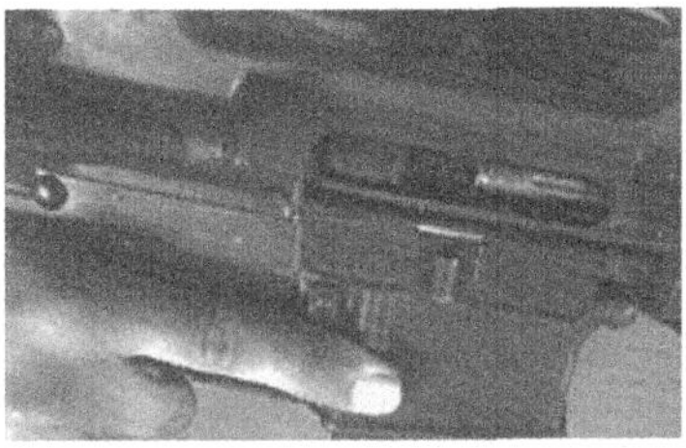

Figure 3-15. Brass Obstructing the Chamber.

To return the weapon to operation—

- Seek cover if the tactical situation permits.
- Attempt to remove the magazine.
- Attempt to lock the bolt to the rear

If the bolt will not lock to the rear, rotate the rifle so the ejection port is facing down; hold the charging handle to the rear as far as it will go and shake the rifle to free the round(s). If the rounds do not shake free, hold the charging handle to the rear and strike the butt of the rifle on the ground or manually clear the round. Conduct a reload. Sight in and attempt to fire.

Indicator: The bolt is locked to the rear. See figure 3-16. To clear return the weapon to operation—

Figure 3-16. Bolt Locked to the Rear.

Note

Although a dry weapon is not considered a true stoppage or mechanical failure, the Marine must take action to return the weapon to operation.

- Seek cover if the tactical situation permits.
- Conduct a dry reload.
- Sight in and attempt to fire.

Audible Pop or Reduced Recoil

An audible pop occurs when only a portion of the propellant is ignited. It is normally identifiable by reduced recoil and is sometimes accompanied by excessive smoke escaping from the chamber area. To clear the rifle in a combat environment:

- Place the rifle in Condition 4.
- Move take down pin from left to right as far as it will go to allow the lower receiver to pivot.
- Remove the bolt carrier group.
- Inspect the bore for an obstruction from the chamber end.
- Insert a cleaning rod into the bore from the muzzle end and clear the obstruction.
- Reassemble the rifle.
- Conduct a reload.
- Sight in and attempt to fire.

3008. Weapons Carries

Weapons carries provide an effective way to handle the rifle while remaining alert to enemy engagement. Weapons carries are tied to threat conditions and are assumed in response to a specific threat situation. The weapons carry assumed prepares the Marine, both mentally and physically, for target engagement. The sling provides additional support for the weapon when firing; therefore, the hasty sling should be used in conjunction with the carries.

Tactical Carry

A Marine carries the rifle at the tactical carry if no immediate threat is present. The tactical carry permits control of the rifle while a Marine is moving, yet it still allows quick engagement of the enemy. A Marine performs the following steps to assume the tactical carry:

- Place left hand on the handguards, right hand around the pistol grip, trigger finger straight along the receiver (see fig. 3-17), and right thumb on top of the selector lever (see fig. 3-18).

Figure 3-17. Straight Trigger Finger.

Figure 3-18. Thumb on Selector Lever.

- Place the buttstock along the side of the body at approximately hip level.
- Angle the muzzle upward about 45 degrees in a safe direction.
- Position the muzzle in front of the eyes, slightly below eye level (see fig. 3-19).
- Move the head and the eyes with the muzzle as it moves.

Figure 3-19. Tactical Carry.

Alert Carry

A Marine carries the rifle at the alert if enemy contact is likely. The alert is also used for moving in close terrain (e.g., urban, jungle). A Marine can engage the enemy faster from the alert than from the tactical carry. However, the alert is more tiring than the tactical carry and its use can be physically demanding. A Marine performs the following steps to assume the alert:

- Place the left hand on the handguards, the right hand around the pistol grip, the trigger finger straight along the receiver (see fig. 3-17), and the right thumb on top of the selector lever (see fig. 3-18).
- Place the buttstock in the shoulder.
- Angle the muzzle downward about 45 degrees and point it in a safe direction or the general direction of likely enemy contact (see fig. 3-20).

Figure 3-20. Alert Carry.

Ready Carry

A Marine carries the rifle at the ready if contact with the enemy is imminent. The ready allows immediate target engagement, but it is very tiring to maintain over a long period of time. A Marine performs the following steps to assume the ready:

- Place left hand on handguards, right hand around the pistol grip, the trigger finger straight along the receiver (see fig. 3-17), and the right thumb on top of the selector lever (see fig. 3-18).
- Place the buttstock in the shoulder.
- Point the muzzle in the direction of the enemy.

- Lower the sights to just below eye level so that a clear field of view is maintained so that a target may be identified (see fig. 3-21).

Figure 3-21. Ready Carry.

3009. Weapons Transports

Weapons transports are used to carry the rifle over the back or shoulders when moving for long periods; they provide a more relaxed position for walking. Weapons transports are used if no immediate threat is present. They are also used whenever one or both hands are needed for other work.

Strong Side Sling Arms Transport (Muzzle Up)

To assume the strong side sling arms (muzzle up) transport from the tactical carry, a Marine performs the following steps (see fig. 3-22):

- Release the hold on the pistol grip.
- Lower the buttstock and bring the rifle to a vertical position.
- With the right hand, grasp the sling above the left forearm.

Figure 3-22. Strong Side Sling Arms Transport (Muzzle Up).

- With the left hand, guide the rifle around the right shoulder.
- With the right hand, apply downward pressure on the sling. This stabilizes the rifle on the shoulder.

Weak Side Sling Arms Transport (Muzzle Down)

The weak side sling arms (muzzle down) transport can be used in inclement weather to keep moisture out of the rifle's bore. To assume this transport from the tactical carry, a Marine performs the following steps (see fig. 3-23):

- Release the hold on the pistol grip.
- With the left hand, rotate muzzle down and bring the rifle to a vertical position on the left side of the body. The pistol grip is pointed outboard.
- With right hand, place sling on left shoulder.
- Grasp sling above the waist with the left hand.
- With the left hand, apply downward pressure on the sling. This stabilizes the rifle on the shoulder.

Figure 3-23. Weak Side Sling Arms Transport (Muzzle Down.)

Cross Body Sling Arms Transport

A Marine uses the cross body sling arms transport if he requires both hands for work. The rifle is slung across the back with the muzzle up or down. Normally, the rifle is carried with the muzzle down to prevent pointing the muzzle in an unsafe direction.

To assume a cross body sling transport, a Marine performs the following steps from weak side sling arms (muzzle down) (see fig. 3-24):

Figure 3-24. Cross Body Sling Arms Transport (Muzzle Down).

- With the right hand, grasp the sling.
- With the left hand, grasp the handguards.
- Pull up on the rifle with both hands.
- Slide the sling over the head.
- Position the rifle so that it rests comfortably across the back.

To assume a cross body sling transport, a Marine performs the following steps from strong side sling arms (muzzle up) (see fig. 3-25):

- With the left hand, grasp the sling.
- With the right hand, grasp the pistol grip.
- Pull up on the rifle with both hands.
- Slide sling over head.
- Position the rifle so that it rests comfortably across the back.

Figure 3-25. Cross Body Sling Arms. Transport (Muzzle Up).

Note

Ensure the muzzle of the rifle is maintained in a safe direction when assuming this transport.

3010. Transferring the Rifle

Proper weapons handling is required every time a Marine picks up a weapon, passes a weapon to another Marine, or receives a weapon from another Marine. It is the responsibility of the Marine receiving or taking charge of a weapon to determine its condition. Depending on the situation, there are two procedures that can be used to transfer a rifle from one Marine to another Marine: show clear transfer and condition unknown transfer.

Show Clear Transfer

When time and the tactical situation permit, the Marine should transfer the rifle using the show clear transfer. To properly pass a rifle between Marines, the Marine handing off the rifle must perform the following procedures:

- Ensure the rifle is on safe.
- Remove the magazine if it is present.
- Lock the bolt to the rear.
- Visually inspect the chamber to ensure there is no ammunition present.
- Leave the bolt locked to the rear and hand the weapon to the other Marine.

The Marine receiving the weapon must:

- Ensure the rifle is on safe.
- Visually inspect the chamber to ensure there is no ammunition present.
- Release the bolt catch and observe the bolt going forward on an empty chamber.
- Close the ejection port cover.

Condition Unknown Transfer

There are times when time or the tactical situation does not permit a show clear transfer of the rifle. The procedures for the condition unknown transfer are conducted by a Marine when he takes charge of a rifle in any situation when the condition of the rifle is unknown (e.g., an unmanned rifle from a casualty, a rifle stored in a rifle rack). To properly take charge of a rifle when its condition is unknown, the Marine must perform the following procedures:

- Ensure the rifle is on safe.
- Conduct a chamber check to determine the condition of the weapon (see para. 3003).
- Remove the magazine and observe if ammunition is present. If time permits, count the rounds.
- Insert the magazine into the magazine well.
- Close the ejection port cover.

3011. Clearing Barrel Procedures

Purpose of a Clearing Barrel

The sole purpose of a clearing barrel is to provide a safe direction in which to point a weapon when loading; unloading; and unloading and showing clear. See figure 3-26 on page 3-12. Clearing barrel procedures are identical to the weapons handling procedures for the rifle for the loading; unloading; and unloading and showing clear.

Figure 3-26. Clearing Barrel.

Procedures for "Load"

On the command "Load," the Marine will perform the following steps to take the rifle from Condition 4 to Condition 3:

- With a straight trigger finger, point the rifle in the clearing barrel.
- Ensure the rifle is on safe.
- Withdraw a magazine from the magazine pouch.
- Observe the magazine to ensure it is filled.
- Fully insert the magazine into the magazine well.
- Without releasing the magazine, tug downward on the magazine to ensure it is seated.
- Fasten the magazine pouch.
- Close the ejection port cover.

Procedures for "Make Ready"

If standard operating procedures (SOP) or rules of engagement (ROE) require the rifle to be carried in Condition 1, the Marine will "Make Ready" at the clearing barrel. On the command "Make Ready," the Marine will perform the following steps to take the rifle from Condition 3 to Condition 1:

- Pull the charging handle to the rear and release. There are two methods of doing this:
  - Grip the pistol grip firmly with the right hand and pull the charging handle with the left hand to its rearmost position and release.
  - Or grip the handguards firmly with the left hand and pull the charging handle with the right hand to its rearmost position and release.
- To ensure ammunition has been chambered, conduct a chamber check.
- Check the sights (to ensure proper BZO setting, correct rear sight aperture, etc.).
- Close the ejection port cover.

Procedures for "Unload"

On the command "Unload," the Marine will perform the following steps to take the rifle from any condition to Condition 4:

- With a straight trigger finger, point the rifle in the clearing barrel.
- Ensure the weapon is on safe.
- Remove the magazine from the rifle and retain it on your person.
- While cupping the left hand under the ejection port, rotate weapon until the ejection port is facing down.
- Pull the charging handle to the rear and catch the round in the left hand (see fig. 3-17).
- Lock the bolt to the rear.
- Put weapon on safe if not already on safe.
- Ensure that the chamber is empty and that no ammunition is present.
- Release the charging handle and observe the bolt moving forward on an empty chamber.
- Close the ejection port cover.
- Check the sights (for proper BZO setting, correct rear sight aperture, etc.).
- Place any ejected round into the magazine and return the magazine to the magazine pouch and close the magazine pouch.

Procedures for "Unload and Show Clear"

On the command "Unload and Show Clear," the Marine will perform the following steps to take the rifle from any condition to Condition 4:

- With a straight trigger finger, point the rifle in the clearing barrel.
- Ensure the weapon is on safe.
- Remove the magazine from the rifle and retain it on your person.
- While cupping the left hand under the ejection port, rotate the weapon until the ejection port is facing down.
- Pull the charging handle to the rear and catch the round in the left hand.
- Lock the bolt to the rear.

- Ensure the chamber is empty and no ammunition is present.
- Have an observer inspect the weapon to ensure no ammunition is present. The observer:
  - Visually inspects chamber to ensure it is empty, no ammunition is present, and the magazine is removed.
  - Ensures the weapon is on safe.
  - Acknowledges the rifle is clear.
- After receiving acknowledgment from the observer that the rifle is clear, the Marine releases the bolt catch and observes the bolt moving forward on an empty chamber.
- Close the ejection port cover.
- Check the sights (for proper BZO setting, correct rear sight aperture, etc.).
- Place any ejected round into the magazine and return the magazine to the magazine pouch and close the magazine pouch.

CHAPTER 4. FUNDAMENTALS OF MARKSMANSHIP

The fundamentals of marksmanship are aiming, breathing, and trigger control. These techniques provide the foundation for all marksmanship principles and skills. For rifle fire to be effective, it must be accurate. A rifleman who merely sprays shots in the vicinity of the enemy produces little effect. The fundamentals of marksmanship, when applied correctly, form the basis for delivering accurate fire on enemy targets. These skills must be developed so that they are applied instinctively. During combat, the fundamentals of marksmanship must be applied in a time frame consistent with the size and the distance of the target. At longer ranges, the target appears to be smaller and a more precise shot is required to accurately engage the target. The fundamentals are more critical to accurate engagement as the range to the target increases. To be accurate at longer ranges, the Marine must take the time to slow down and accurately apply the fundamentals. At shorter ranges, the enemy must be engaged quickly before he engages the Marine. As the size of the target increases, and the distance to the target decreases, the fundamentals, while still necessary, become less critical to accuracy.

Note

☞The procedures in this manual are written for right-handed Marines; left-handed Marines should reverse instructions as necessary.

4001. Aiming

Sight Alignment

Sight alignment is the relationship between the front sight post and rear sight aperture and the aiming eye. This relationship is the most critical to aiming and must remain consistent from shot to shot. To achieve correct sight alignment (see fig. 4-1):

- Center the tip of the front sight post vertically and horizontally in the rear sight aperture.
- Imagine a horizontal line drawn through the center of the rear sight aperture. The top of the front sight post will appear to touch this line.
- Imagine a vertical line drawn through the center of the rear sight aperture. The line will appear to bisect the front sight post.

Sight Picture

Sight picture is the placement of the tip of the front sight post in relation to the target while maintaining sight alignment. Correct sight alignment but improper sight placement on the target will cause the bullet to impact the target incorrectly on the spot where the sights were aimed when the bullet exited the muzzle.

To achieve correct sight picture, place the tip of the front sight post at the center of the target while maintaining sight alignment (see fig. 4-2). Center mass is

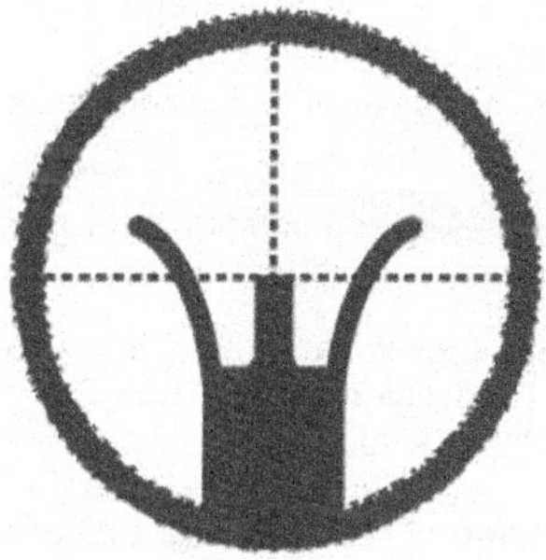

Figure 4-1. Correct Sight Alignment.

Figure 4-2. Correct Sight Picture.

Figure 4-3. Examples of Correct Sight Picture.

the correct aiming point so that point of aim/point of impact is achieved.

The sighting system for the M16A2 rifle is designed to work using a center mass sight picture.

In combat, targets are often indistinct and oddly shaped. The center mass hold provides a consistent aiming point (see fig. 4-3).

Importance of Correct Sight Alignment

A sight alignment error results in a misplaced shot. The error grows proportionately greater as the distance to the target increases. An error in sight picture, however, remains constant regardless of the distance to the target. See figure 4-4.

Factors Affecting Sight Alignment and Sight Picture

Stock Weld

Stock weld is the point of firm contact between the cheek and the stock of the rifle. See figure 4-5. The head should be as erect as possible to enable the aiming eye to look straight through the rear sight aperture. If the position of the Marine's head causes him to look across the bridge of his nose or out from under his eyebrow, the eye will be strained. The eye functions best in its natural forward position. Changing the placement of the cheek up or down on the stock from shot to shot may affect the zero on the rifle due to the perception of the rear sight aperture. A consistent and proper stock weld is critical to the aiming process because it provides consistency in eye relief, which affects the ability to align the sights.

Eye Relief

Eye relief is the distance between the rear sight aperture and the aiming eye. See figure 4-6. Normal eye relief is two to six inches from the rear sight aperture. The distance between the aiming eye and the rear sight aperture depends on the size of the Marine and the firing position. While eye relief varies slightly from one position to another, it is important to have the same eye relief for all shots fired from a particular position.

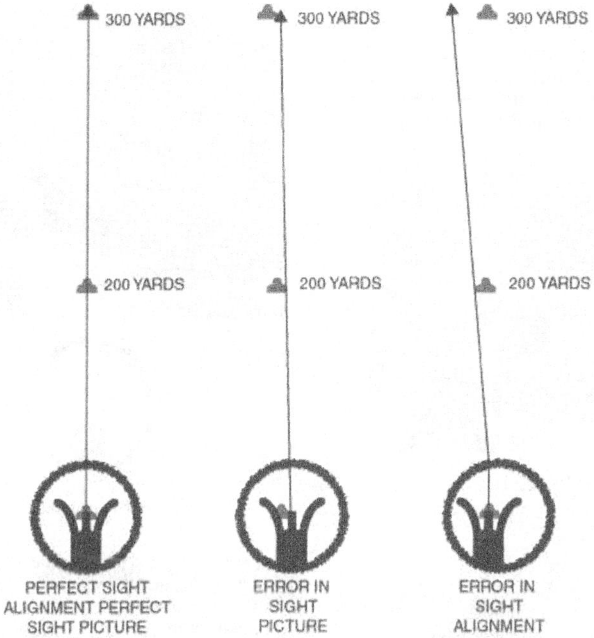

Figure 4-4. Importance of Correct Sight Alignment.

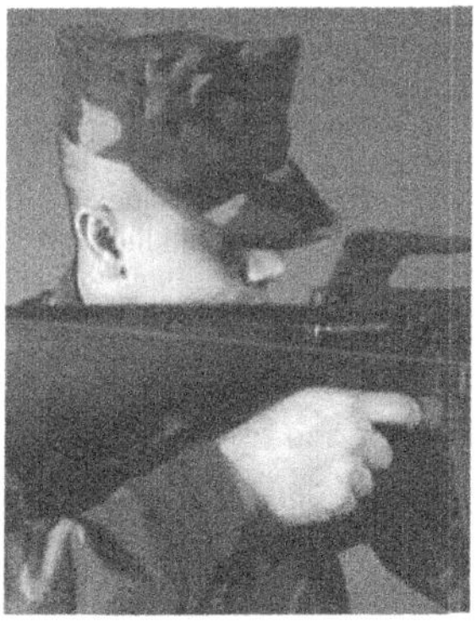

Figure 4-5. Stock Weld.

Figure 4-6. Proper Eye Relief.

If the eye is too close to the rear sight aperture, it will be difficult to line up the front sight post in the rear sight aperture (see fig. 4-7). Moving the eye back from the rear sight aperture will make the aperture appear smaller and allow the tip of the front sight post to be easily lined up inside the rear sight aperture.

Figure 4-7. Shortened Eye Relief.

If the eye is too far from the rear sight aperture, it will be difficult to acquire the target and to maintain a precise aiming point (see fig. 4-8).

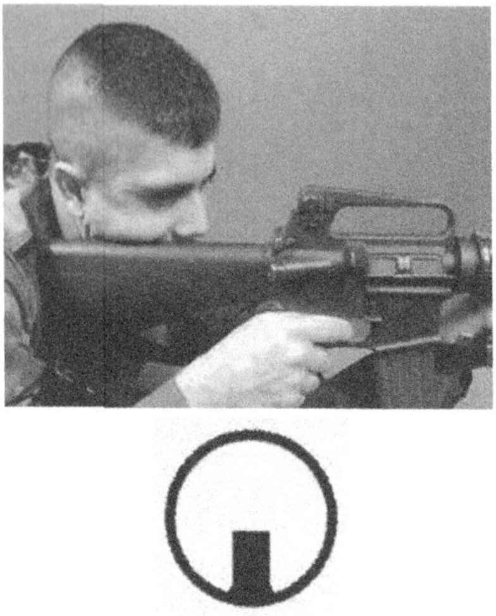

Figure 4-8. Extended Eye Relief.

Wearing of Glasses

Wearing glasses can alter the perception of sight alignment and sight picture. If wearing glasses, it is critical to look through the optic center of the lens.

Acquiring and Maintaining Sight Alignment and Sight Picture

The human eye can focus clearly on only one object at a time. For accurate shooting, it is important to focus on the tip of the front sight post. When the shot is fired, focus must be on the tip of the front sight post; peripheral vision will include the rear sight and the target. The rear sight and the target will appear blurry. Staring or fixing the vision on the front sight post for longer than a few seconds can distort the image, making it difficult to detect minute errors in sight alignment.

Proper stock weld and placement of the rifle butt in the shoulder aid in establishing sight alignment quickly. The rifle butt's placement in the shoulder serves as the pivot point for presenting the rifle up to a fixed point on the cheek (stock weld). During combat, a Marine will look at the target as the rifle is presented. As rifle sights become level with the aiming eye, a Marine visually locates the target through the rear sight aperture. As the rifle settles, a Marine's focus shifts back to the front sight post to place the tip of the post on the target and obtain sight alignment and sight picture. To maintain sight alignment and sight picture, the Marine's focus should shift repeatedly from the front sight post to the target until correct sight alignment and sight picture are obtained. This enables the detection of minute errors in sight alignment and sight picture.

Size and Distance to the Target

During combat, the fundamentals of marksmanship must be applied in a time frame consistent with the size and the distance to the target.

Long-range Engagements

At longer ranges, the target appears smaller and a more precise shot is required to accurately engage the target. Sight alignment and sight picture are more critical to accurate engagement as the range to the target increases. To be accurate at longer ranges, the Marine must take the time to slow down and accurately apply the fundamentals.

As the distance to the target increases, the front sight post covers more of the target, making it difficult to establish a center of mass hold (see fig. 4-9). Since the Marine must see the target to engage it, there is a tendency to look at the target by lowering the tip of the front sight post. This causes shots to impact low or miss the target completely. A Marine must consciously aim at the center of mass and attempt to maintain a center mass sight picture.

Short-range Engagements

At shorter ranges, the enemy must be engaged quickly before he engages the Marine. As distance to the target decreases, the size of the target appears to increase, and sight alignment becomes less critical to accuracy. At very short ranges, a deviation in sight alignment can still produce accurate results as long as the tip of the front sight post is in the rear sight aperture and on the target (see fig. 4-10). The time required to engage a target is unique to each individual. Although a Marine must engage the target rapidly, some semblance of sight alignment is still required to be accurate.

4002. Breath Control

Proper breath control is critical to the aiming process. Breathing causes the body to move. This movement transfers to the rifle making it impossible to maintain proper sight picture. Breath control allows the Marine to fire the rifle at the moment of least movement.

Breath Control During Long-range or Precision Fire (Slow Fire)

It is critical that Marines interrupt their breathing at a point of natural respiratory pause before firing a long-range shot or a precision shot from any distance. A respiratory cycle lasts 4 to 5 seconds. Inhaling and exhaling each require about 2 seconds. A natural pause of 2 to 3 seconds occurs between each respiratory cycle. The pause can be extended up to 10 seconds. During the pause, breathing muscles are relaxed and the sights settle at their natural point of aim. To minimize movement, Marines must fire the shot during the natural respiratory pause. The basic technique is as follows:

- Breathe naturally until the sight picture begins to settle.
- Take a slightly deeper breath.
- Exhale and stop at the natural respiratory pause.
- Fire the shot during the natural respiratory pause.

Note

If the sight picture does not sufficiently settle to allow the shot to be fired, resume normal breathing and repeat the process.

Breath Control During All Other Combat Situations

A Marine in a combat environment may not have the time to fire a shot during the natural respiratory pause.

Figure 4-9. Sight Picture at Long-range Engagements.

Figure 4-10. Sight Picture at Short-range Engagements.

It may be necessary to take several deep breaths before holding the breath. A Marine should not make an exaggerated effort to perform breath control. A natural respiratory pause will help stabilize the shooter's sight picture. The basic technique is as follows:

- Take a deep breath filling the lungs with oxygen.

Note

It may be necessary to take several deep breaths quickly before holding the breath.

- Hold the breath and apply pressure to the trigger.
- Fire the shots.

4003. Trigger Control

Trigger control is the skillful manipulation of the trigger that causes the rifle to fire without disturbing sight alignment or sight picture. Controlling the trigger is a mental process, while pulling the trigger is a physical process.

Grip

A firm grip is essential for effective trigger control. The grip is established before starting the application of trigger control and it is maintained through the duration of the shot. To establish a firm grip on the rifle, position the "V" formed between the thumb and index finger on the pistol grip behind the trigger. The fingers and the thumb are placed around the pistol grip in a location that allows the trigger finger to be placed naturally on the trigger and the thumb in a position to operate the safety. Once established, the grip should be firm enough to allow manipulation of the trigger straight to the rear without disturbing the sights. See figure 4-11.

Trigger Finger Placement

Correct trigger finger placement allows the trigger to be pulled straight to the rear without disturbing sight alignment. The trigger finger should contact the trigger naturally. The trigger finger should not contact the rifle receiver or trigger guard.

Figure 4-11. Grip.

Types of Trigger Control

There are two techniques of trigger control: uninterrupted and interrupted

Uninterrupted Trigger Control

The preferred method of trigger control in a combat environment is uninterrupted trigger control. After obtaining sight picture, the Marine applies smooth, continuous pressure rearward on the trigger until the shot is fired.

Interrupted Trigger Control

Interrupted trigger control is used at any time the sight alignment is interrupted or the target is temporarily obscured. An example of this is extremely windy conditions when the weapon will not settle, forcing the Marine to pause until the sights return to his aiming point. To perform interrupted trigger control:

- Move the trigger to the rear until an error is detected in the aiming process.
- When this occurs, stop the rearward motion on the trigger, but maintain the pressure on the trigger, until sight picture is achieved.
- When the sight picture settles, continue the rearward motion on the trigger until the shot is fired.

Resetting the Trigger

During recovery, release the pressure on the trigger slightly to reset the trigger after the first shot is delivered (indicated by an audible click). Do not remove the finger from the trigger. This places the trigger in position to fire the next shot without having to reestablish trigger finger placement.

4004. Follow-Through/Recovery

Follow-Through

Follow-through is the continued application of the fundamentals until the round has exited the barrel. In combat, follow-through is important to avoid altering the impact of the round by keeping the rifle as still as possible until the round exits the barrel.

Recovery

It is important to get the rifle sights back on the target for another shot. This is known as recovery. Shot recovery starts immediately after the round leaves the barrel. To recover quickly, a Marine must physically bring the sights back on target as quickly as possible.

CHAPTER 5. RIFLE FIRING POSITIONS

In a combat environment, a Marine must be prepared to engage the enemy under any circumstance. There are four basic firing positions: prone, sitting, kneeling, and standing. These positions provide a stable foundation for effective shooting. Any firing position must provide stability, mobility, and observation of the enemy. During training, a Marine learns positions in a step-by-step process, guided by a series of precise movements until the Marine assumes a correct position. The purpose of this is to ensure that the Marine correctly applies all of the factors that assist him in holding the rifle steady. The Marine will gradually become accustomed to the feel of the positions through practice and eventually will be able to know instinctively whether his position is correct. In combat, it may not be possible to assume a textbook firing position due to terrain, available cover, engagement time, dispersion of targets, and other limiting factors. Modifications to the basic positions may have to be made to adjust to the combat environment. The Marine must strive to assume a position that offers stability for firing, maximum cover and concealment from the enemy, and maximum observation of the target.

Note

+The procedures in this manual are written for right-handed Marines; left-handed Marines should reverse instructions as necessary.

5001. Selecting a Firing Position

The selection of a firing position (prone, sitting, kneeling, standing) is based on terrain, available cover, dispersion of targets, and other limiting factors. A Marine must select a position that offers stability, mobility, and observation.

Stability

A firing position must provide a stable platform for accurate and consistent shooting. If the position is solid, the front sight can be held steady and the rifle sights should recover after recoil to the same position on the target. This allows for rapid reengagement of the enemy. The prone position provides the most stability for firing, while the standing position provides the least stability.

Mobility

A firing position must provide a Marine with the mobility required to move to new cover or to another area. The standing position permits maximum mobility, and it also allows the most lateral movement for engagement of widely dispersed targets. The prone position allows the least mobility and allows limited lateral movement.

Observation of the Enemy

A firing position must limit a Marine's exposure to the enemy, yet allow observation of the enemy. Manmade structures and terrain features (e.g., vegetation, earth contours) often dictate the shooting position. The standing position normally provides the best field of view, but it usually allows the most exposure to the enemy. The prone position normally allows the least exposure, but it usually provides a limited field of view.

5002. Types and Uses of the Rifle Web Sling

The rifle sling, when adjusted properly, provides maximum stability for the weapon, and helps hold the front sight still and reduce the effects of the rifle's recoil. Once a sling adjustment is found that provides maximum control of the weapon, the same sling adjustment should be maintained. Varying the sling tension extensively will affect the strike of the bullet, which will make maintaining a BZO difficult. Using the same sling adjustment will ensure the accuracy of rounds on target. There are two basic types of rifle sling adjustments: the hasty sling and loop sling.

Note

In training situations, the parade sling may be used. See paragraph 5008.

Hasty Sling

Application

The hasty sling is used in all firing positions. The hasty sling is advantageous in combat because it can be acquired quickly and it provides added stability to the rifle. The same sling setting can be used for all firing positions. If properly adjusted, the hasty sling supports the weight of the weapon, provides stability for the rifle, and reduces the effects of the rifle's recoil. When using the hasty sling, controlled muscle tension is applied to offer resistance against the sling, enabling the rifle sights to be held steady.

Donning the Hasty Sling

A Marine performs the following steps to form a hasty sling:

- Hold the rifle vertical with the barrel pointing upward.
- Unhook the J-hook from the lower sling swivel.
- Loosen the sling keeper.
- Adjust the sling until the J-hook hangs below the rifle butt. (The distance will vary based on the individual Marine, but the J-hook will usually hang approximately 3 to 10 inches below the rifle butt.) Secure the sling keeper. See figure 5-1.

Figure 5-1. J-hook Location.

- Turn the sling a half turn outboard to allow the sling to lay flat against the arm.
- Attach the J-hook to the lower sling swivel so the open end of the J-hook faces outboard, away from the rifle. See figure 5-2.
- While holding the rifle with the right hand, place the left arm through the sling near the lower sling swivel. Slide the arm up through the sling below the half twist. The sling makes contact low on the arm just below the triceps, above the elbow. The sling lies flat on the back of the arm.

Figure 5-2. J-hook Turned Outboard.

- With the left hand, grasp the handguard by pinching it in the "V" formed by the thumb and forefinger. The sling lies flat against the back or side of the wrist or on the arm near the wrist. See figure 5-3.

Figure 5-3. Sling Against the Wrist.

- Move the left hand as required to level the rifle with the line of sight. Placement of the forward hand controls the tension on the sling between the back of the wrist or arm and the upper sling swivel (see fig. 5-4). This hand placement, with a straight locked wrist, will cause the sling to pull straight under the handguards and serves to stabilize the front sight of the rifle.

- Move the feed end of the sling in or out of the sling keeper to adjust the hasty sling. Sling tension is in-

Figure 5-4. Position of Forward Hand.

creased by pushing the elbow outboard. See figure 5-5. (This enables one sling setting to fit all positions.) It is important for the hasty sling to be adjusted so it supports the rifle. The sling setting must allow the left elbow to push outboard against the sling so the elbow is not inverted under the rifle.

Figure 5-5. Position of Left Elbow.

- Locate the sling keeper near the feed end of the sling and secure so the backside or flat end of the sling keeper is against the arm.

Loop Sling

Application

The loop sling provides the greatest amount of stability during firing. This stability allows the Marine to perfect marksmanship fundamentals. A loop sling takes longer to don or remove than a hasty sling. Therefore, it has limited combat application and is best used where stability of hold is needed for a precision or long-range shot. The loop sling is used in the prone, sitting, and kneeling positions.

Donning the Loop Sling

- Place the rifle butt on the right hip and cradle the rifle in the right arm.
- Disconnect the J-hook from the lower sling swivel.
- With the M-buckle near the hook, feed the sling through the top of the M-buckle to form a loop large enough to slip over the arm. See figure 5-6.
- Give the loop a half turn outboard and insert the left arm through the loop, positioning the loop above the biceps. The loop is high on the left arm above the biceps muscle in such a position that it does not transmit pulse beat to the rifle.
- Position the M-buckle on the outside of the left arm. See figure 5-7.
- Tighten the loop on the left arm, ensuring the M-buckle moves toward the center of the arm as the loop tightens. The sling must pull from the center of the arm to be properly positioned. In this way, as tension is applied to the sling in the firing position, the loop will tighten.

To adjust the sling for the proper length, loosen the sling keeper and pull up or down (toward or away) from the loop. This adjustment varies with every individual and every firing position. The loop should not be tightened excessively on the arm. If blood flow is restricted, excessive pulse beat is transmitted through the rifle sling to the rifle and causes a noticeable, rhythmic movement of the rifle sights. When this occurs, a stable hold at the desired aiming point is impossible to achieve.

Figure 5-6. Forming a Loop.

Figure 5-7. Position of M-buckle.

Tension on the rifle sling is correct when it causes the rifle butt to be forced rearward into the pocket of the shoulder. This serves to keep the butt plate in the shoulder pocket during recoil. To increase tension on the rifle sling, the sling must be shortened. To lessen the tension, the rifle sling must be lengthened.

Move the sling keeper toward the left arm and secure it. The sling keeper should be positioned near the feed end of the sling.

Place the left hand over the sling from the left side and under the rifle. The rifle handguard should rest in the "V" formed between the thumb and forefinger and across the palm of the hand.

Move the left hand as required to achieve desired sight picture. Adjust the length of the sling for proper sling tension and support. See figure 5-8.

Figure 5-8. Loop Sling Donned.

5003. Factors Common to All Shooting Positions

There are seven factors that are common to all shooting positions that affect the ability to hold the rifle steady, maintain sight alignment, and control the trigger. The way these factors are applied differs slightly for each position, but the principles of each factor remain the same.

Left Hand

Placement of the left hand affects placement of left elbow, eye relief, stock weld, and sling tension.

Hasty Sling

In a hasty sling configuration, the sling is attached to the upper and lower sling swivels of the rifle. When the left arm is placed in the hasty sling, tension created by the sling travels from side to side. The tension created by the sling affects how the position is established. There are fundamental differences between the application of the seven factors when using the hasty sling. The most obvious of these is placement of the left hand and the left elbow.

To maximize the support provided by the hasty sling, the left elbow should not be inverted and under the rifle. Instead, the elbow should be pushed outboard against the sling. To achieve this, the position of the shooter's body must almost face the target as opposed to being perpendicular to the target. In addition, the hasty sling must be loosened to allow the elbow to push out against the sling far enough so that the elbow is not under the rifle.

The tension on the sling created by the hasty sling causes the center of balance to change on the rifle. When the elbow is under the rifle with the hasty sling donned, the sling pulls down on the sling swivel disrupting the center of balance and causing the muzzle to drop. Therefore, the elbow must be pushed outboard.

Outboard tension on the sling by the elbow drives the buttstock into the pocket of the shoulder. To enable this, the sling must make contact on the arm just below the triceps, above the elbow. See figure 5-9.

Figure 5-9. Position of Left Elbow with Hasty Sling.

To stabilize the front sight of the rifle, the forward hand, wrist, and forearm should be straight with the wrist locked in place; the hand should be rotated up so the rifle rests in the "V" formed by the thumb and index finger; the fingers will not curl around the handguards. Instead, they will pinch the handguard slightly to keep the hand from slipping on the handguard during recoil. See figure 5-10.

Figure 5-10. Position of Left Hand with Hasty Sling.

When the wrist of the left hand is straight and locked, it creates resistance on the sling close to the muzzle. The sling is in contact with the back or side of the wrist or on the arm near the wrist. This resistance allows the front sight to be stabilized.

In contrast, when the rifle rests across the palm of the hand, the only resistance created is where the sling meets the triceps. Since the resistance is further from the muzzle of the rifle, it makes stabilizing the front sight more difficult.

Loop Sling

With the loop sling donned, the handguard of the rifle rests in the "V" formed by the thumb and index finger of the left hand. The rifle rests across the heel of the hand. The left elbow should be positioned under the weapon to create bone support and a consistent resistance to recoil. The fingers can curl around the handguard, but should apply only the minimum amount of pressure to prevent the hand from slipping on the handguard. See figure 5-11.

Figure 5-11. Position of Left Hand with Loop Sling.

Rifle Butt in the Pocket of the Shoulder

The rifle butt placed firmly in the pocket formed in the right shoulder provides resistance to recoil, helps steady the rifle, and prevents the rifle butt from slipping in the shoulder during firing. Consistent placement of the rifle butt in the shoulder pocket is essential to maintaining a BZO and firing tight shot groups.

Hasty Sling

With the hasty sling donned, the butt of the rifle is placed in the pocket of the shoulder. The body is squared to the target to provide a pocket for the butt of the weapon.

Outboard tension on the sling by the left elbow drives the buttstock into the pocket of the shoulder.

The rifle butt should be placed high in the shoulder to achieve a proper stock weld, allowing the Marine to bring the stock up to his head, rather than lower his head to the stock, which can degrade acquisition of sight alignment and sight picture. See figure 5-12.

Figure 5-12. Buttstock in the Shoulder with Hasty Sling.

Loop Sling

With the loop sling donned, the toe of the rifle butt is placed in the pocket of the shoulder. See figure 5-13.

Figure 5-13. Buttstock in the Shoulder with Loop Sling.

Grip of the Right Hand

Grasp the pistol grip with the right hand and place the forefinger on the trigger, with the thumb and remaining fingers wrapped around the pistol grip. See figure 5-14.

Figure 5-14. Grip of the Right Hand.

Firm rearward pressure should be exerted to help keep the rifle butt firmly in the shoulder, reducing the effects of recoil (this type of pressure is particularly true with the loop sling).

The trigger finger should be placed naturally on the trigger and care should be taken to ensure that the trigger finger can move independently without dragging on the side of the receiver. Proper placement of the right hand high on the pistol grip allows the trigger to be moved straight to the rear without disturbing sight alignment.

Right Elbow

The right elbow should be positioned naturally to provide balance to the position and create a pocket in the shoulder for the rifle butt. If the elbow is correctly positioned, it helps to form a pocket in the right shoulder where the rifle butt rests. The exact placement of the elbow varies with each shooting position but should remain consistent from shot to shot, ensuring the resistance to recoil remains constant. See figure 5-15.

Figure 5-15. Right Elbow.

Stock Weld

The placement of the shooter's cheek against the stock should remain firm and consistent from shot to shot. Consistency of stock weld is achieved through proper placement of the rifle butt in the pocket of the shoulder. A firm contact between the cheek and the stock enables consistent eye relief and enables the head and rifle to recoil as a single unit. Stock weld provides quick recovery between rapid fire shots, keeps the aiming eye centered in the rear sight aperture, and prevents the head from bouncing off the stock during recoil. The head should remain erect to allow the aiming eye to look straight through the rear sight aperture. See figure 5-16.

Figure 5-16. Stock Weld.

Breathing

Breathing causes chest movement and a corresponding movement in the rifle and its sights. Applying breath control will minimize this movement and the effect it has on aiming.

Muscular Tension/Relaxation

Muscular Tension–Hasty Sling

With the hasty sling donned, the shooter must apply an amount of controlled muscular tension in the left arm to keep the sling taut and stabilize the weapon sights. Resistance against the hasty sling controls the point at which the rifle sights will settle. The muscular tension is applied outward against the sling rather than in an effort to hold the rifle up. However, muscular tension should not be excessive to cause the shooter to shake, tremble, or experience fatigue.

Muscular Relaxation–Loop Sling

When using the loop sling, the muscles should be relaxed. Relaxation prevents undue muscle strain and reduces excessive movement. If proper relaxation is achieved, natural point of aim and sight alignment are more easily maintained.

5004. Elements of a Good Shooting Position

There are three elements of a good shooting position that apply when using a loop sling: bone support, muscular relaxation, and natural point of aim. The three elements of a shooting position applied with the loop sling do not apply in the same way as when firing with a hasty sling. While some degree of bone support is still achieved with the hasty sling, muscular **tension** is applied rather than muscular **relaxation**. Natural point of aim, however, applies to both the loop sling and the hasty sling.

Bone Support

The body's skeletal structure provides a stable foundation to support the rifle's weight. A weak shooting position will not withstand a rifle's repeated recoil when firing at the sustained rate or buffeting from wind. To attain a correct shooting position, the body's bones must support as much of the rifle's weight as possible. Proper use of the sling provides additional support.

The weight of the weapon should be supported by bone rather than muscle because muscles fatigue whereas bones do not.

By establishing a strong foundation for the rifle utilizing bone support, the Marine can relax as much as possible while minimizing weapon movement due to muscle tension.

Muscular Relaxation

Once bone support is achieved, muscles are relaxed. Muscular relaxation helps to hold the rifle steady and increase the accuracy of the aim. Muscular relaxation also permits the use of maximum bone support to create a minimum arc of movement and consistency in resistance to recoil. Muscular relaxation cannot be achieved without bone support. During the shooting process, the muscles of the body must be relaxed as much as possible. Muscles that are tense will cause excessive movement of the rifle, disturbing the aim. When proper bone support and muscular relaxation are achieved, the rifle will settle onto the aiming point, making it possible to apply trigger control and deliver a well-aimed shot.

Natural Point of Aim

The point at which the rifle sights settle when in a firing position is called the natural point of aim.

Since the rifle becomes an extension of the body, it may be necessary to adjust the position of the body until the rifle sights settle naturally on the desired aiming point on the target.

When in a shooting position with proper sight alignment, the position of the tip of the front sight post will indicate the natural point of aim. When completely relaxed, the tip of the front sight post should rest on the desired aiming point.

One method of checking for natural point of aim is to aim in on the target, close the eyes, take a couple of breaths, and relax as much as possible. When the eyes are opened, the tip of the front sight post should be positioned on the desired aiming point while maintaining sight alignment.

For each shooting position, specific adjustments will cause the rifle sights to settle center mass, achieving a natural point of aim.

In all positions, the natural point of aim can be adjusted by—

- Varying the placement of the left hand in relation to the handguards.
- Moving the left hand forward on the handguards to lower the muzzle of the weapon, causing the sights to settle lower on the target.
- Moving the left hand back on the handguards to raise the muzzle of the weapon, causing the sights to settle higher on the target.
- Varying the placement of the stock in the shoulder.
- Moving the stock higher in the shoulder to lower the muzzle of the weapon, causing the sights to settle lower on the target.
- Moving the stock lower in the shoulder to raise the muzzle of the weapon, causing the sights to settle higher on the target.
- Natural point of aim can be adjusted right or left by adjusting body alignment in relation to the target.

In the prone position, if the natural point of aim is above or below the desired aiming point, move the body slightly forward or back using the left elbow as a pivot and by digging the toes in.

- Pushing the body forward causes the sights to settle lower on the target.
- Pulling the body backward causes the sights to settle higher on the target.

In the kneeling and sitting positions, natural point of aim can be adjusted by varying the placement of the left elbow on the knee.

- Moving the left elbow forward on the knee lowers the muzzle of the weapon, causing the sights to settle lower on the target.
- Moving the left elbow back on the knee raises the muzzle of the weapon, causing the sights to settle higher on the target.

5005. Prone Position

Application

The prone position provides a very steady foundation for shooting and presents a low profile for maximum concealment. However, the prone position is the least mobile of the shooting positions and may restrict a Marine's field of view for observation. In this position, the Marine's weight is evenly distributed on the elbows, providing maximum support and good stability for the rifle.

Assuming the Prone Position

The prone position can be assumed by either moving forward or dropping backward into position, depending on the combat situation.

Moving Forward into Position

To move forward into the prone position, the Marine performs the following steps:

- With the hasty sling donned, stand erect and face the target, keeping the feet a comfortable distance apart (approximately shoulder width).
- Place the left hand on the handguard, the right hand on the pistol grip.
- Lower yourself into position by dropping to both knees (see fig. 5-17).

Figure 5-17. Dropping to Both Knees.

- Then shift the weight forward to lower the upper body to the ground using the right hand to break the forward motion. See figure 5-18.

Figure 5-18. Moving Forward Into Position.

Dropping Back into Position

It may be necessary to drop backward into position to avoid crowding cover, or to avoid covering uncleared terrain. To drop back into the prone position, the Marine performs the following steps:

- Face the target.
- Place the left hand on the handguard, the right hand on the pistol grip.
- Squat to the ground and break the fall with either hand (see fig. 5-19).
- Kick both legs straight to the rear (see fig. 5-20).
- If the fall was broken using the left hand, reestablish the hasty sling.

Figure 5-19. Breaking the Fall.

Figure 5-20. Kicking Back Into Position.

Straight Leg Prone Position with the Hasty Sling

Apply the seven factors to this position (para. 5003). To assume the straight leg prone position with the hasty sling, either move forward or drop back into position (see figs. 5-21, 5-22, and 5-23):

- Once on the ground, extend your left elbow in front of you. Stretch your legs out behind you. Spread the feet a comfortable distance apart with the toes pointing outboard and the inner portion of the feet in contact with the ground.
- As much of the body mass should be aligned directly behind the rifle as possible.
- If body alignment is correct, weapon recoil is absorbed by the whole body and not just the shoulder.
- Grasp the pistol grip with the right hand and place the rifle butt in the right shoulder pocket.
- Lower the head and place the cheek firmly against the stock to allow the aiming eye to look through the rear sight aperture.
- Rotate the left hand up, pinching the handguard between the thumb and forefinger.
- Slide both elbows outboard on the ground so there is outboard tension against the sling (moving the elbows out tightens the sling) and both shoulders are level. The elbows should provide a tripod of support with the body.
- Adjust the position of the left hand on the handguard to allow the sling to support the weapon and the front sight to be centered in the rear sight aperture. To adjust for elevation:
  - Move the left hand rearward or forward on the handguards (moving the hand rearward elevates the muzzle).
  - Open or close "V" of the left hand for small adjustments. (closing the "V" elevate the muzzle).
- To adjust for a minor cant in the rifle, rotate the handguard left or right in the "V" formed by the thumb and forefinger by rotating the pistol grip left or right.

Straight Leg Prone Position with the Hasty Sling

Figure 5-21. Front View.

Figure 5-22. Left View.

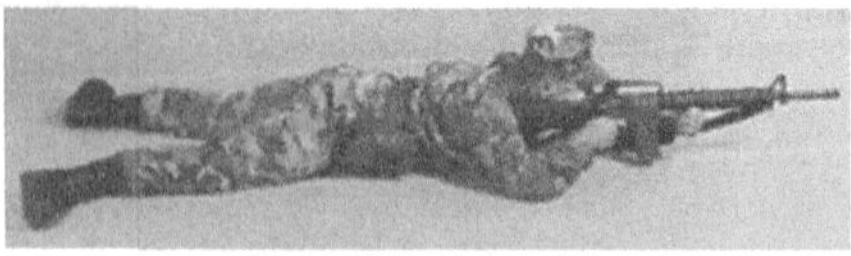

Figure 5-23. Right View.

Straight Leg Prone Position with the Loop Sling

Figure 5-24. Front View.

Figure 5-25. Left View.

Figure 5-26. Right View.

Straight Leg Prone Position with the Loop Sling

Apply the three elements and seven factors to this position (para. 5004). To assume straight leg prone position with the loop sling, either move forward or drop back into position (see figs. 5-24, 5-25, and 5-26):

- Once on the ground, roll the body to the left side as you extend and invert the left elbow on the ground. Stretch your legs out behind you. Spread the feet a comfortable distance apart with the toes pointing outboard and the inner portion of the feet in contact with the ground.
- As much of the body mass should be aligned directly behind the rifle as possible.
- If body alignment is correct, weapon recoil is absorbed by the whole body and not just the shoulder.
- Grasp the rifle butt with the right hand and place the rifle butt into the right shoulder pocket.
- Grasp the pistol grip with the right hand.
- Rotate the body to the right while the elbow is lowered to the ground so the shoulders are level. The right hand pulls and holds the rifle in the shoulder.
- Lower the head and place the cheek firmly against the stock to allow the aiming eye to look through the rear sight aperture.
- Move the left hand to a location under the handguard, which provides maximum bone support and stability for the weapon. This may require that you remove the rifle from the shoulder to reposition the left hand.

Cocked Leg Prone Position with the Hasty Sling

Figure 5-27. Front View.

Figure 5-28. Left View.

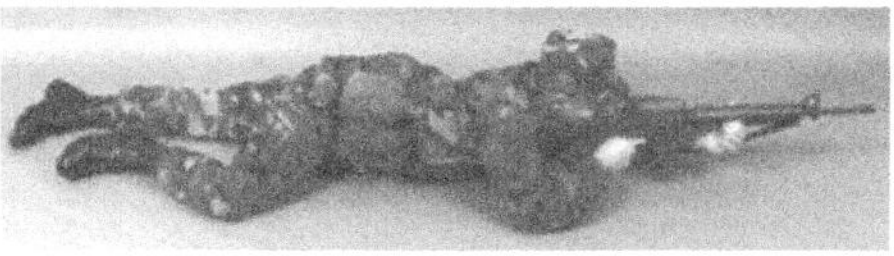

Figure 5-29. Right View.

Cocked Leg Prone Position with the Hasty Sling

Apply the seven factors to this position (para. 5003). To assume the cocked leg prone position with the hasty sling, either move forward or drop back into position (see figs. 5-27, 5-28, and 5-29):

- Once on the ground, roll the body to the left side. The left leg is stretched out behind you, almost in a straight line. This allows the mass of the body to be placed behind the rifle to aid in absorbing recoil.
- Turn the toe of the left foot inboard so the outside of the left leg and foot are in contact with the ground. Bend the right leg and draw it up toward the body to a comfortable position. Turn the right leg and foot outboard so the inside of the right boot is in contact with the ground. Cocking the leg will raise the diaphragm, making breathing easier.
- Grasp the pistol grip with the right hand and place the rifle butt in the right shoulder pocket.
- Lower the head and place the cheek firmly against the stock to allow the aiming eye to look through the rear sight aperture.
- Rotate the left hand up, pinching the handguard between the thumb and forefinger.
- Roll the body to the right while lowering the right elbow to the ground. Slide both elbows outboard on the ground so there is outboard tension against the sling (moving the elbows out tightens the sling). The right shoulder is higher than the left shoulder in the cocked leg position.
- Adjust the position of the left hand on the handguard to allow the sling to support the weapon and the front sight to be centered in the rear sight aperture. To adjust for elevation—
  - Move the left hand rearward or forward on the handguards (moving the hand rearward elevates the muzzle).
  - Open or close "V" of the left hand for small adjustments (closing the "V" elevates the muzzle).

To adjust for a minor cant in the rifle, rotate the handguard left or right in the "V" formed by the thumb and forefinger by rotating the pistol grip left or right.

Cocked Leg Prone Position with the Loop Sling

Apply the three elements and seven factors to this position (para. 5004). To assume the cocked leg prone position with the loop sling, either move forward or drop back into position (see figs 5-30, 5-31, and 5-32):

- Once on the ground, roll the body to the left side and extend and invert the left elbow on the ground. The left leg is stretched out behind you, almost in a straight line. This allows the mass of the body to be placed behind the rifle to aid in absorbing recoil.
- Turn the toe of the left foot inboard so the outside of the left leg and foot are in contact with the ground. Then bend the right leg and draw it up toward the body to a comfortable position. Turn the right leg and foot outboard so the inside of the right boot is in contact with the ground. Cocking the leg will raise the diaphragm, making breathing easier.
- Grasp the rifle butt with the right hand and place the rifle butt into the right shoulder pocket.

Cocked Leg Prone Position with the Loop Sling

Figure 5-30. Front View.

Figure 5-31. Left View.

Figure 5-32. Right View.

- Grasp the pistol grip with the right hand.
- Roll the body to the right while lowering the right elbow to the ground. The right shoulder is higher than the left shoulder in the cocked leg position.
- Lower the head and place the cheek firmly against the stock to allow the aiming eye to look through the rear sight aperture.
- Move the left hand to a location under the handguard which provides maximum bone support and stability for the weapon.

5006. Sitting Position

There are three variations of the sitting position: crossed ankle, crossed leg, and open leg. Experiment with all the variations and select the position that provides the most stability for firing.

Although the sitting position provides an extremely stable base, it limits lateral movement and maneuver ability. It has several variations that can be adapted to the individual Marine. The sitting position provides greater elevation than the prone position while still having a fairly low profile.

Crossed Ankle Sitting Position with the Hasty Sling

The crossed ankle sitting position is an extremely stable shooting position. This position places most of the body's weight behind the weapon and aids in quick shot recovery. Apply the seven factors to this position (para. 5003). To assume the crossed ankle sitting position with the hasty sling (see figs. 5-33, 5-34, 5-35)—

- Square the body to the target.
- Grasp the handguard with the left hand.
- Bend at knees and break the fall with the right hand.
- Push backward with the feet to extend the legs and place the buttocks on the ground.
- Cross the left ankle over the right ankle.
- Grasp the pistol grip with the right hand and place the rifle butt in the right shoulder pocket.
- Lower the head and place the cheek firmly against the stock to allow the aiming eye to look through the rear sight aperture.
- Rotate the left hand up, pinching the handguard between the thumb and forefinger.
- Bend forward at the waist and place the flat portion of the back of the left arm, just above the elbow, on the left leg just below the knee.
- Place the right elbow on the inside of the right knee.
- Adjust the position to adjust sling tension. To tighten the sling: square the body more to the target, or move the left elbow out, or draw the feet up slightly toward the body.
- Adjust the position of the left hand to allow the sling to support the weapon and the front sight to be centered in the rear sight aperture.

To adjust for elevation:

- Move left hand rearward or forward on handguards (moving the hand rearward elevates the muzzle).
- Open or close the "V" of the left hand for small adjustments (closing the "V" elevates the muzzle).

To adjust for a minor cant in the rifle, rotate the handguard left or right in the "V" formed by the thumb and forefinger by rotating the pistol grip left or right.

Crossed Ankle Sitting Position with the Hasty Sling

Figure 5-33. Left View.

Figure 5-34. Front View.

Figure 5-35. Right View.

Crossed Ankle Sitting Position with the Loop Sling

Figure 5-36. Right View.

Figure 5-37. Front View.

Crossed Ankle Sitting Position with the Loop Sling

Apply the three elements and seven factors to this position (para. 5004). To assume crossed ankle sitting position with the loop sling (see figs. 5-36 and 5-37)—

- Position the body at approximately a 30-degree angle to the target.
- Place the left hand under the handguard.
- Bend at knees and break the fall with the right hand.
- Push backward with the feet to extend the legs and place the buttocks on the ground.
- Cross the left ankle over the right ankle.
- Bend forward at the waist and place the left elbow on the left leg below the knee.
- Grasp the rifle butt with the right hand and place the rifle butt into the right shoulder pocket.
- Grasp the pistol grip with the right hand.
- Lower right elbow to the inside of the right knee.
- Lower the head and place the cheek firmly against the stock to allow the aiming eye to look through the rear sight aperture.
- Move the left hand to a location under the handguard, which provides maximum bone support and stability of the weapon.

Crossed Leg Sitting Position with the Hasty Sling

The crossed leg sitting position provides a medium base of support and places some of the body's weight behind the weapon for quick recovery after each shot. Apply the seven factors to this position (para. 5003). To assume the crossed leg sitting position with the hasty sling (see figures 5-38, 5-39, and 5-40)—

- Square the body to the target.
- Grasp the handguard with the left hand.
- Bend at the knees while breaking the fall with the right hand.
- Place the buttocks on the ground as close to the feet as you comfortably can.
- Cross the left leg over the right leg.
- Grasp the pistol grip with the right hand and place the rifle butt in the right shoulder pocket.

Crossed Leg Sitting Position with the Hasty Sling

Figure 5-38. Left View.

Figure 5-39. Front

Figure 5-40. Right View.

- Lower the head and place the cheek firmly against the stock to allow the aiming eye to look through the rear sight aperture.
- Rotate the left hand up, pinching the handguard between the thumb and forefinger.
- Bend forward at the waist while placing the left elbow into the bend of the knee or placing the flat portion of the back of the left arm, just above the elbow, in front of the knee.
- Place the right elbow on the inside of the right knee.
- Adjust the position to adjust sling tension. Squaring the body more to the target or drawing the feet closer together tightens the sling by forcing the left elbow outboard.
- Adjust the position of the left hand to allow the sling to support the weapon and the front sight to be centered in the rear sight aperture.

To adjust for elevation:

- Move left hand rearward or forward on handguards (moving the hand rearward elevates the muzzle).
- Open or close the "V" of the left hand for small adjustments (closing the "V" elevates the muzzle).

To adjust for a minor cant in the rifle, rotate the handguard left or right in the "V" formed by the thumb and forefinger by rotating the pistol grip left or right.

Crossed Leg Sitting Position with the Loop Sling

Apply the three elements and seven factors to this position (para. 5004). To assume crossed leg sitting position with loop sling (see figs. 5-41, 5-42, and 5-43)—

- Position body at a 45- to 60-degree angle to target.
- Place the left hand under the handguard.
- Cross the left leg over the right leg.
- Bend at the knees while breaking the fall with the right hand.
- Place the buttocks on the ground as close to the crossed legs as you comfortably can.
- Bend forward at the waist while placing the left elbow on the left leg into the bend of the knee.
- Grasp the rifle butt with the right hand and place the rifle butt into the right shoulder pocket.
- Grasp the pistol grip with the right hand.
- Lower right elbow to the inside of the right knee.
- Lower the head and place the cheek firmly against the stock to allow the aiming eye to look through the rear sight aperture.
- Move the left hand to a location under the handguard which provides maximum bone support and stability of the weapon.

Open Leg Sitting Position with the Hasty Sling

The open leg sitting position provides a medium base of support and is most commonly used when firing from a forward slope. Apply the seven factors to this position (para. 5003). To assume the open leg sitting position with the hasty sling (see figs. 5-44, 5-45, and 5-46)—

- Square the body to the target.
- Place the feet approximately shoulder width apart.
- Grasp the handguard with the left hand.
- Bend at the knees while breaking the fall with the right hand.
- Push backward with the feet to extend the legs and place the buttocks on the ground.
- Grasp the pistol grip with the right hand and place the rifle butt in the right shoulder pocket.
- Lower the head and place the cheek firmly against the stock to allow the aiming eye to look through the rear sight aperture.

Crossed Leg Sitting Position with the Loop Sling

Figure 5-41. Left View.

Figure 5-42. Front View.

Figure 5-43. Right View.

Open Leg Sitting Position with the Hasty Sling

Figure 5-44. Left View.

Figure 5-45. Front View.

Figure 5-46. Right View.

- Rotate the left hand up, pinching the handguard between the thumb and forefinger.
- Place the flat portion of the back of the left arm, just above the elbow, in front of the left knee.
- Place the right elbow on the inside of the right knee or place the flat portion of the back of the right arm, just above the elbow, in front of the knee.
- Adjust the position to adjust sling tension. Widening the stance tightens the sling by forcing the left elbow outboard. There must be some controlled muscular tension in the legs to hold them up and offer resistance to recoil.
- The position of the left hand to allow the sling to support the weapon and the front sight to be centered in the rear sight aperture.

To adjust for elevation:

- Move left hand rearward or forward on handguards (moving the hand rearward elevates the muzzle).
- Open or close the "V" of the left hand for small adjustments (closing the "V" elevates the muzzle).

To adjust for a minor cant in the rifle, rotate the handguard left or right in the "V" formed by the thumb and forefinger by rotating the pistol grip left or right.

Open Leg Sitting Position with the Loop Sling

Apply the three elements and seven factors to this position (para. 5004). To assume the open leg sitting position with the loop sling (see figs. 5-47, 5-48, and 5-49 on page 5-16)—

- Position the body at approximately a 30-degree angle to the target.
- Place the feet approximately shoulder width apart.
- Place the left hand under the handguard.
- Bend at the knees while breaking the fall with the right hand.
- Push backward with the feet to extend the legs and place the buttocks on the ground.
- Place the left elbow on the inside of the left knee.
- Grasp the rifle butt with the right hand and place the rifle butt into the right shoulder pocket.
- Lower right elbow to the inside of the right knee.
- Lower the head and place the cheek firmly against the stock to allow the aiming eye to look through the rear sight aperture.
- Move the left hand to a location under the handguard which provides maximum bone support and stability of the weapon.

5007. Kneeling Position

Description

The kneeling position is quick to assume and easy to maneuver from. It is usually assumed after initial engagement has been made from a standing position. It can easily be adapted to available cover. A tripod is formed by the left foot, right foot, and right knee when the Marine assumes the position, providing a stable foundation for shooting. The kneeling position also presents a higher profile to facilitate a better field of view as compared to the prone and sitting positions.

Assuming the Kneeling Position

The kneeling position can be assumed by either moving forward or dropping back into position, depending on the combat situation. For example, it may be necessary to drop back into position to avoid crowding cover, or to avoid covering uncleared terrain.

Open Leg Sitting Position with the Loop Sling

Figure 5-47. Left View.

Figure 5-48. Front View.

Figure 5-49. Right View.

Moving Forward into Position

To move forward into the kneeling position, the Marine steps forward toward the target with his left foot and kneels down on his right knee.

Dropping Back into Position

To drop back into the kneeling position, the Marine leaves his left foot in place and steps backward with his right foot and kneels down on his right knee.

High Kneeling Position with the Hasty Sling

Apply the seven factors to this position (para. 5003). To assume the high kneeling position with the hasty sling, either move forward or drop back into position (see figs. 5-50, 5-51, and 5-52):

- Square the body to the target.
- Keep the right ankle straight, with the toe of the boot in contact with the ground and curled under by the weight of the body.
- Place the right portion of the buttocks on or over the right heel. Contact with the heel provides more stability to the position, however, it is not mandatory that the buttocks make contact.
- Place the left foot forward to a point that allows the shin to be vertically straight. For the shin to be vertical, the heel should be directly under the knee. The left foot must be flat on the ground. To provide a wider base of support, slide the right knee and left foot outboard to form a tripod with the right foot.
- Grasp the pistol grip with the right hand and place the rifle butt in the right shoulder pocket.
- Lower the head and place the cheek firmly against the stock to allow the aiming eye to look through the rear sight aperture.
- Rotate the left hand up, pinching the handguard between the thumb and forefinger. The left hand will not grasp the slip ring or the magazine. The magazine must be on the inside of the left arm.
- Place the flat portion of the back of the upper left arm, just above the elbow, on the left knee or against the inside of the left knee so it is in firm

High Kneeling Position with the Hasty Sling

Figure 5-50. Left View.

Figure 5-51. Front View.

Figure 5-52. Right View.

contact. The upper portion of the triceps or the armpit will not rest on the knee.

- Bend the right elbow to provide the least muscular tension possible, lowering it to a natural position.
- Adjust the position to adjust sling tension. Widening the stance by moving the left foot and knee outboard will allow the sling to be tightened.
- Adjust the position of the left hand to allow the sling to support the weapon and the front sight to be centered in the rear sight aperture.

To adjust for elevation:

- Move left hand rearward or forward on handguards (moving the hand rearward elevates the muzzle).
- Open or close the "V" of the left hand for small adjustments (closing the "V" elevates the muzzle).

To adjust for a minor cant in the rifle, rotate the handguard left or right in the "V" formed by the thumb and forefinger by rotating the pistol grip left or right.

High Kneeling Position with the Loop Sling

Apply the three elements and seven factors to this position (para. 5004). To assume the high kneeling position with the loop sling, either move forward or drop back into position (see figs. 5-53, 5-54, and 5-55):

- Position the body at a 45-degree angle to the target.
- Place the left hand under the handguard.
- Kneel down on right knee so right lower leg is approximately parallel to the target (45 to 90 degrees).
- Keep the right ankle straight, with the toe of the boot in contact with the ground and curled under by the weight of the body.
- Place the right portion of the buttocks on or over the right heel; contact with the heel provides more stability to the position.
- Place the left foot forward to a point that allows the shin to be vertically straight. For the shin to be vertical, the heel should be directly under the knee. The left foot must be flat on the ground since it will be supporting the majority of the weight.
- Place the flat portion of the back of upper left arm, just above the elbow, on the left knee so it is in firm contact with the flat surface formed on top of the bent knee. The point of the left elbow will extend just slightly past the left knee. The upper portion of the triceps or the armpit will not rest on the knee.
- Lean slightly forward into the sling for support.
- Grasp the rifle butt with the right hand and place the rifle butt into the right shoulder pocket.
- Grasp the pistol grip with the right hand.
- Bend the right elbow to provide the least muscular tension possible and lower it to a natural position.
- Lower the head and place the cheek firmly against the stock to allow the aiming eye to look through the rear sight aperture.
- Move the left hand to a location under the handguard, which provides maximum bone support and stability for the weapon.

Medium Kneeling Position

This is also referred to as the bootless kneeling position. Assume the medium kneeling position in the same way as the high kneeling position with the ex-

High Kneeling Position with the Loop Sling

Figure 5-53. Left View.

Figure 5-54. Front View.

Figure 5-55. Right View.

ception of the right foot. The right ankle is straight and the foot is stretched out with the bootlaces in contact with the ground. The buttocks are in contact with the heel of the right foot. See figure 5-56.

Figure 5-56. Medium Kneeling Position.

Low Kneeling Position

The low kneeling position is most commonly used when firing from a forward slope. Assume the low kneeling position in the same way as the high kneeling position with the exception of the placement of the right foot. Turn the right ankle so the outside of the foot is in contact with the ground and the buttocks are in contact with the inside of the foot. See figure 5-57.

Figure 5-57. Low Kneeling Position.

5008. Standing Position

Description

The standing position is the quickest position to assume and the easiest to maneuver from. It allows greater mobility than other positions. The standing position is often used for immediate combat engagement. The standing position is supported by the shooter's legs and feet and provides a small area of contact with the ground. In addition, the body's center of gravity is high above the ground. Therefore, maintaining balance is critical in this position.

Standing Position with the Hasty Sling

Apply the seven factors to this position (para. 5003). To assume the standing position with the hasty sling (see figs. 5-58, 5-59, and 5-60)—

- Square the body to the target.
- Spread the feet apart to a comfortable distance with the left foot slightly in front of the right foot. This distance may be wider than shoulder width.
- Distribute the weight evenly over both feet and hips. Balance will shift forward slightly to reduce recovery time and increase the stability of the hold. The legs should be slightly bent for balance.
- Grasp the pistol grip with the right hand and place the rifle butt in the right shoulder pocket.
- Bring the rifle sights up to eye level instead of lowering the head to the sights and place the cheek firmly against the stock. Ensure the head is erect so the aiming eye can look through rear sight aperture.
- Rotate the left hand up, pinching the handguard between the thumb and forefinger.
- The left hand will be under the handguard with the thumb on the outboard side of the handguard.The left hand will not grasp the slip ring or magazine. The magazine must be on the inside of the left arm.
- Hold the right elbow in a natural position.
- Adjust the position of the left hand on the handguard to allow the sling to support the weapon and front sight to be centered in the rear sight aperture.

To adjust for elevation:

- Move left hand rearward or forward on handguards (moving the hand rearward elevates the muzzle).
- Open or close the "V" of the left hand for small adjustments (closing the "V" elevates the muzzle).
- Adjust the position to adjust sling tension. Moving the left elbow out or squaring the body more to the target tightens the sling.

To adjust for a minor cant in the rifle, rotate the handguard left or right in the "V" formed by the thumb and forefinger by rotating the pistol grip left or right.

Standing Position with the Hasty Sling

Figure 5-58. Left View.

Figure 5-59. Front View.

Figure 5-60. Right View.

Standing Position with the Parade Sling

The parade sling is used to emphasize marksmanship fundamentals while firing from the standing position on KD courses during entry-level training. To achieve proper positioning of the parade sling, perform the following steps:

- Attach sling to rifle by placing feed end of sling down through the upper sling swivel.
- Place feed end of sling through sling keeper and lock into place.
- Attach J-hook to lower sling swivel.
- Pull feed end of sling through sling keeper until sling is taut.
- Move sling keeper down near feed end of sling.
- Lock sling keeper into place.

Perform the following steps to assume the standing position with a parade sling (figure 5-61):

- Stand erect.
- Face approximately 90 degrees to the right of the line of fire.
- Place feet approximately shoulder width apart.
- Place left hand under handguard.
- Grasp pistol grip with right hand.
- Place rifle butt into right shoulder pocket.
- Invert left elbow across rib cage.
- Rest left arm naturally against rib cage.
- Lower right elbow to a natural position.
- Place cheek firmly against stock to obtain a firm stock weld.

Figure 5-61. Standing Position with the Parade Sling.

CHAPTER 6. USE OF COVER AND CONCEALMENT

Note

+The procedures in this manual are written for right-handed Marines; left-handed Marines should reverse instructions as necessary.

6001. Cover and Concealment

In a combat environment, a Marine must be prepared to fire from any type of cover or concealment. Cover is anything that protects a Marine from enemy fire. Cover may be an existing hole, a hastily dug shelter, or a well-prepared fighting position with overhead protection. Concealment is anything that hides a Marine from enemy view, but it may not afford protection. Concealment can be obtained from buildings, trees, crops, and skillful use of ground contours. A Marine can use any object or terrain feature that protects him from enemy fire, hides him from enemy view, allows him to observe the enemy, and provides support for a firing position.

Types of Cover

Frontal Cover

A firing position should have frontal cover that provides protection from small arms fire and indirect fire fragments. Ideally, frontal cover should be thick enough to stop small arms fire, high enough to provide protection from enemy fire, and wide enough to provide cover when firing to the left or right edge of a sector of fire.

Ideal Cover

The ideal cover provides:

- Overhead, flank, and rear protection from direct and indirect fire.
- Free use of personal weapons.
- Concealment from enemy observation.
- A concealed route in and out.
- Unobstructed view of the sector of fire.

Common Cover Materials

Any material that protects a Marine from small arms fire can be used for cover. Some common materials include sandbags, trees, logs, and building debris. Table 6-1 presents some common materials and their minimum thickness required for protection from small arms fire.

Table 6-1. Minimum Thickness for Protection Against Small Arms.

Material	Minimum Thickness (in inches)
Concrete	7
Broken stone (rubble)	20
Dry sand	24
Wet sand	35
Logs (oak)	40
Earth (packed)	48

Sandbags

Cover can be improved and positions can be fortified by filling sandbags with dirt/sand and placing them around the position. Sandbags should be tightly packed because bullets can easily penetrate moist or loosely packed sandbags. Overlapping sandbags increase protection and decrease the bullet's ability to penetrate the sandbag. A minimum thickness of three sandbags is required to stop small arms fire.

Trees/Logs

Wood is a relatively dense material and offers good cover and protection. Bullets have a tendency to fragment when they penetrate wood. Live trees have a greater resistance to bullet penetration than dead trees. Wood that has been treated with creosote, such as telephone poles and railroad ties, offers better protection from projectiles than untreated wood, but it still does not ensure protection from small arms fire.

Cinder Bocks

Cinderblocks are not impenetrable cover. Although they are made of a dense material, the composition of a cinderblock is so brittle that a bullet can shatter the block upon impact. This can cause injury to a Marine by secondary fragmentation.

Firing From Specific Types of Cover

Effective cover allows a Marine to engage enemy targets while protecting himself from enemy fire. Several types of cover provide support, protection, and concealment and do not interfere with target engagement. A Marine must adapt firing positions to the type of cover available.

Fighting Hole

A Marine should use fighting holes if available. See figure 6-1. After a Marine enters the fighting hole, he adds or removes dirt, sandbags or other supports to fit his height. To assume a firing position, a Marine performs the following steps:

- Place the right foot to the rear as a brace.
- Lean forward until the chest is against the forward wall of the fighting hole.
- Extend the left arm and elbow over the forward side of the fighting hole so the left forearm rests against the back of the parapet.
- Place the rifle butt into the pocket of the right shoulder and grasp the pistol grip with the right hand.
- Place the right elbow on solid support using the elbow rest of the fighting hole or sandbags placed around the fighting hole.

Figure 6-1. Fighting Hole Position.

Rooftop

If possible, a Marine's entire body should be positioned behind the apex of the rooftop, using the apex to support the rifle. See figure 6-2. If the body cannot be positioned behind the apex, place the left arm over the apex of the roof to hold the weight of the body. Expose as little of the head and shoulders as possible. See figure 6-3.

Figure 6-2. Rooftop Position Behind the Apex.

Figure 6-3. Rooftop Position Supported by the Apex.

Window

The Marine can establish a supported or unsupported position from a window.

Unsupported. A Marine can establish an unsupported position back from the opening of the window so that the muzzle does not protrude and interior shadows provide concealment so as not to provide a silhouette to the enemy. See figure 6-4.

Figure 6-4. Unsupported Window Position.

Supported. When additional stability is needed, a Marine can establish a supported position by placing the rifle handguards or his forearm in the "V" formed by the side and bottom of the windowsill. A drawback to this technique is that the muzzle of the weapon and the Marine may be exposed to view. See figures 6-5 and 6-6.

Vehicle

In many combat situations, particularly in urban environments, a vehicle may be the best form of cover. When using a vehicle for cover, the engine block provides the most protection from small arms fire. The Marine should establish a position behind the front wheel so the engine block is between him and the target (see figs. 6-7 and 6-8). From this position, the Marine may fire over, under or around the vehicle. This is a particularly effective position for larger vehicles that are high off the ground.

The Marine can establish additional support for the rifle by positioning himself behind the doorjamb (frame of door) and placing the rifle against the "V" formed by the open door and doorframe (see figure 6-9). From this position, the Marine may fire over the hood of the vehicle while using the engine block for protection.

Figure 6-5. Supported Window Position (Handguards on Support).

Figure 6-6. Supported Window Position (Forearm on Support).

Figure 6-7. Firing Around Front of Vehicle.

Figure 6-8. Firing Over Front of Vehicle.

Figure 6-9. Establishing a Supported Position in a Vehicle.

However, this position limits lateral mobility and it is more difficult to maneuver from.

At the back of the vehicle, only the axle and the wheel provide cover. If the Marine must shoot from the back of the vehicle, he must position himself directly behind the wheel as much as possible (see fig. 6-10).

Figure 6-10. Using the Back of a Vehicle for Cover.

6002. Supported Firing Positions

Supports are foundations for positions; positions are foundations for the rifle. To maximize the support the position provides, the firing position should be adjusted to fit or conform to the shape of the cover. Elements of a sound firing position, such as balance and stability, must be incorporated and adjusted to fit the situation and type of cover. A supported firing position should minimize exposure to the enemy, maximize the stability of the rifle, and provide protection from enemy observation and fires. A Marine can use any available support (e.g., logs, rocks, sandbags or walls) to stabilize his firing position. The surrounding combat environment dictates the type of support and position used.

Considerations Using Cover and Concealment

Cover and concealment considerations are similar, regardless of the combat environment (e.g., urban, desert, jungle).

Adjusting the Shooting Position

The type of cover can dictate which shooting position (e.g., standing, kneeling, sitting, and prone) will be the most effective. For example, a Marine's height in relation to the height of the cover aids in the selection of a firing position.

The firing position selected should be adjusted to fit the type of cover to:

- Provide stability. The position should be adjusted to stabilize the rifle sights and allow the management of recoil to recover on target.
- Permit mobility. The position should be adjusted to permit lateral engagement of dispersed targets and movement to other cover.
- Allow observation of the area/enemy while minimizing exposure to the enemy.

The firing position is adjusted to fit the type of cover by adjusting the seven factors (i.e., left hand, pocket of shoulder, right elbow, stock weld, grip of right hand) to support the rifle or the position.

Keeping the Entire Body Behind Cover

A Marine should minimize exposure of any part of his body to fire. Be especially aware of the head, right elbow, knees or any other body part that may extend beyond the cover.

Firing From the Right or Left Side of Cover

To minimize exposure and maximize the cover's protection, a right-handed Marine should fire from the right side of cover and a left-handed Marine should fire from the left side, if possible (see fig. 6-11).

Figure 6-11. Firing from the Right Side of Cover.

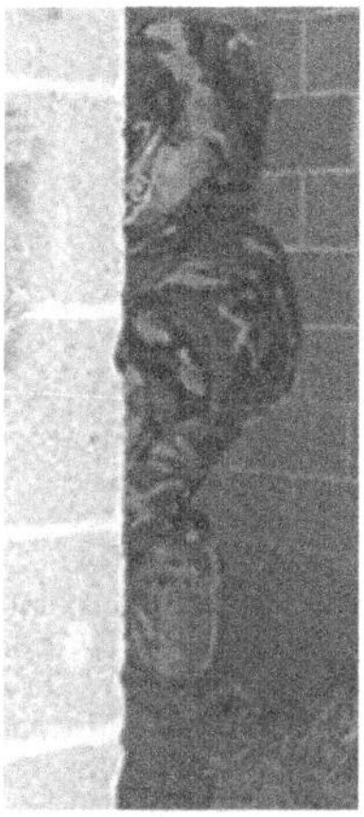

Figure 6-12. Firing from the Left Side of Cover.

If, however, a right-handed Marine must fire from the left side of cover, he fires right-handed but adjusts his position behind cover and uses the rollout technique (see para. 6003) to engage the target. See figure 6-12.

Firing Over the Top of Cover

Firing over the top of cover provides a wider field of view and lateral movement. When firing over the top of cover, the position may be supported and stabilized by resting the handguard or the left forearm on the cover (see fig. 6-13). The Marine should keep as low a profile as possible; the rifle should be as close to the top of cover as possible.

Figure 6-13. Firing Over the Top of Cover.

Maintaining Muzzle Awareness

When firing over the top of cover, a Marine must remember that the sights are higher than the barrel and remain aware of the location of his muzzle. Therefore, a Marine must maintain a position that ensures the muzzle is high enough to clear the cover (e.g., window sill, top of wall) as he obtains sight alignment/sight picture on the target (see figs. 6-14a and 6-14b).

Clearing the Ejection Port

Ensure the cover does not obstruct the ejection port. If the ejection port is blocked, the obstruction can interfere with the ejection of the spent cartridge case and cause a stoppage.

Figures 6-14a and 6-14b. Clearing Cover with the Muzzle.

Resting the Magazine

The bottom, front or side of the rifle magazine can rest on or against support to provide additional stability (see figs. 6-15, 6-16, and 6-17).

CAUTION

The back of the magazine should not be pulled back against support because it can cause a stoppage by not allowing a round to feed from the magazine.

Seven Factors

The seven factors (left hand, rifle in shoulder pocket, stock weld, right elbow, grip of the right hand, breathing, muscular tension) are applied when firing from cover, however, some may have to be modified slightly to accommodate the artificial support provided the rifle and position.

Left Hand

The support should be used to help stabilize both the firing position and the rifle to enable the Marine to maintain sight alignment and sight picture.

The forearm or left hand can contact the support to stabilize the weapon. Rest the forearm or the meaty portion of the bottom of the left hand on the support and rest the rifle in the "V" formed by the thumb and forefinger of the left hand (see fig. 6-18).

The rifle's handguards may rest on the support, but the barrel may not (see fig. 6-19). Placement of left hand on the handguard may have to be adjusted forward or back to accommodate the cover and the additional support provided by the rifle resting on the cover. If the handguards are resting on the cover, the left hand

Figure 6-15. Bottom of Magazine on Support.

Figure 6-16. Front of Magazine on Support.

Figure 6-17. Side of Magazine on Support.

can pull down on the handguards to further stabilize the weapon.

Rifle in Shoulder Pocket

Regardless of the combat situation or the height of the cover, the rifle butt must remain in the shoulder pocket to manage the effects of recoil, stabilize the rifle, and maintain the rifle's battlesight zero. The firing position must be adjusted behind cover to enable the rifle to be placed in the shoulder.

Stock Weld

Regardless of the cover, the firing position must be adjusted to allow stock weld to be achieved. Proper stock weld provides quick recovery between shots and keeps the aiming eye centered in the rear sight aperture.

Right Elbow

The right elbow can be placed on or against the support to stabilize the weapon and the position. Ensure the elbow is not extended beyond cover revealing the position to the enemy.

Grip of the Right Hand

If rifle handguards, forward hand, or forearm rest on cover for support, the grip of the right hand should pull back and down on the pistol grip to further stabilize the weapon in the shoulder and on the support.

Breathing

Breathing does not change when firing from a supported position.

Muscular Tension

To create balance and support for the position, the Marine may shift his body weight into or against the support (see fig. 6-20). This enables the Marine to use cover to support his body weight, reducing the need for muscular tension.

Figure 6-18. Forearm Resting on Cover.

Figure 6-19. Handguards Resting on Cover.

Figure 6-20. Shifting Body Weight into Cover.

Types of Supported Positions

Supported Prone

If possible, a Marine should use the supported prone position when firing from behind cover. This position is the steadiest, provides the lowest silhouette, and provides maximum protection from enemy fire.

- Support the position by placing the handguards, the forearm, or the magazine on or against support (see fig. 6-21).

Figure 6-21. Supported Prone.

- The prone position can be assumed behind a tree, a wall, a log, or almost any type of cover. It is flexible and allows shooting from all sides of cover and from cover of various sizes.
- The body must be adjusted to conform to the cover. For example, if the cover is narrow, keep the legs together. The body should be in line with the rifle and directly behind the rifle (see figs. 6-22 and 6-23). This presents a smaller target to the enemy and provides more body mass to absorb recoil.

Figure 6-22. Supported Prone Behind Narrow Cover—Rear View.

Supported Kneeling

When the prone position cannot be used because of the height of the support, the supported kneeling position may be appropriate. The kneeling position provides additional mobility over the prone position.

- The kneeling position allows shooting from all sides and from cover of varying sizes. This position may be altered to maximize the use of cover or support by assuming a variation of the kneeling position (high, medium or low).
- In the kneeling position, the Marine must not telegraph his position behind the cover with his knee. When shooting around the sides of cover, the Marine should strive to keep his right knee in line with his left foot so as not to reveal the position to the enemy. See figure 6-24.
- Support the position by placing the handguards, the forearm, or the magazine on or against support. In addition, the position (e.g., a knee, the side of the body) may rest against support (see fig. 6-25).
- If the rifle is resting on support, the Marine may not need to stabilize the weapon by placing his left elbow on his knee (see fig. 6-26).

Supported Sitting

A supported sitting position may be used to fire over the top of cover when mobility is not as critical. A sitting position can be comfortably assumed for a longer period of time than a kneeling position and it can conform to higher cover when a prone position cannot be used.

Figure 6-23. Supported Prone Behind Narrow Cover—Front View.

Figure 6-24. Supported Kneeling.

Figure 6-25. Supported Kneeling with Body Against Support.

- Support the position by placing the handguards, the forearm, or the magazine on or against support (see fig. 6-27).

- If the rifle is resting on support, the Marine may not need to stabilize the weapon by placing his left or right elbows on his legs (see fig. 6-28).

Supported Standing

The supported standing position provides greater mobility than the other positions and usually provides greater observation of the enemy. The supported standing position is effectively used behind high cover (e.g., window, over a wall) or narrow cover (e.g., tree, telephone pole).

- To assume the supported standing position, the Marine leans his body forward or against support to stabilize the weapon and the position.

Figure 6-26. Supported Kneeling with Rifle Resting on Support.

Figure 6-27. Supported Sitting.

Figure 6-28. Supported Sitting with Rifle on Support.

- Support the position by placing the handguards, the forearm, or the magazine on or against support. In addition, the position (e.g., the side of the body) may rest against support (see fig. 6-29).

Figure 6-29. Supported Standing.

6003. Searching for and Engaging Targets From Behind Cover

To locate targets when behind cover or to ensure the area is clear before moving, the Marine must expose as little of himself as possible to the enemy. Additionally, the Marine must be ready to fire if a target is located. Two techniques that can be used to locate and engage targets from behind cover are: the pie and rollout. These techniques minimize the Marine's exposure to enemy fire while placing the Marine in a position to engage targets or to move to another location if necessary. These techniques are also used to enter a building or structure. Both techniques are used in the kneeling and standing positions. To be accurate in engaging targets using either technique, the seven factors must be applied. See page 6-6 for a discussion of seven factors.

Pie Technique

To perform the pie technique:

- Staying behind cover, move back and away from the leading edge of the cover. The surroundings and situation will dictate the distance you should move back and away from the cover. Generally, the further back the Marine is from cover, the greater his area of observation; staying too close to cover decreases the area of observation. However, if the Marine is too far back from the leading edge of cover, he may unknowingly expose himself to the enemy.
- Assume a firing position and lower the rifle sights enough to have a clear field of view, orienting the muzzle on the leading edge of the cover. (In a building, the baseboards serve as a reference point for the muzzle of the rifle when searching for targets.)
- Taking small side steps, slowly move out from behind the cover, covering the field of view with the aiming eye and muzzle of the weapon. Wherever the eyes move, the muzzle should move (eyes, muzzle, target). The muzzle should remain on the leading edge of cover, serving as a pivot point when moving out.
- Continue taking small side steps and moving out from cover until a target is identified or the area is found to be clear.
- When a target is identified, sweep the safety, place the finger on the trigger, and engage the target.

Rollout Technique

To perform the rollout technique:

- Staying behind cover, move back and position the body so it is in line with the leading edge of the cover, ensuring that no part of the body extends beyond the cover.
- Assume a firing position and come to the Ready, ensuring the muzzle is just behind the cover.
- Canting the head and weapon slightly, roll the upper body out to the side just enough to have a clear field of view and allow the muzzle to clear the cover. Keeping the feet in place, push up on the ball of one foot to facilitate rolling out.
- Continue rolling out from cover until a target is identified or the area is found to be clear.
- When a target is identified, sweep the safety, place the finger on the trigger, and engage the target.
- If a target is identified before moving out from cover, the rifle should be taken off safe before moving out using the rollout technique.

Combining Techniques

In some situations, it may be necessary to combine the pie and rollout techniques to search an entire area for targets (e.g., corner of a building, a doorway). Changing from one technique to another may permit the Marine to minimize his exposure to enemy observation and fires.

6004. Moving Out From Behind Cover

A Marine must be constantly aware of his surroundings and available cover, should a threat appear. He should avoid obvious danger areas and move quickly through danger areas that cannot be avoided.

When moving from cover to cover, the Marine should select the next covered location and plan his route before moving from his present position. This is done by quickly looking from behind cover to ensure the area is clear, ensuring the head and eyes are exposed for as short a time as possible.

If necessary, the Marine should conduct a Condition 1 reload before moving from cover.

Once the Marine is committed to moving, all focus should be on moving until cover is reassumed.

CHAPTER 7. RIFLE PRESENTATION

Note

☞The procedures in this manual are written for right-handed Marines; left-handed Marines should reverse instructions as necessary.

7001. Presentation of the Rifle

In a combat environment, targets may present themselves with little or no warning. To maintain an advantage, the Marine carries his weapon in a position appropriate to the threat level that permits the rifle to be both easily carried and presented as quickly as possible. A carry is also established based on the situation such as moving in a close quarter environment, moving over or under objects, etc.

Presenting the Rifle From the Tactical Carry

The Marine uses the Tactical Carry when no immediate threat is present. This carry permits the rifle to be easily carried for long periods of time, but it does not permit the quickest presentation to a target. If the situation changes and a target presents itself, a Marine performs the following steps to present the rifle from the Tactical Carry once a target appears:

- Extend the rifle toward the target keeping the muzzle slightly up so the buttstock clears all personal equipment. Continue to look at the target.
- At the same time, place the rifle in Condition 1. Two methods can be used to place rifle in Condition 1 if it is in Condition 3:
  - Grip the pistol grip firmly with the right hand. Pull the charging handle with the left hand to its rearmost position and release (see fig. 7-1).

Figure 7-1. Pulling Charging Handle with Left Hand.

  - Grip the handguards firmly with the left hand. Pull the charging handle with the right hand to its rearmost position and release (see fig. 7-2).

Figure 7-2. Pulling Charging Handle with Right Hand.

- As the rifle is being presented, take the rifle off safe and place the trigger finger on the trigger (see figs. 7-3a, 7-3b, and 7-3c on page 7-2).
- Level the rifle while pulling it firmly into the pocket of the shoulder to obtain proper stock weld. Do not move the head down to meet the stock of the rifle.

Note

If the rifle is in the shoulder properly, the aiming eye will be able to look through the rear sight as soon as the stock makes contact with the cheek.

- As the sights become level with the aiming eye, visually locate the target through the rear sight aperture. As the rifle sights settle, shift the focus back to the front sight post to obtain sight alignment, and place the tip of the post center mass on the target to obtain sight picture.

Presenting the Rifle From the Alert Carry and From the Ready Carry

The Marine uses the Alert Carry when enemy contact is likely. The Alert is also used for moving in close terrain (e.g., urban, jungle).

The Marine uses the Ready Carry when enemy contact is imminent.

Figures 7-3a, 7-3b, and 7-3c. Sweeping the Safety.

To present the rifle from the Alert and from the Ready, a Marine performs the following steps once a target appears:

- While looking at the target, bring the muzzle up by raising the left hand, allowing the rifle butt to pivot in the shoulder. At the same time, pull the rifle firmly into the pocket of the shoulder.
- As the rifle is being presented, take the rifle off safe and place the trigger finger on the trigger (see figs. 7-3a, 7-3b, and 7-3c).
- As the stock makes contact with the cheek, level the rifle to obtain proper stock weld.
- Do not move the head down to meet rifle stock.

Note

If the rifle is in the shoulder properly, the aiming eye will be able to look through the rear sight as soon as the stock makes contact with the cheek.

- As the sights become level with the aiming eye, visually locate the target through the rear sight aperture. As the rifle sights settle, shift the focus back to the front sight post to obtain sight alignment, and place the tip of the post center mass on the target to obtain sight picture.

Presenting the Rifle From the Strong Side Sling Arms Transport

Once a target appears, a Marine performs the following to present the rifle from Strong Side Sling Arms:

- While looking at the target, lean forward slightly to facilitate removal of the rifle from the shoulder.
- Reach under the right arm with the left hand between the sling and the body and grasp the handguards. See figure 7-4). At the same time, pull down on the sling and raise right elbow out and parallel to the deck.
- Roll the right shoulder forward and release the sling from the right hand once the handguards have cleared the elbow. At the same time, pull rifle forward off the shoulder with left hand.
- Continue pulling the rifle forward with the left hand while rotating the rifle parallel to the deck; when the right arm is free of the sling and the rifle clears all personal gear, grasp the charging handle with the right hand (see fig. 7-5) and pull it to its rearmost position and release.
- Establish a firing grip with right hand while keeping the trigger finger straight along the receiver.
- Take rifle off safe and place the trigger finger on the trigger (see figs. 7-3a, 7-3b, and 7-3c).
- Level the rifle while pulling it firmly into the pocket of the shoulder to obtain proper stock weld. Do not move the head down to meet the stock of the rifle.

Figure 7-4. Grasping the Handguards.

Figure 7-5. Clearing Gear and Grasping the Charging Handle.

Note

If the rifle is in the shoulder properly, the aiming eye will be able to look through the rear sight as soon as the stock makes contact with the cheek.

- As the sights become level with the aiming eye, visually locate the target through the rear sight aperture. As the rifle sights settle, shift the focus back to the front sight post to obtain sight alignment, and place the tip of the post center mass on the target to obtain sight picture.

Presenting the Rifle From the Weak Side Sling Arms Transport

The hasty sling should be maintained while presenting the rifle from this transport. To present the rifle from Weak Side Sling Arms, a Marine performs the following steps once a target appears:

- While looking at the target, lean forward slightly to facilitate removal of rifle from the shoulder.
- Grasp the sling with the right hand to prevent the rifle from falling off the shoulder.
- Grasp handguards with left hand (the index finger points toward the muzzle). See figure 7-6.
- Rotate the rifle counterclockwise while extending the muzzle toward the target.
- Continue extending the rifle toward the target to ensure the rifle clears all personal gear.
- Grasp the charging handle with the right hand and pull it to its rearmost position and release (see fig. 7-7).
- Establish a firing grip with right hand while keeping the trigger finger straight along the receiver.
- Take the rifle off safe and place the trigger finger on the trigger (see figures 7-3a, 7-3b, and 7-3c).
- Level the rifle while pulling it firmly into the pocket of the shoulder to obtain proper stock weld. Do not move the head down to meet the stock of the rifle.

Figure 7-6. Grasping the Handguards.

Note

If the rifle is in the shoulder properly, the aiming eye will be able to look through the rear sight as soon as the stock makes contact with the cheek.

Figure 7-7. Clearing Gear and Grasping the Charging Handle.

- As the sights become level with the aiming eye, visually locate the target through the rear sight aperture. As the rifle sights settle, shift the focus back to the front sight post to obtain sight alignment, and place the tip of the post center mass on the target to obtain sight picture.

7002. Search and Assess

After a Marine engages a target, he must immediately search the area and assess the results of his engagement. Searching and assessing enables the Marine to avoid tunnel vision that can restrict the focus so that an indication of other targets is overlooked.

Purpose

The Marine searches the area for additional targets or for cover. The Marine assesses the situation to determine if he needs to re-engage a target, engage a new target, take cover, assume a more stable position, cease engagement, etc.

Technique

To search and assess, a Marine performs the following steps:

- Keeping the buttstock in the shoulder, lower the muzzle of the rifle slightly to look over the sights.
- Place the trigger finger straight along the receiver (see fig. 7-8).
- Search the area and assess the situation/threat by moving the head, eyes, and rifle left and right (approximately 45 degrees from center) to cover the immediate area. The muzzle moves with the head and eyes in one fluid motion while searching. Keep both eyes open to increase the field of view.

Figure 7-8. Straight Trigger Finger.

- Once a Marine determines the area is clear of enemy threat, he places the rifle on safe.

Searching and Assessing to a Higher Profile

Depending on the tactical situation, the Marine may choose to increase his area of observation by searching and assessing to a higher profile position.

Prone to Kneeling

After searching and assessing at the prone position, move to a kneeling position by performing the following steps:

- While maintaining control of the pistol grip, lower the rifle butt out of the shoulder.
- Drop the left hand to the deck and, bringing it back, push up off the deck to both knees (see figs. 7-9a and 7-9b).
- Grasp the handguard with the left hand and place the rifle butt in the pocket of the shoulder.
- Assume a kneeling position and search and assess (see fig. 7-10).

Sitting to Kneeling

After searching and assessing at the sitting position, move to a kneeling position by performing the following steps:

- Maintain control of the rifle with the rifle butt in the pocket of the shoulder.

Figures 7-9a and 7-9b. Pushing Up Off the Deck to Both Knees.

- Uncross the legs to an open leg position.
- Tuck the right foot underneath the left thigh, as close to the buttocks as possible (see fig. 7-11).
- Lean forward and to the right and roll on to the right knee to a kneeling position and search and assess (see fig. 7-12).

Note

It may be necessary to release the rifle with the right hand and push off the deck with the right hand to assist in rolling up to a kneeling position.

Kneeling to Standing

After searching and assessing at the kneeling position, maintain control of the rifle with the rifle butt in the pocket of the shoulder, and stand while continuing to search and assess.

Figure 7-10. Kneeling Search and Assess.

Figure 7-11. Tucking the Right Foot.

Figure 7-12. Rolling Up to a Kneeling Position.

Chapter 8. Effects of Weather

Wind, temperature, and precipitation can affect the trajectory of the bullet. In addition, all weather conditions have a physical and psychological effect on Marines. Marines must use techniques to offset the effects of wind, light, temperature, and precipitation (snow, sleet, and rain). Through proper training, Marines can develop the confidence required to reduce the physical and psychological effects of weather during combat situations.

8001. Physical Effects of Wind on the Bullet

Physical Effects

The weather condition that presents the greatest problem to shooting is the wind. Wind affects a bullet's trajectory. The effect of wind on the bullet as it travels down range is referred to as deflection. The wind deflects the bullet laterally in its flight to the target (see fig. 8-1).

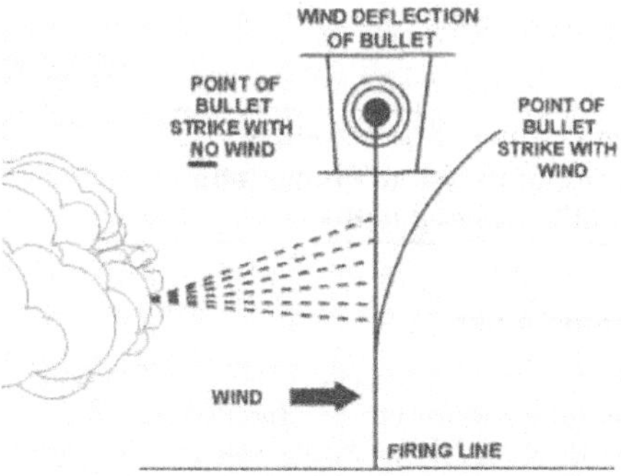

Figure 8-1. Deflection of a Bullet.

The bullet's exposure time to the wind determines the amount the bullet is deflected from its original trajectory. Deflection increases as the distance to the target increases. There are three factors that affect the amount of deflection of the bullet:

- Velocity of the wind—The greater the velocity of the wind, the more the bullet will be deflected.
- Range to the target—As the distance to the target increases, the speed of the bullet slows allowing the wind to have a greater effect on shot placement.
- Velocity of the bullet—A bullet with a high muzzle velocity will not be affected by the wind as much as a bullet with a low muzzle velocity.

Determining Windage Adjustments to Offset Wind Effects

The velocity and direction of the wind in relationship to the bullet must be determined to offset the wind's effects. If Marines can classify wind values and determine velocity within 5 mph, they can effectively engage targets in windy conditions.

Wind Direction

Determine wind direction by observing direction vegetation is moving, by feeling the wind blow against the body, or by observing direction of a flag (in training).

Wind Value Classifications

Winds are classified according to the direction from which they are blowing in relation to the direction of fire. The clock system indicates wind direction and value (see fig. 8-2). Winds can be classified as half

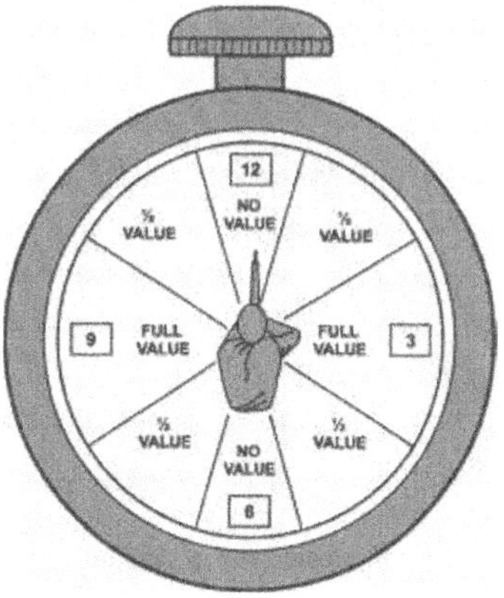

Figure 8-2. Clock System.

value, full value, or no value. The target is always located at 12 o'clock.

Wind Velocity

There are two methods used to determine wind velocity: observation and flag. The flag method is used as a training tool on the known distance (KD) range to learn the observation method. This method teaches Marines to relate the effect a given wind condition has on the natural surroundings in order to develop the base of knowledge used during the observation method. The observation method is the primary method used to estimate wind velocity and direction in a tactical situation. The following are guidelines used during the observation method:

- Under 3 miles per hour (mph) the wind can hardly be felt on the face. The presence of a slight wind can be detected by drifting smoke.
- 3 to 5 mph winds can be felt lightly on the face.
- 5 to 8 mph winds keep tree leaves in a constant motion.
- 8 to 12 mph winds raise dust and loose paper.
- 12 to 15 mph winds cause small trees to sway.
- 16 to 25 mph winds cause large trees to sway.

Flag Method

The flag method is primary method used on the KD range. To estimate wind velocity in miles per hour:

Estimate the angle created between the flagpole and the flag in degrees.

- Divide the angle by four to estimate wind velocity in miles per hour. See figure 8-3.

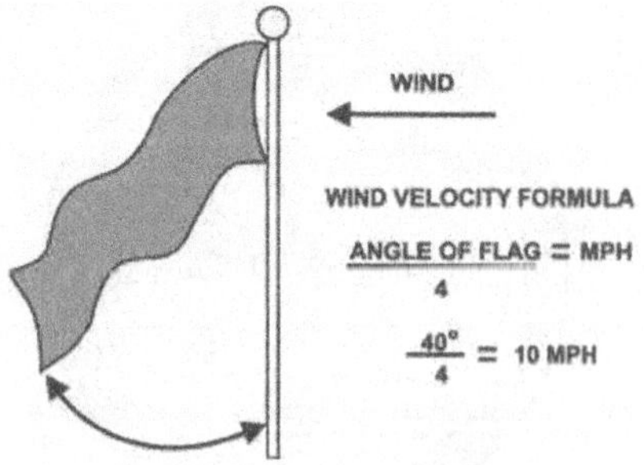

Figure 8-3. Flag Method.

Note

Information given is based on a dry flag. A wet flag is heavy and gives a false reading.

Windage Adjustments

After identifying wind direction, wind classification, and wind velocity, windage adjustments needed to enable the bullet to strike the target are estimated in the following ways:

Observation Method. Using the windage chart provided in figure 8-4, match the wind velocity, wind direction, and range to the target to the information in the chart to estimate the correct number of clicks to apply to the windage knob.

Flag Method. Using the windage chart provided in figure 8-5, match the wind velocity, wind direction, and range to the target to the information in the chart to determine the correct number of clicks to apply to the windage knob.

Once the number of windage clicks is determined, turn the windage knob causing the rear sight aperture to move into the direction of the wind. (See chapter 9.)

8002. Physical Effects of Temperature and Precipitation on the Bullet and the Rifle

Temperature

Extreme changes in temperature cause fluctuation in the rifle's chamber pressure. This fluctuation is caused by changes in the propellant's temperature. In cold weather, as rifle chamber pressure decreases, the bullet exits the muzzle at a lower velocity, and the bullet impacts the target below the point of aim. In extreme heat, the rifle's chamber pressure increases causing the bullet to exit the muzzle at a higher velocity and impact the target above the point of aim. Hot air is less dense than cool air and provides less resistance to the bullet; this allows the bullet to travel faster and experience less deflection from the wind. Cold air is dense and provides the bullet with more resistance; this causes the bullet to travel slower and experience greater deflection from the wind.

	WINDS CAN BE FELT LIGHTLY ON FACE AND LEAVES ARE IN CONSTANT MOTION		WINDS RAISE DUST AND LOOSE PAPER 8-12 MPH		WINDS CAUSE SMALL TREE TO SWAY 12-15 MPH		WINDS CAUSE LARGE TREES TO SWAY 16-20 MPH		21-25 MPH	
RANGE YARDS	WIND VALUE		WIND VALUE		WIND VALUE		WIND VALUE		WIND VALUE	
	FULL	HALF	FULL	HALF	FULL	HALF	FULL	HALF	FULL	HALF
200	2	1	3	1	5	2	6	3	8	4
300	3	1	6	3	10	5	13	6	16	8
400	4	2	9	4	13	6	18	9	22	11
500	6	3	12	6	18	9	24	12	30	15

Figure 8-4. Windage Click Chart for the Observation Method.

RANGE FLAG ANGLES	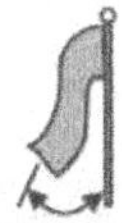 5 MPH		 10 MPH		 15 MPH		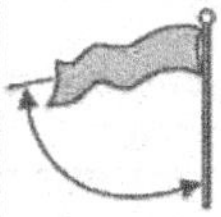 20 MPH		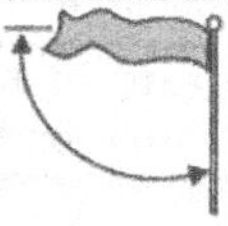 25 MPH	
RANGE YARDS	WIND VALUE		WIND VALUE		WIND VALUE		WIND VALUE		WIND VALUE	
	FULL	HALF	FULL	HALF	FULL	HALF	FULL	HALF	FULL	HALF
200	2	1	3	1	5	2	6	3	8	4
300	3	1	6	3	10	5	13	6	16	8
400	4	2	9	4	13	6	18	9	22	11
500	6	3	12	6	18	9	24	12	30	15

Figure 8-5. Windage Click Chart for the Flag Method.

Once the rifle is zeroed, a change in temperature of 20 degrees or more can cause the bullet to strike above or below the point of aim. Therefore, if the temperature changes 20 degrees or more, a Marine should re-zero the rifle.

If the rifle is exposed to below freezing temperatures, it should not be brought immediately into a warm location. Condensation may form on and in the rifle, and it may freeze if re-exposed to the cold. Ice that forms inside the rifle may cause it to malfunction. Ice can form on the rear sight aperture due to condensation, making it impossible to acquire sight alignment.

Precipitation

Freezing rain and other types of precipitation may make the rifle difficult to handle, foul the rifle and cause stoppages, or build up in the barrel or compensator and cause erratic shots. Care should be taken to keep the barrel and muzzle free of water. If the rifle has been submerged, ensure the bore is drained before firing. To drain the bore, pull the charging handle slightly to the rear and hold for a few seconds while the muzzle points down. Once the barrel has been drained, turn the rifle muzzle up to allow the water to drain out of the stock.

8003. Physical and Psychological Effects of Weather on Marines

Wind

Marines can shoot effectively in windy conditions if they apply a few basic techniques and develop the proper mental attitude. The Marine can combat the wind in a number of ways:

- Make subtle changes to the basic shooting positions, such as increasing muscular tension, to reduce movement of the rifle sights.
- Select a more stable firing position.
- Seek support to stabilize the rifle.
- Hold the shot and apply the fundamentals during a lull in the wind.

Temperature

Extreme Heat

In extreme heat, a Marine may experience rapid fatigue. Heat can cause muscle cramps, heat exhaustion, heat stroke, blurred vision, and reduced concentration levels that result in inaccurate shooting. Increased fluid intake and good physical condition can offset the effects of extreme heat. Sweat running into the eyes can cause irritation and make it difficult to see the sights. Extreme heat also can create ground mirages that cause a target to appear indistinct and to drift from side to side. Heat waves or mirages may also distort the target shape or the appearance of the front sight post. A mirage created by the heat of the barrel reduces a Marine's ability to see the sight clearly. To overcome the effects of heat and accurately engage a target, a Marine should maintain a center of mass hold.

Extreme Cold

Extreme cold may affect a Marine's ability to concentrate. If a Marine's hands are numb, he will have difficulty holding a frigid rifle and executing effective trigger control. To protect the hands in a cold environment, a Marine should wear arctic mittens or gloves. To operate the rifle while wearing arctic mittens or gloves, a Marine depresses the trigger guard plunger to open the trigger guard. This allows easier access to the trigger. See figure 8-6. The hasty sling can assist in holding the hand in place on the hand guards so the hand does not slip while wearing mittens.

Precipitation

Precipitation (rain, snow, hail, sleet) can affect target engagement, a Marine's comfort level, and a Marine's ability to concentrate. The amount and type of precipitation may obscure or completely hide the target and it may reduce a Marine's ability to establish an accurate sight picture. Precipitation collecting on rear sight aperture can make it difficult to establish sight alignment and sight picture. Protect sights as much as possible

Figure 8-6. Open Trigger Guard.

during periods of precipitation. It is easy to lose concentration when wet and uncomfortable. Proper dress reduces the effects of precipitation on the Marine.

Light

Light conditions can change the appearance of a target. Light affects each Marine differently. Light can affect range estimation, visual acuity, or the placement of the tip of the front sight on the target. By maintaining a center of mass hold, the effects of light can be reduced.

Bright Light

Bright light conditions exist under a clear blue sky with no fog or haze present to filter the sunlight. Bright light can make a target appear smaller and farther away. As a result, it is easy to overestimate range. Maintaining a center of mass hold, regardless of how indistinct the target appears, ensures the best chances for an effective shot.

Overcast

An overcast condition exists when a solid layer of clouds blocks the sun. The amount of available light changes as the overcast thickens. Overcast conditions can make a target appear larger and closer. As a result, it is easy to underestimate range. During a light overcast, the target appears very distinct and the rifle sights appear very distinct, making it easy to establish sight alignment. As the overcast thickens, it becomes difficult to identify the target from its surroundings.

Haze

Hazy conditions exist when fog, dust, humidity or smoke is present. Hazy conditions can make a target appear indistinct making it difficult to establish sight picture.

CHAPTER 9. ZEROING

To be combat effective, it is essential for the Marine to know how to zero his rifle. Zeroing is adjusting the sights on the weapon to cause the shots to impact where the Marine aims. This must be done while compensating for the effects of weather and the range to the target. It is critical that Marines can zero their rifles and make the sight adjustments required to engage targets accurately.

See the appropriate technical manual for procedures on boresighting and zeroing with supplemental aiming devices (e.g., laser, night vision devices).

Note

☞The procedures in this manual are written for right-handed Marines; left-handed Marines should reverse instructions as necessary.

9001. Elements of Zeroing

There are five basic elements involved in zeroing a rifle: line of sight, aiming point, centerline of the bore, trajectory, and range. See figure 9-1.

Line of Sight

The line of sight is a straight line, which begins with the shooter's eye, proceeds through the center of the rear sight aperture, and passes across the tip of the front sight post to a point of aim on a target.

Aiming Point

The aiming point is the precise point where the tip of the front sight post is placed in relationship to target.

Centerline of the Bore

Centerline of the bore is an imaginary straight-line beginning at the chamber end of the barrel, proceeding out of the muzzle, and continuing indefinitely.

Trajectory

In flight, a bullet does not follow a straight line but travels in a curve or arc, called trajectory. Trajectory is the path a bullet travels to the target. As the bullet exits the muzzle, it travels on an upward path, intersecting the line of sight (because the sights are above the muzzle). As the bullet travels farther, it begins to drop and intersects the line of sight again.

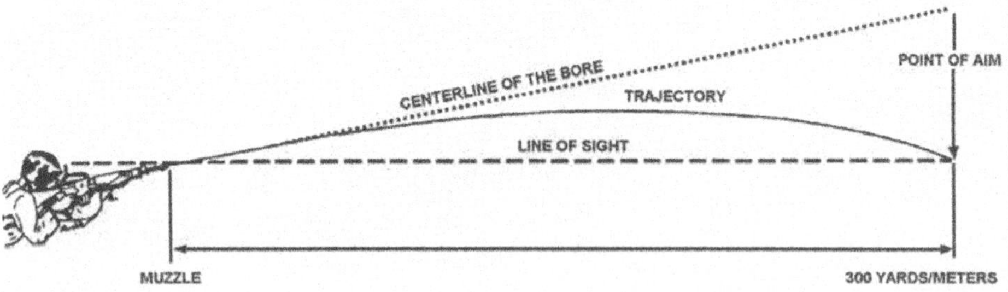

Figure 9-1. Elements of Zeroing.

Range

Range is the KD from the rifle muzzle to the target.

9002. Types of Zeros

Battlesight Zero (BZO)

A BZO is the elevation and windage settings required to place a single shot, or the center of a shot group, in a predesignated location on a target at 300 yards/meters, under ideal weather conditions (i.e., no wind). A BZO is the sight settings placed on your rifle for combat. In combat, your rifle's BZO setting will enable engagement of point targets from 0–300 yards/meters in a no-wind condition.

Zero

A zero is the elevation and windage settings required to place a single shot, or the center of a shot group, in a predesignated location on a target at a specific range, from a specific firing position, under specific weather conditions.

True Zero

A true zero is the elevation and windage settings required to place a single shot, or the center of a shot group, in a predesignated location on a target at a specific range other than 300 yards/meters, from a specific firing position, under ideal weather conditions (i.e., no wind).

9003. M16A2 Sighting System

The sighting system of the M16A2 service rifle consists of a front sight post and two rear sight apertures windage and elevation knob. Scales of the sighting system may be applied accurately to both yard and meter measurements. For example, a rear sight elevation setting of 8/3 may be used for 300 yards/meters.

Front Sight

The front sight post is used to adjust for elevation. The front sight consists of a square, rotating sight post with a four-position, spring-loaded detent (see fig. 9-2). To adjust for elevation, use a pointed instrument (or the tip of a cartridge) to depress the detent and rotate the front sight post (see fig. 9-3). To raise the strike of the bullet, rotate the post clockwise (in the direction of the arrow marked UP) or to the right. To lower the strike of the bullet, rotate the the post counter-clockwise (in the opposite direction of the arrow) or to the left.

Figure 9-2. Front Sight.

Figure 9-3. Front Sight Adjustment.

Rear Sight

The rear sight consists of two sight apertures, a windage knob, and an elevation knob. See figure 9-4. The large aperture marked 0-2 is used for target engagement during limited visibility, when a greater field of view is desired, or for engagements of targets closer than 200 yards/meters. The unmarked aperture (small aperture) is used for zeroing and normal firing situations.

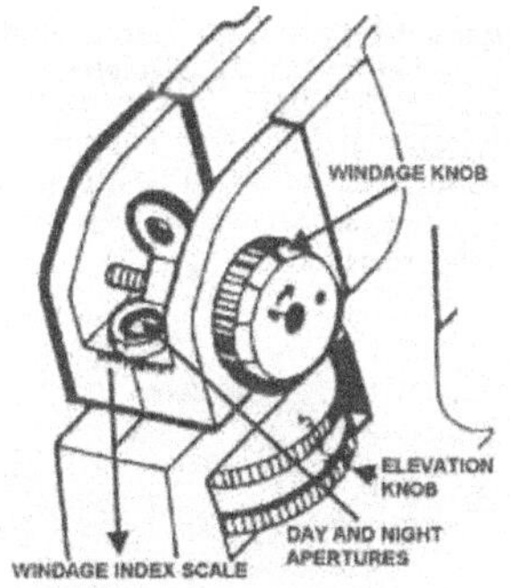

Figure 9-4. Rear Sight.

Elevation Knob

The rear sight elevation knob is used to adjust the sight for a specific range to the target. The elevation knob is indexed as shown in figure 9-5. Each number on the knob represents a distance from the target in 100-yard/meter increments. To adjust for range to the target, rotate the elevation knob so the desired setting is aligned with the index on the left side of the receiver.

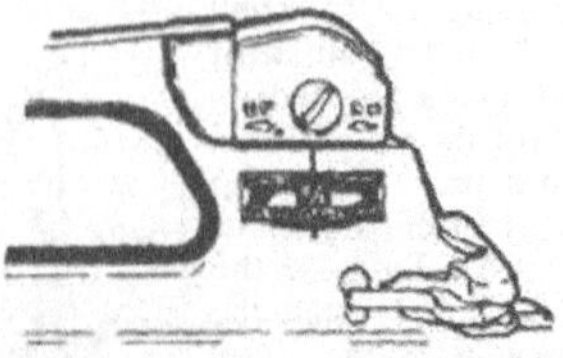

Figure 9-5. Elevation Knob.

If the elevation knob is turned so the number 8/3 aligns with the elevation index line, the 3 indicates 300 yards/meters (see fig. 9-6).

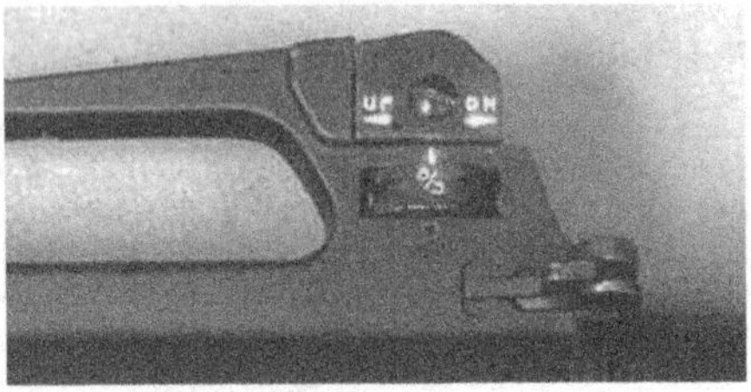

Figure 9-6. Rear Sight Elevation Knob Set for 300 Yards/Meters.

When the rear sight elevation knob is set on 8/3 for 800 yards/meters, there will be a considerable gap (about a quarter inch) between the rear sight housing and the upper receiver (see fig. 9-7).

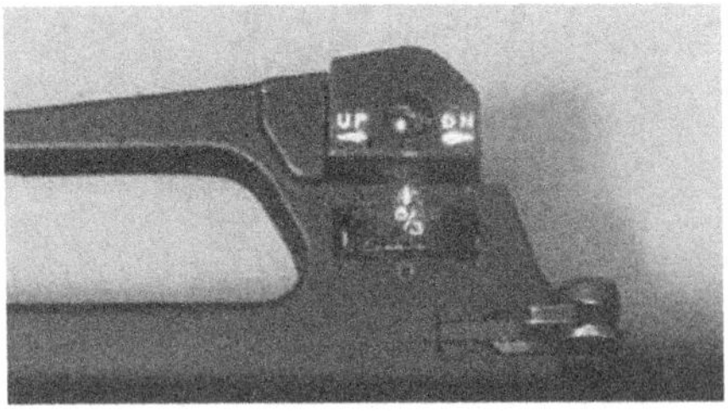

Figure 9-7. Rear Sight Elevation Knob Set for 800 Yards/Meters.

A hasty sight setting is the setting placed on the rear sight elevation knob to engage targets beyond 300 yards/meters. Hasty sight settings for ranges of 400 to 800 yards/meters are applied by rotating the rear sight elevation knob to the number that corresponds with the engagement distance of the enemy. Aligning the number 4, 5, 6 or 7 with the elevation index line places the elevation at 400, 500, 600 or 700 yards/meters, respectively. If a clockwise rotation is continued, the number 8/3 appears for the second time on the elevation index line and indicates an 800 yard/meter elevation.

Windage Knob

The windage knob is used to adjust the strike of the round right or left. The windage knob is marked with an arrow and the letter R that shows the direction the strike of the round is being moved. See figure 9-8.

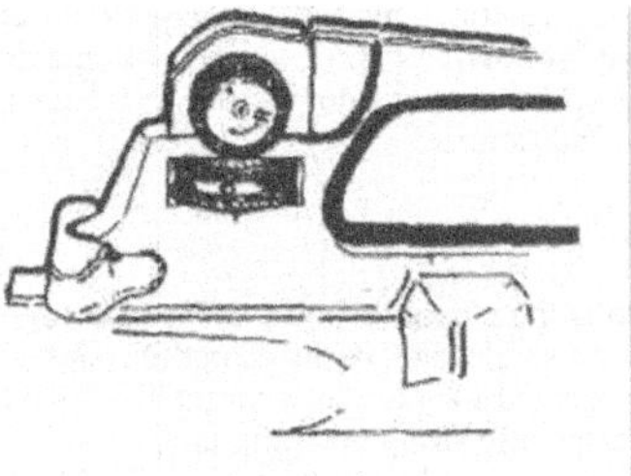

Figure 9-8. Windage Knob.

To move the strike of the round to the right, rotate the windage knob clockwise (in the direction of the arrow marked R).

To move the strike of the round to the left, rotate the windage knob counterclockwise.

9004. Windage and Elevation Rules

Moving the front sight post, elevation knob or windage knob one graduation or notch is referred to as moving one "click" on the sight. The windage and elevation rules define how far the strike of the round will move on the target for each click of front and rear sight elevation or rear sight windage for each 100 yards/meters of range to the target.

Front Sight Elevation Rule

One click of front sight elevation adjustment will move the strike of the round on target approximately 1.25 inches for every 100 yards of range to the target or 3.5 centimeters for every 100 meters of range to the target.

Rear Sight Elevation Rule

One click of rear sight elevation adjustment will move the strike of the round on the target approximately 1 inch for every 100 yards of range to the target or 2.5 centimeters for every 100 meters of range to the target.

Windage Rule

One click of windage adjustment will move the strike of the round on the target approximately 0.5 inch for every 100 yards of range to the target or 1.25 centimeters for every 100 meters of range to the target.

9005. Initial Sight Settings

Initial sight settings are those settings that serve as the starting point for initial zeroing from which all sight adjustments are made. (If the Marine already has a BZO established on his rifle, he may begin the zeroing process by using the previously established BZO sight settings.) To set the sights to initial sight settings:

Front Sight Post

To set the front sight post to initial sight setting, depress the front sight detent and rotate the front sight post until the base of the front sight post is flush with the front sight housing.

Rear Sight Elevation Knob

To set the elevation knob at the initial sight setting, perform the following:

- Rotate the rear sight elevation knob counterclockwise until the moveable rear sight housing is bottomed out on the upper receiver. See figure 9-9.

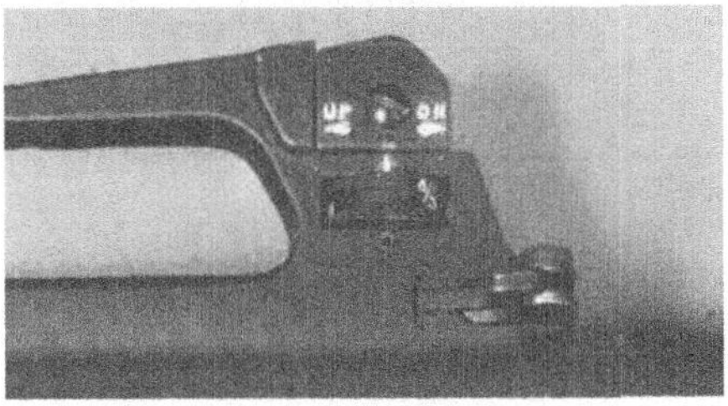

Figure 9-9. Bottoming Out Elevation Knob.

Note

Once bottomed out, the rear sight elevation knob should be three clicks counterclockwise from 8/3. If the sight fails to move three clicks counterclockwise from 8/3, it must be adjusted by a qualified armorer.

- Rotate rear sight elevation knob clockwise until the number 8/3 aligns with the index mark located on the left side of the upper receiver. See figure 9-10.

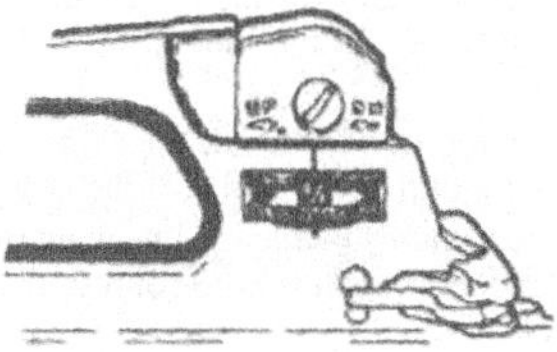

Figure 9-10. Elevation Knob Set at 8/3.

Windage Knob

To set the windage knob to initial sight setting, rotate windage knob until the index line located on the top of the large rear sight aperture aligns with the centering on the windage index scale located on the moveable base of the rear sight assembly. See figure 9-11.

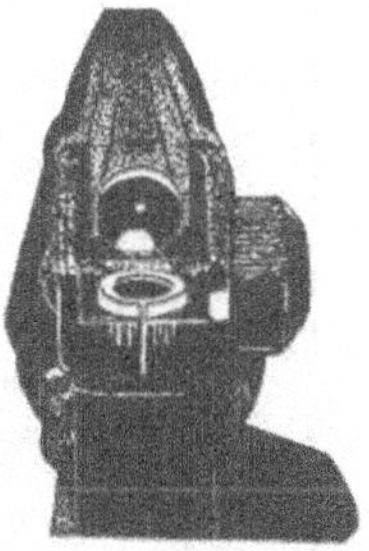

Figure 9-11. Aligning Index Line.

9006. Zeroing Process

During the zeroing process, all elevation adjustments are made on the front sight post. Once a BZO is established, the front sight post should never be moved, except when rezeroing the rifle. (The rear sight elevation knob is used for dialing in the range to the target.) Zeroing is conducted at a range of 300 yards/meters. To prepare a rifle for zeroing, the rifle sights must be adjusted to the initial sight settings as outlined in paragraph 9005. Perform the following steps to zero the rifle:

- Fire a 3-shot group.
- Triangulate the shot group to find the center. See figure 9-12.
- Determine the vertical distance in inches from the center of the shot group to the center of the target.
- Make elevation adjustments on the front sight post to move the center of the shot group to the center of the target.
- Determine the horizontal distance from the center of the shot group to the center of the target.
- Make lateral adjustments on the windage knob to move the center of the shot group to the center of the target.
- Repeat preceding steps until shot group is centered.
- Fire a 4-round shot group to confirm sight setting.

Once the sight setting is confirmed, determine the value and direction of the wind and remove the number of clicks added to the windage knob (if necessary) to compensate for current wind conditions. This becomes the BZO setting for the rifle.

9007. Battlesight Zero

Zeroing is conducted at a range of 300 yards/meters. If a 300-yard/-meter range is not available, a field expedient BZO can be established at a reduced range of 36 yards/30 meters. When a rifle is zeroed for 300 yards/meters, the bullet crosses the line of sight twice. It first

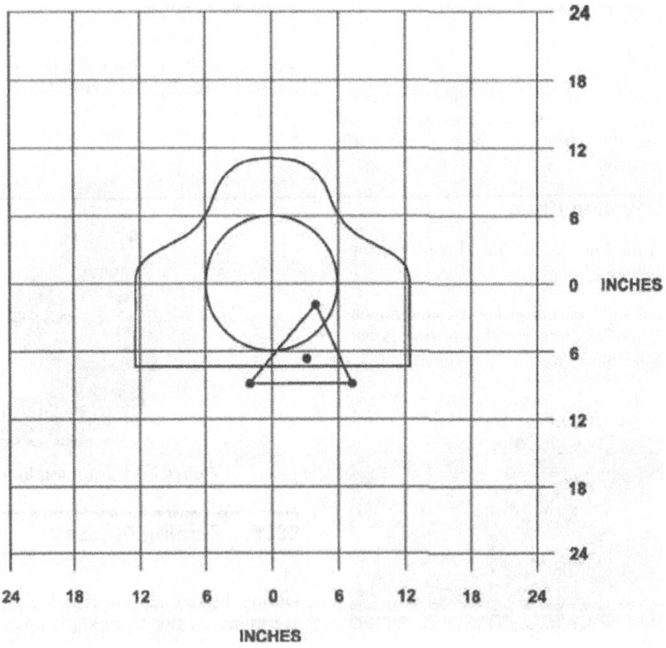

Figure 9-12. Triangulating Shot Group.

crosses the line of sight on its upward path of trajectory at 36 yards/30 meters, and again farther down range at 300 yards/meters (see fig. 9-13). Therefore a rifle's BZO may be established at a distance of 36 yards/30 meters and the same BZO will be effective at 300 yards/meters.

To establish a field expedient BZO at 36 yards or 30 meters when a 300-yard/-meter range is not available, a Marine performs the same steps as the zeroing process outlined in paragraph 9006. However, since wind does not affect the round at 36 yards/30 meters, windage is not added nor is it removed from the windage knob after confirming the BZO. To be accurate, the target must be placed exactly 36 yards (or 30 meters) from the muzzle of the rifle.

9008. Factors Causing a BZO to be Reconfirmed

Marines are responsible for maintaining a BZO on their rifles at all times. Many factors influence the BZO of a rifle. Atmospheric conditions, humidity, and temperature can cause BZOs to change on a daily basis. If operating in a combat environment, Marines should confirm their BZO as often as possible. To confirm a BZO, a Marine may begin the zeroing process by using the previously established BZO sight settings rather than placing the sights at initial sight setting. The following factors cause a BZO to be reconfirmed.

Maintenance

It is possible for the BZO to change if ordnance personnel perform maintenance on a rifle. If maintenance was performed, it is critical that the rifle be rezeroed as soon as possible.

Temperature

An extreme change in temperature (i.e., 20 degrees or more) will cause the elevation BZO to change. Changes in temperature cause chamber pressure to increase when hot and decrease when cold. This causes shots to impact the target high in hot temperatures and low in cold temperatures.

Climate

Changing climates (i.e., moving from a dry climate to a tropical climate) can mean changes in air density, moisture content, temperature or barometric pressure. Any of these elements can affect the rifle's BZO.

Ammunition

Inconsistencies in the production of ammunition lots can change a rifle's BZO.

Ground Elevation

Drastic changes in ground elevation can create changes in air density, moisture content, temperature or barometric pressure. Any of these elements can affect the rifle's BZO.

Uniform

If Marines zero their rifles in utility uniform and fire in full battle gear, their BZOs will change. The wearing of full battle gear changes eye relief, placement of the rifle in the shoulder pocket, and the way the rifle is supported on the handguard. Marines must establish their BZOs while wearing the uniform and equipment they will be wearing while engaging targets.

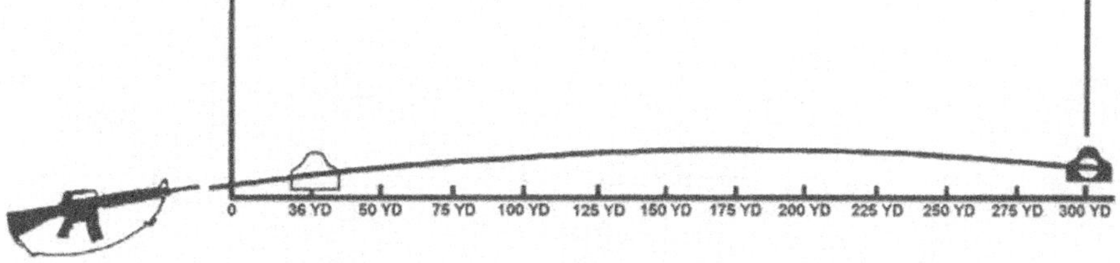

Figure 9-13. Bullet Crossing the Line of Sight Twice.

9009. Factors Affecting the Accuracy of a BZO

Anything the Marine changes from shot to shot affects the accuracy of his BZO. It further affects the accuracy of shot placement. The following factors, when applied inconsistently, diminish the accuracy of a BZO:

- Any of the seven factors (left hand, rifle butt in the pocket of the shoulder, grip of the right hand, right elbow, stock weld, breathing, and muscular relaxation/tension).
- Stability of hold.
- Sling tension.
- Trigger control.
- Sight picture.

CHAPTER 10. ENGAGEMENT TECHNIQUES

A Marine must maintain the ability to react instinctively in a combat environment—day or night. He must possess a combat mindset that eliminates hesitation, fear, or uncertainty of action. A Marine's combat mindset allows him to engage the enemy rapidly and focus on the actions required to fire well-aimed shots. He must remember that speed alone does not equate to effective target engagement. He should fire only as fast as he can fire accurately. He should never exceed his physical capability to engage a target effectively. To be effective in combat, a Marine must train to perfect the physical skills of target engagement (such as presenting the weapon and assuming a shooting position) until they become instinctive. A Marine must also employ effective engagement techniques that enable acquisition and engagement of a variety of targets in diverse combat conditions.

10001. Target Detection

To be proficient, a Marine rifleman must be able to detect targets, determine the range to targets, and accurately engage the targets. There are many variables affecting a Marine's ability to detect and determine the range to combat targets. Enemy targets on the battlefield may be single or multiple, stationary or moving, or completely hidden from view. Success in locating an enemy target will depend upon the observer's position, his skill in searching an area, and his ability to recognize target indicators.

Target Indicators

Most combat targets are detected at close range by smoke, flash, dust, noise or movement, and are usually seen only momentarily. Target indicators are anything that reveals an individual's position to the enemy. These indicators are grouped into three general areas: movement, sound, and improper camouflage.

Movcmont

The human eye is attracted to movement, especially sudden movement. The Marine need not be looking directly at an object to notice movement. The degree of difficulty in locating moving targets depends primarily on the speed of movement. A slowly moving target will be harder to detect than one with quick, jerky movements.

Sound

Sound can also be used to detect an enemy position. Sound may be made by movement, rattling equipment, or talking. Sound provides only a general location of the enemy, making it difficult to pinpoint a target by sound alone. However, sound can alert the Marine to the presence of a target and increase his probability of locating it through other indicators.

Improper Camouflage

There are three indicators caused by improper camouflage: shine, outline, and contrast with the background. Most targets on the battlefield are detected due to improper camouflage. However, many times an observation post or enemy firing position will blend almost perfectly with the natural background. Only through extremely careful, detailed searching will these positions be revealed.

Shine. Shine is created from reflective objects such as metal or glass. It may also come from pools of water and even the natural oils from the skin. Shine acts as a beacon to the target's position.

Outline. Most enemy soldiers will camouflage themselves, their equipment, and their positions. The outline of objects such as the body, head and shoulders, weapons, and web gear are recognizable even from a distance. The human eye will often pick up a recognizable shape and concentrate on it even if the object cannot be identified immediately. The reliability of this indicator depends upon visibility and the experience of the observer.

Contrast With the Background. Indicators in this category include objects that stand out against (contrast with) a background because of differences in color, surface, and shape. For instance, a target wearing a dark uniform would be clearly visible in an area of snow or sand. Geometric shapes, such as helmets or rifle barrels, can be easy to detect in a wooded area. Fresh soil around a fighting hole contrasts with the otherwise unbroken ground surface. While observing an area, take note of anything that looks out of place or unusual and study it in more detail. This will greatly increase your chances of spotting a hidden enemy.

Identifying Target Location

Observation Position

A good position is one that offers maximum visibility of the area while affording cover and concealment. The optimal observation position should allow the Marine to scan all the areas of observation and offer enough concealment to prevent his position from being detected.

The Marine should avoid positions that are obvious or stand out, such as a lone tree in a field or a pile of rocks on a hill. These positions may be ideal points for easy observation, but they will also make it easier for the enemy to locate the Marine.

Methods for Searching an Area

In searching an area, the Marine will be looking for target indicators. There are two techniques for searching an area: the hasty search and the detailed search.

Hasty Search. When a Marine moves into a new area, he must quickly check for enemy activity that may pose an immediate danger. This search is known as the hasty search and should take about 30 seconds, depending on the terrain.

Quickly glance at various points throughout the area rather than sweeping the eyes across the terrain in one continuous movement. The Marine should search the area nearest him first since it poses the greatest potential for danger.

This method of search is effective because it takes advantage of peripheral vision. Peripheral vision enables the detection of any movement in a wide area around the object being observed. For this technique to be effective, the eyes must be focused briefly on specific points (i.e., areas that may provide cover or concealment for the enemy).

Detailed Search. A detailed search is a systematic examination of a specific target indicator or of the entire observation area. A detailed search should be conducted immediately on target indicators located during the hasty search. The detailed search should be made from top to bottom or side to side, observing the entire object in exact detail. If multiple indicators were observed during the hasty search, the detailed search should begin with the indicator that appears to pose the greatest threat.

After a thorough search of target indicators, or if no indicators were located during the hasty search, a detailed search should be made of the entire observation area. The 50-meter overlapping strip method is normally used.

Normally, the area nearest the observer offers the greatest potential danger and should be searched first. Begin the search at one flank, systematically searching the terrain at the front in 180-degree arcs, searching everything in exacting detail, 50 meters in depth. See figure 10-1.

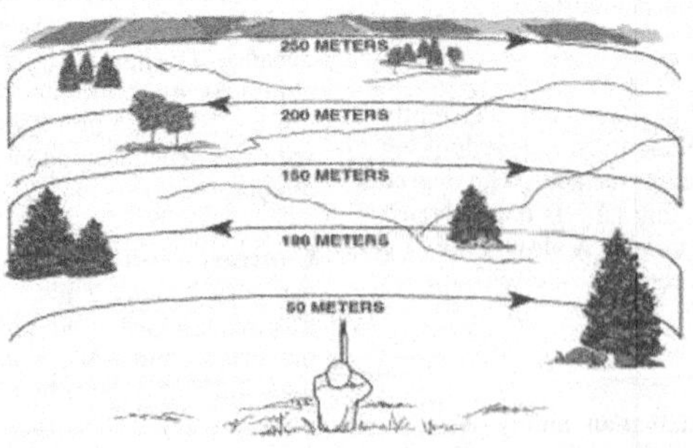

Figure 10-1. Searching the Terrain in Overlapping Strips.

After reaching the opposite flank, systematically cover the area between 40 and 90 meters from your position. The second search of the terrain includes about 10 meters of the area examined during the first search. This technique ensures complete coverage of the area.

Continue the overlapping strip search method for as far as you can see.

Maintaining Observation

Method

The combat situation will dictate the method of maintaining observation of an area. Generally, the observation method will include a combination of hasty and detailed searches.

Sequence of Observation

Observation is often conducted by a two-man team. One team member should constantly observe the entire area using the hasty search technique and the other team member should conduct a detailed overlapping strip search. If you are observing as an individual, devise a plan to ensure that the area of observation is completely covered. When entering a new area, immediately conduct a hasty search. Since a hasty search may fail to detect some indicators, periodically conduct a detailed search of the area. A detailed search should also be conducted any time your attention has been diverted from the search area.

Remembering Target Location

Most targets are seen only briefly and most areas contain multiple targets. Once you have located a target indicator, you will need to remember its location to engage it successfully. To help remember the location of a target, select a known feature and use it as a reference point to determine the distance and general direction to the target.

10002. Range Estimation

To engage targets at unknown distances, a Marine must determine the distance from his location to a known point. This is known as range estimation. The ability to determine range is a skill that must be developed if a Marine is to successfully engage targets at unknown distances. Precise range estimation enhances accuracy, enhances the chance of survival, and determines if a target can be effectively engaged using the rifle's existing BZO or if a new sight setting or point of aim is required.

Range Estimation Methods

Unit of Measure Method

To use this method, a Marine visualizes a distance of 100 meters on the ground, and then estimates how many of these units will fit between him and the target. This determines the total distance to the target. See figure 10-2.

Figure 10-2. Unit of Measure Method.

The greatest limitation of this method is that its accuracy is related to the amount of visible terrain. For example, if a target appears at a range of 500 meters or more and only a portion of the ground between a Marine and the target can be seen, it becomes difficult to use the unit of measure method to estimate range accurately. A Marine must practice this method frequently to be proficient. Whenever possible, a Marine should select an object, estimate the range, and then verify the actual range by either pacing or using another accurate measurement.

Rifle Front Sight Post Method

The area of the target covered by the rifle's front sight post can be used to estimate range to a target. A Marine notes the appearance of the front sight post on a known-distance target. A Marine then uses this as a guide to determine range over an unknown distance. Because the apparent size of the target changes as the distance to the target changes, the amount of the target covered by the front sight post varies based on the range. In addition, a Marine's eye relief and perception of the front sight post affect the amount of the target that is visible. To use this method, a Marine must apply the following guidelines:

- The front sight post covers the width of a man's chest or body at approximately 300 meters.
- If the target is less than the width of the front sight post, the target is in excess of 300 meters and the rifle's BZO cannot be used effectively.
- If the target is wider than the front sight post, the target is less than 300 meters and can be engaged point of aim/point of impact using the rifle's BZO.

See figure 10-3.

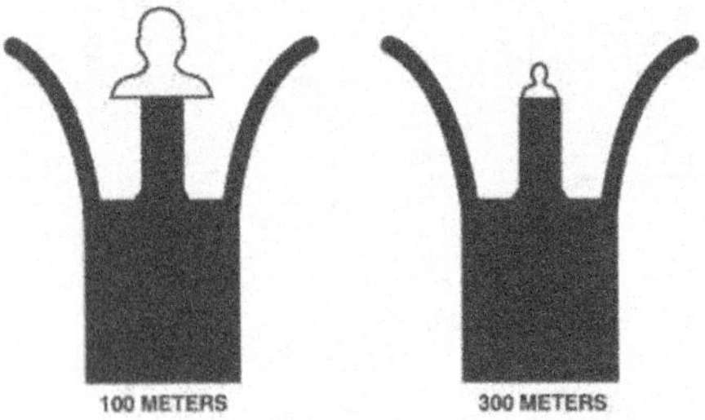

Figure 10-3. Front Sight Post Method.

Visible Detail Method

The amount of detail seen at various ranges can provide a Marine with an estimate of the target's distance. To use this method, a Marine must be familiar with the size and various details of personnel and equipment at known distances. Visibility (such as weather, smoke or darkness) limits the effectiveness of this method. A Marine should observe a man while he is standing, kneeling, and in the prone position at known ranges of 100 to 500 meters. He should note the man's size, characteristics/size of his uniform and equipment, and any other pertinent details. The Marine then uses this as a guide to determine range over an unknown distance. A Marine also should study the appearance of other familiar objects such as rifles and vehicles. To use this method, a Marine applies the following general guidelines:

- At 100 meters, the target is clearly observed in detail and facial features are distinguished
- At 200 meters, the target is clearly observed. There is a loss of facial detail. The color of the skin and equipment are still identifiable.
- At 300 meters, the target has a clear body outline, face color usually remains accurate, but remaining details are blurred.
- At 400 meters, the body outline is clear, but remaining detail is blurred.
- At 500 meters, the body shape begins to taper at the ends, and the head becomes indistinct from the shoulders.
- At 600 meters, the body appears wedge-shaped and headless.

Bracketing Method

This method of range estimation estimates the shortest possible distance and the greatest possible distance to the target. For example, a Marine estimates that a target may be as close as 300 meters but it could be as far away as 500 meters. The estimated distances are averaged to determine the estimated range to the target. For example, the average of 300 meters and 500 meters is 400 meters.

Halving Method

To use the halving method, a Marine estimates the distance halfway between him and the target, then doubles that distance to get the total distance to the target. A Marine must take care when judging the distance to the halfway point; any error made in judging the halfway distance is doubled when estimating the total distance.

Combination Method

The methods previously discussed require optimal conditions with regard to the target, terrain, and visibility in order to obtain an accurate range estimation. A Marine should estimate the range using two methods and then compare the estimates, or two Marines can compare their estimates. The average of the two estimates should be close to the actual range to target.

Factors Affecting Range Estimation

The following specific factors will affect the accuracy of estimation. A Marine must be aware of these factors and attempt to compensate for their effects.

Nature of the Target

- An object with a regular outline such as a steel helmet, rifle, or vehicle on a clear day will appear to be closer than one with an irregular outline such as a camouflaged object.

- A target that contrasts with its background will appear to be closer than a target that blends in with its background.
- A partially exposed object will appear to be farther away than it is.
- A target will appear to be farther away if the target is smaller than the objects surrounding it.

Nature of the Terrain

- Terrain that slopes upward gives the illusion of shorter distance.
- Terrain that slopes downward gives the illusion of greater distance.
- Terrain with dead space makes the target appear to be closer.
- Smooth terrain such as sand, water, or snow gives the illusion of greater distance.

Limited Visibility

The position of the light source significantly affects a Marine's ability to estimate range. Other factors that affect range estimation include smoke, fog, rain, angled light, or anything that obscures the battlefield. Generally, when the sun is bright, a target appears further away. When the sky is overcast, a target generally appears closer. Target contrast is another factor to consider when estimating range as well as obstacles located between the shooter and the target. If there is contrast (e.g., color variation) between the target and the background, the target will appear closer. If there is little or no contrast between the target and the background, the target will appear farther away. If there is an object between the target and the shooter that will distract the shooter (e.g., a tree), the target will appear farther away.

10003. Offset Aiming

The conditions of rifle fire in combat may not permit mechanical adjustments of the sights. To engage a target during combat, a Marine may be required to aim his rifle at a point on the target other than center mass. This is known as offset aiming. Offset aiming is used to compensate for the distance and size of the target, wind, and speed and angle of a moving target. There are two primary techniques for offset aiming: point of aim technique and known strike of the round.

Point of Aim Technique

The point of aim technique is the shifting of the point of aim (sight picture) to a predetermined location on or off the target to compensate for a known condition (i.e., wind, distance, and movement). Each predetermined location is known as a point of aim. Figure 10-4 illustrates points of aim for elevation.

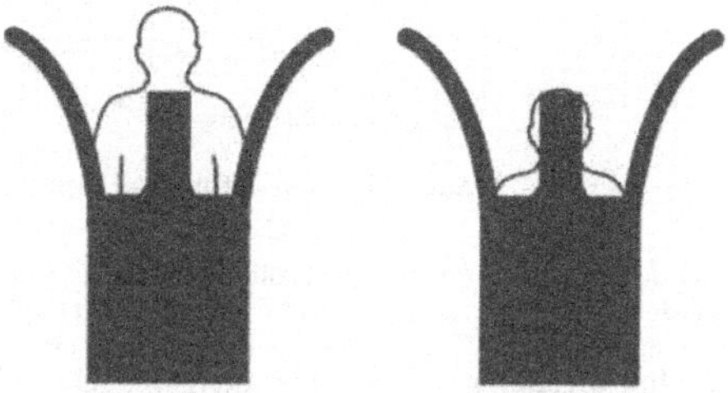

Figure 10-4. Points of Aim for Elevation.

Elevation

Predetermined points of aim sector the target horizontally. The tip of the front sight post held at shoulder level is considered one point of aim; the tip of the front sight post held at the top of the target's head is considered two points of aim. A Marine uses these points of aim to compensate for the elevation required to engage a target beyond the BZO capability of the rifle or to engage a small target (e.g., head shot) inside the BZO of the weapon.

Beyond the BZO. To use the point of aim technique to engage a target beyond the BZO of the rifle, a Marine must apply the following guidelines:

- When range to the target is estimated to be beyond 300 meters out to 400 meters, hold one point of aim.
- When the range to the target is estimated to be beyond 400 meters out to 500 meters, hold two points of aim.

Note

It is better to apply a hasty sight setting at ranges beyond the rifle's BZO. Points of aim are only guidelines at these distances because the front sight will mask the target when the front sight is held above center mass, making it difficult to acquire sight picture.

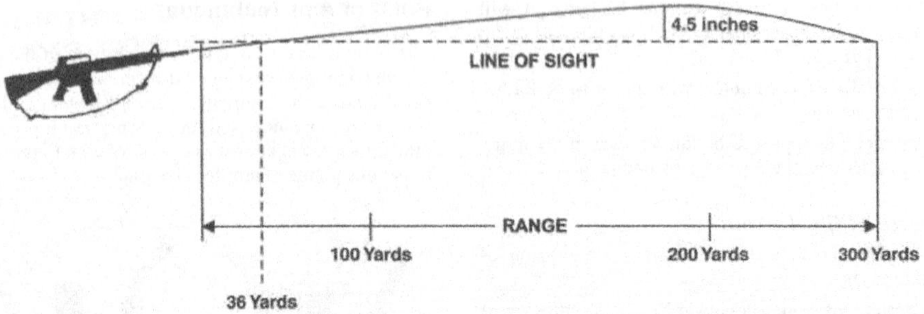

Figure 10-5. Trajectory and Point of Aim/Point of Impact for 300 Yard BZO.

Inside the BZO. If the rifle is properly zeroed for 300 yards/meters, the trajectory (path of the bullet) will rise approximately 4.5 inches (11 centimeters) above the line of sight at a distance of approximately 175 yards (160 meters). At other distances, the strike of the bullet will be less than 4.5 inches above the point of aim. Only at 36 yards/30 meters and 300 yards/meters does the point of impact coincide with the point of aim. If only a portion of the target is visible (e.g., the head of an enemy soldier), the trajectory of the bullet may have to be taken into consideration when firing at a distance less than 300 yards/meters. If a Marine does not consider trajectory, he may shoot over the top of the target if the target is small and at a distance less than 300 yards/meters. See figure 10-5.

Windage

Predetermined points of aim sector the target vertically (see fig. 10-6). The tip of the front sight post centered on the edge of the target into the wind is considered one point of aim; the trailing edge of the front sight post held on the edge of the target into the wind is considered two points of aim. The same units of measure are applied off the target for holds of additional points of aim. These points of aim are used to compensate for wind affecting the strike of the round and when there is no time to adjust the rifle's sights, or when a lead is required to engage a moving target (points of aim for moving targets are discussed in paragraph 10007). Table 10-1 provides points of aim for full value winds.

Known Strike of the Round

This offset aiming technique shifts the aiming point (sight picture) to compensate for rounds that strike off target center. The known strike of the round method is used if the strike of the round is known. To engage a target using this method, a Marine aims an equal distance from center mass opposite the known strike of the round. For example, if the round strikes high and left, a Marine aims an equal and opposite distance low and right.

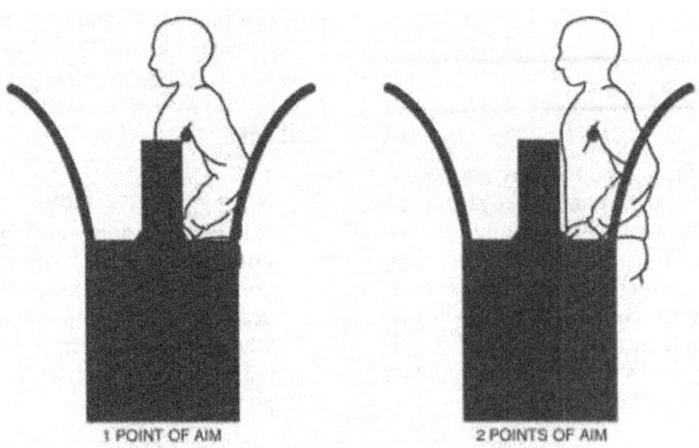

Figure 10-6. Points of Aim.

Table 10-1. Points of Aim for Full Value Winds.

Wind	Distance (yards/meters)	Points of Aim
Light	0-300	0
Medium	200-300	1
Strong	0-200	1
Strong	200-300	2

10004. Techniques of Fire

The size and distance to the target should dictate the technique of fire.

Two-Shot Technique

In combat, an effective technique for eliminating a threat is to rapidly fire more than one shot on the target. Two shots fired in rapid succession will increase the trauma (i.e., shock, blood loss) on the target, increasing the Marine's chances of quickly eliminating the threat. Firing two shots enables the Marine to break out of the tunnel vision often associated with firing in combat and then assess the situation to determine follow-on action. To execute the two-shot technique, the Marine acquires sight picture for each shot fired (each pull of the trigger).

Single Shot Technique

If the target is at a long range or if the target is small (i.e., partially exposed), it can best be engaged with a single, precision shot. A Marine's stability of hold and sight alignment are more critical to accurate engagement of long-range or small targets. To engage a target with the single shot technique, the Marine must slow down the application of the fundamentals and place one well-aimed shot on target.

Sustained Rate of Fire

An effective method for delivering suppressive fire is to fire at the sustained rate of 12 to 15 rounds per minute. Management of recoil is critical to bring the sights back on target after the shot is fired.

Three-Round Burst Technique

When set on burst, the design of the rifle permits three shots to be fired from a single trigger pull. The rounds fire as fast as the weapon will function and cause the muzzle to climb during recoil. The ability to manage recoil is extremely important when firing the rifle on burst. To achieve the desired effect (i.e., 3 rounds on target), the Marine must control the jump angle of the weapon to maintain the sights on target. At short ranges (i.e., 25 meters or less), firing on three-round burst can be an effective technique to place rounds on a man-sized target quickly to increase trauma on the target. To execute the three-round burst technique, the Marine places the selector lever on burst, aims center mass, and acquires sight picture once for the single trigger pull.

10005. Engaging Immediate Threat Targets

Immediate threat target engagement is characterized by short-range engagement (i.e., less than 50 meters) with little or no warning that requires an immediate response to engage an enemy. This type of engagement is likely in close terrain (e.g., urban, jungle). If this type of engagement is likely, the large rear sight aperture (0-2) could be placed up to provide a wider field of view and detection of targets. Marksmanship skills include quick presentation and compression of the fundamentals (i.e., quick acquisition of sight picture, uninterrupted trigger control). At close ranges, perfect sight alignment is not as critical to accuracy on target. However, the front sight post must be in the rear sight aperture; proper sight alignment is always the goal.

10006. Engaging Multiple Targets

When engaging multiple targets, a Marine must prioritize each target and carefully plan his shots to ensure successful target engagement. Mental preparedness and the ability to make split-second decisions are the key to successful engagement of multiple targets. The proper mindset allows a Marine to react instinctively and to control the pace of the battle rather than just reacting to the threat.

After the first target is engaged, a Marine must immediately engage the next target and continue to engage targets until they are eliminated. While engaging multiple targets, a Marine must be aware of his surroundings and not fixate on just one target. He must rapidly prioritize the targets, establish an engagement sequence, and engage the targets. A Marine also must maintain constant awareness and continuously search the terrain for additional targets.

Prioritizing Targets

The combat situation will usually dictate the order of multiple target engagement. Target priority is based on factors such as proximity, threat, and opportunity; no two situations will be the same. The principal method is to determine the level of threat for each target so all may be engaged in succession from the most threatening to the least threatening. The target that poses the greatest threat (e.g., closest, greatest firepower) should be engaged first. Prioritizing targets is an ongoing process. Changes in threat level, proximity, or the target itself may cause a Marine to revise his priorities. Therefore, a Marine must remain alert to changes in a target's threat level and proximity and other target opportunities as the battle progresses.

Technique of Engagement

To engage multiple targets, the Marine performs the following steps:

- The first target with two rounds.
- The recoil of the rifle can be used to direct the recovery of the weapon on to the next target. As the weapon is coming down in its recovery, the Marine physically brings the sights onto the desired target. Pressure is maintained on the trigger throughout recovery and trigger control is applied at a rate consistent with the Marine's ability to establish sight picture on the desired target.
- When possible, such as when all targets are of equal threat, the Marine should engage targets in a direction that maximizes his ability to support and control the weapon.
- The preceding steps are repeated until all targets have been engaged.

Firing Position

The selection and effective use of a firing position are critical to the successful engagement of multiple targets. A Marine should make a quick observation of the terrain and select a firing position that provides good cover and concealment, as well as the flexibility to engage multiple targets. If enemy targets are widely dispersed, the selected position must provide the Marine with flexibility of movement. The more restrictive the firing position, the longer it will take a Marine to eliminate multiple targets.

Prone

The prone position limits left and right lateral movement and is, therefore, not recommended for engaging short-range dispersed targets. Because the elbows are firmly placed on the ground in the prone position, upper body movement is restricted.

Sitting

Like the prone position, the sitting position allows limited lateral movement. This makes engagement of widely-dispersed multiple targets difficult. To ease engagement, the forward arm can be moved by pivoting on the elbow, but this movement disturbs the stability of the position.

Kneeling

The kneeling position provides a wider, lateral range of motion since only one elbow is used for support. A Marine moves from one target to another by rotating at the waist to move the forward arm in the direction of the target, either right or left.

Standing

The standing position allows maximum lateral movement. Multiple targets are engaged by rotating the upper body to a position where the sights can be aligned on the desired target. If severe or radical adjustments are required to engage widely dispersed targets, a Marine moves his feet to establish a new position rather than give up maximum stability of the rifle. This avoids poorly placed shots that can result from an unstable position.

10007. Engaging Moving Targets

In combat, it is unlikely that a target will remain stationary. The enemy will move quickly from cover to cover, exposing himself for the shortest possible time. Therefore, a Marine must quickly engage a moving target before it disappears.

Types of Moving Targets

There are two types of moving targets: steady moving targets and stop and go targets. Steady moving targets move in a consistent manner and remain in a Marine's field of vision. A walking or running man is an example of a steady moving target. A stop and go target appears and disappears during its movement. A stop and go target will present itself for only a short time before it reestablishes cover. An enemy moving from cover to cover is an example of a stop and go target. This target is most vulnerable to fire at the beginning and end of its movement to new cover because the target must gain momentum to exit its existing cover and then slow down to occupy a new position.

Leads

When a shot is fired at a moving target, the target continues to move during the time the bullet is in flight. Therefore, a Marine must aim in front of the target; otherwise, the shot will fall behind the target. This is called leading a target. Lead is the distance in advance of the target that the rifle sights are placed to accurately engage the target when it is moving.

Amount of Lead Required

Factors that affect the amount of lead are the target's range, speed, and angle of movement.

Range. Lead is determined by the rifle's distance to the target. When a shot is fired at a moving target, the target continues to move during the time the bullet is in flight. This time of flight could allow a target to move out of the bullet's path if the round was fired directly at the target. Time of flight increases as range to the target increases. Therefore, the lead must be increased as the distance to the target increases.

Speed. If a man is running, a greater lead is required because the man will move a greater distance than a walking man will while the bullet is in flight.

Angle of Movement. The angle of movement across the line of sight relative to the flight of the bullet determines the type (amount) of lead.

Three Types of Leads: Full, Half, No

- **Full Lead**. The target is moving straight across a Marine's line of sight with only one arm and half the body visible. This target requires a full lead because it will move the greatest distance across a Marine's line of sight during the flight of the bullet.
- **Half Lead**. The target is moving obliquely across a Marine's line of sight (at a 45-degree angle). One arm and over half the back or chest are visible. This target requires half of a full lead because it will move half as far as a target moving directly across a Marine's line of sight during the flight of the bullet.
- **No Lead**. The target is moving directly toward or away from a Marine and presents a full view of both arms and the entire back or chest. No lead is required. A Marine engages this target as if it were a stationary target because it is not moving across his line of sight.

Point of Aim Technique

Predetermined points of aim sector the target vertically. The tip of the front sight post centered on the leading edge of the target is considered one point of aim; the trailing edge of the front sight post held on the leading edge of the target is considered two points of aim. The same units of measure are applied off the target for holds of additional points of aim. These points of aim are used to compensate when a lead is required to engage a moving target. The following guidelines apply if a Marine uses the point of aim technique to establish a lead for a moving target at various ranges and speeds (see fig. 10-7, on page 10-10). These guidelines do not consider wind or other effects of weather. Body width in these examples is considered to be 12 inches (30 centimeters) (side view of the target).

For a slow walking target (approximately 2 to 2.5 mph/3.2 to 4 kph) moving directly across the line of sight (full lead)—

- At a range of 200 meters or less, no lead is required.
- At a range of 300 meters, hold one point of aim in the direction the target is moving.

For a fast walking target (approximately 4 mph/6.4 kph) moving directly across the line of sight (full lead)—

- At a range of 200 meters or less, hold one point of aim in the direction the target is moving.
- At a range of 300 meters, hold two points of aim in the direction the target is moving.

For a target running (approximately 6 mph/9.7 kph) directly across the line of sight (full lead)—

- At a range of 100 meters or less, hold one point of aim in the direction the target is moving.

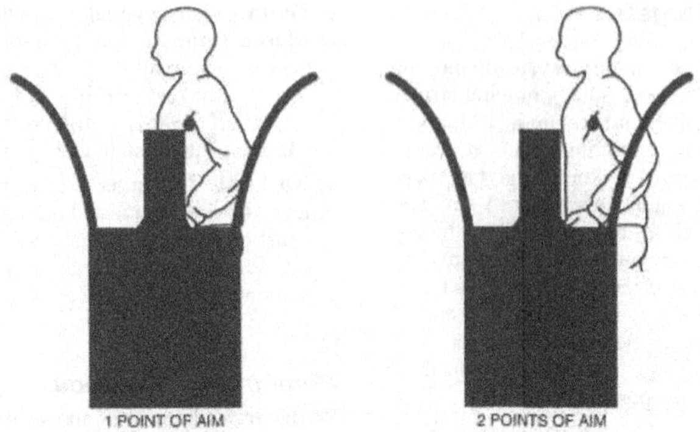

Figure 10-7. Points of Aim.

- At a range of 200 meters, hold two points of aim in the direction the target is moving.

For a target moving at a 45-degree angle (an oblique target) across the line of sight, the lead is one half of the lead that is required for a target moving directly across the line of sight.

Engagement Methods

Moving targets are the most difficult targets to engage. However, they can be engaged successfully by using the tracking or the ambush method.

The Tracking Method

The tracking method is used for a target that is moving at a steady pace over a well-determined route. If a Marine uses the tracking method, he tracks the target with the rifle's front sight post while maintaining sight alignment and a point of aim on or ahead of (leading) the target until the shot is fired. When establishing a lead on a moving target, rifle sights will not be centered on the target and instead will be held on a lead in front of the target. See figure 10-8. A Marine performs the following steps to execute the tracking methods:

- Present the rifle to the target.
- Swing rifle muzzle through the target (from the rear of the target to the front) to the desired lead (point of aim). The point of aim may be on the target or some point in front of the target depending upon the target's range, speed, and angle of movement.
- Track and maintain focus on the front sight post while acquiring the desired sight picture. It may be necessary to shift the focus between the front sight post and the target while acquiring sight picture, but the focus must be on the tip of the front sight post when the shot is fired.
- Engage the target once sight picture is acquired while maintaining the proper lead.
- Follow-through so the lead is maintained as the bullet exits the muzzle.
- Continue to track in case a second shot needs to be fired on the target.

The Ambush Method

The ambush method is used when it is difficult to track the target with the rifle, as in the prone, sitting, or any supported position. The lead required to effectively engaging the target determines the engagement point. With the sights settled, the target moves into the predetermined engagement point and creates the desired sight picture. See figure 10-9.

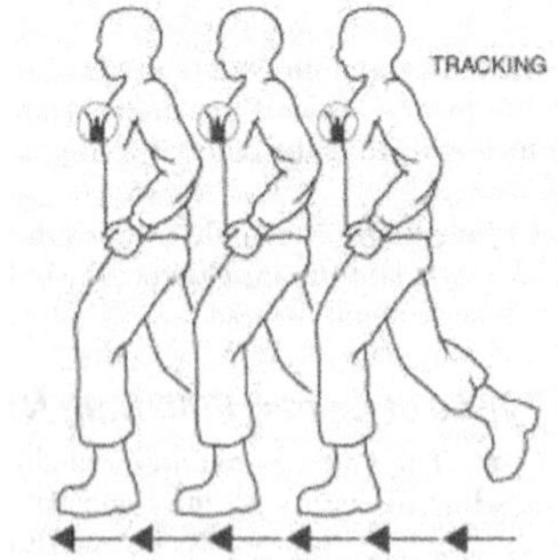

Figure 10-8. Tracking Method.

The trigger is pulled simultaneously with the establishment of sight picture. To execute the ambush method, a Marine performs the following steps:

- Select an aiming point ahead of the target.
- Obtain sight alignment on the aiming point.
- Hold sight alignment until the target moves into vision and the desired sight picture is established.
- Engage the target once sight picture is acquired.
- Follow-through so the rifle sights are not disturbed as the bullet exits the muzzle.

A variation of the ambush method can be used when engaging a stop and go target. A Marine should look for a pattern of exposure (e.g., every 15 seconds). Once a pattern is determined, a Marine establishes a lead by aiming at a point in front of the area in which the target is expected to appear, then he fires the shot at the moment the target appears.

Marksmanship Fundamentals

Engaging moving targets requires concentration and adherence to the fundamentals of marksmanship. The following modifications to the fundamentals of marksmanship are critical to engagement of moving targets.

Sight Picture

Typically, sight picture is the target's center of mass. If a Marine engages a moving target, he bases his sight picture on the target's range, speed, and angle of movement, i.e., sight alignment may be established on a point of aim in front of the target.

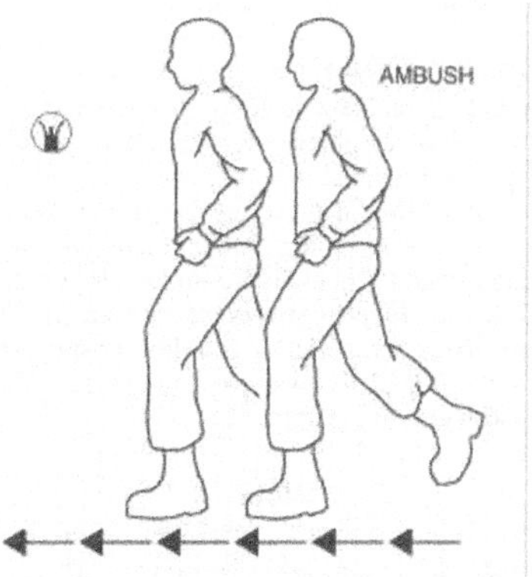

Figure 10-9. Ambush Method.

Trigger Control

As with any target engagement, trigger control is critical to the execution of shots. A Marine can apply pressure on the trigger prior to establishing sight picture, but there should be no rearward movement of the trigger until sight picture is established. Interrupted trigger control is not recommended because the lead will be lost or have to be adjusted to reassume proper sight picture. When using the tracking method, continue tracking as trigger control is applied to ensure the shot does not impact behind the moving target.

Follow-through

If a Marine uses the tracking method to engage moving targets, he continues to track the target during follow-through so the desired lead is maintained as the bullet exits the muzzle. Continuous tracking also enables a second shot to be fired on target if necessary.

Stable Position

To engage moving targets using the tracking method, the rifle must be moved smoothly and steadily as the target moves. A stable position steadies the rifle sights while tracking. Additional rearward pressure may be applied to the pistol grip to help steady the rifle during tracking and trigger control. Elbows may be moved from the support so the target can be tracked smoothly.

10008. Engaging Targets at Unknown Distances

Hasty Sight Setting

While a BZO is considered true for 300 meters, a Marine must be capable of engaging targets beyond this distance. The rifle's sighting system allows sight settings for distances out to 800 meters in 100-meter increments. If a Marine must establish a BZO for extended ranges, it is referred to a hasty sight setting. To achieve a hasty sight setting, a Marine dials the appropriate range numeral on the rear sight elevation knob that corresponds to the range to the target. For example, if the rear sight elevation knob is set at 8/3 and a target appears at 500 meters, rotate the knob to the 5 setting.

Note

Upon completion of firing with a hasty sight setting for extended ranges, return the rear sight to the BZO setting.

Point of Aim Technique

If the distance to the target is beyond the BZO capability of the rifle and time does not permit adjustment of the sights, a Marine can use offset aiming techniques to apply a point of aim for elevation to engage the target. See paragraph 10003.

10009. Engaging Targets During Low Light and Darkness

Combat targets are frequently engaged during periods of darkness or under low-light conditions. Although basic marksmanship fundamentals do not change, the principles of night vision must be applied and target detection is applied differently. During periods of darkness or low light, a Marine's vision is extremely limited. A Marine must apply the techniques of night observation in order to detect potential targets, and he must develop skills that allow him to engage targets under these conditions.

Night Vision

A Marine can improve his ability to see during periods of darkness or low light by obtaining and maintaining night vision. Since adapting to night vision is a slow and gradual process, steps should be taken to protect night vision once it is obtained.

Obtaining Night Vision

There are two methods used to obtain night vision. The first method is to remain in an area of darkness for about 30 minutes. This area can be indoors or outdoors. The major disadvantage of this approach is that an individual is not able to perform any tasks while acquiring night vision in total darkness. The second method is to remain in a darkened area under low intensity red light (similar to the light used in a photographer's darkroom) for about 20 minutes, followed by about 10 minutes in darkness without the red light. This method produces almost complete night vision adaptation while permitting the performance of some tasks during the adjustment period.

Maintaining Night Vision

Because the eyes take a long time to adjust to darkness, it is important to protect night vision once it is acquired. To maintain night vision—

- Avoid looking at any bright light. Bright light will eliminate night vision and require readaptation.
- Shield eyes from parachute flares, from spotlights or from headlights.
- When using a flashlight to read a map or any other written material:
  - Put one hand over the glass to limit the area illuminated and the intensity of the light. Keeping one eye shut will reduce the amount of night vision lost.
  - Cover the light with a red filter to help reduce the loss of night vision.
  - Minimize the time spent using a flashlight.

Factors Affecting Night Vision

Some physical factors may affect your night vision and reduce your ability to see as clearly as possible in low light or darkness. These factors include—

- Fatigue.
- Lack of oxygen.
- Long exposure to sunlight.
- Heavy smoking.
- Drugs.
- Headaches.
- Illness.
- Consumption of alcohol within the past 48 hours.
- Improper diet.

Searching Methods

Once night vision has been acquired, the Marine can locate targets. Some daylight observation techniques (e.g., searching for target indicators) also apply during periods of darkness or low light.

Off-center Vision

Off-center vision is the technique of keeping the attention focused on an object without looking directly at it (see fig. 10-10). To search for targets using off-center vision, never look directly at the object you are observing. You will see the object much better by using off-center vision. Look slightly to the left, right, above or below the object. Experiment and practice to find the best off-center angle for you. For most people, it is about 6 to 10 degrees away from the object, or about a fist's width at arm's length.

Note

Staring at a stationary object in the dark may make it appear to be moving. This occurs because the eye has nothing on which

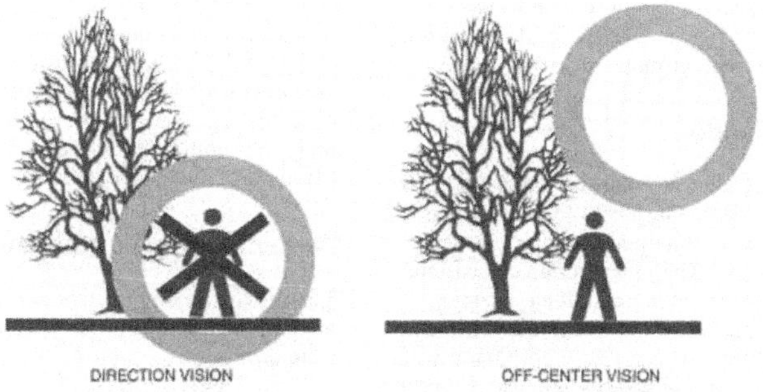

Figure 10-10. Off-center Vision.

to reference the exact position of the object. This illusion can be prevented by visually aligning the object against something else, such as a finger at arm's length.

Scanning/Figure Eight Scan

Scanning is the use of off-center vision to observe an area or object and involves moving the eyes in a series of separate movements across the objective area.

A common method of scanning is to move the eyes in a figure eight pattern (see fig. 10-11). A Marine moves the eyes in short, abrupt, irregular movements over and around the area. Once a target indicator is detected, focus is concentrated in that area, but not directly at it. Pause a few seconds at each point of observation since the eyes cannot focus on a still object while in motion. Rest the eyes frequently when scanning.

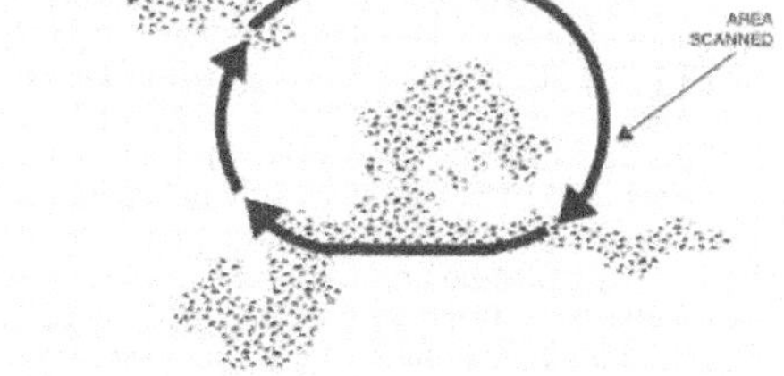

Figure 10-11. Figure Eight Scan.

While you are observing, there may be periodic blackouts of night vision due to simple fatigue. This is normal and is not a cause for alarm. Night vision will quickly return after the eyes are moved and blinked a few times.

It is more effective to scan from a prone position or a position closer to the ground than the object being observed. This creates a silhouetted view of the object.

When scanning an area, look and listen for the same target indicators as in daylight: movement, sound, and improper camouflage.

Objects in bright moonlight/starlight cast shadows just as in sunlight.

Sound seems louder at night than during daylight.

Types of Illumination

There are two types of illumination that assist engagement during low light or darkness: ambient light and artificial illumination. Both ambient light and artificial illumination can affect target perception (distance and size) and night vision capabilities.

- Ambient light is the light produced by natural means (i.e., the sun, moon, and stars). Variations occur in ambient light due to the time of day, time of year, weather conditions, terrain, and vegetation.
- Artificial illumination is the light produced by a process other than natural means. Artificial light can be used to illuminate an area for target detection

or to illuminate a specific target to pinpoint its position. There are two types of artificial illumination used in combat: air and ground.

Effects of Illumination

In some combat situations, ambient light and artificial illumination may assist a Marine in locating targets. However, this light can affect perception of the target and disrupt night vision. The introduction of artificial light requires the eyes to make a sudden, drastic adjustment to the amount of light received. This can cause a temporary blinding because night vision was abruptly interrupted. Ambient light also can cause a blinding effect; e.g., a Marine may experience temporary blindness or reduced night vision if a bright moon suddenly appears from behind the clouds.

- Light behind a Marine or light between the Marine and a target illuminates the front of the target and makes it appear closer than it is.
- Light beyond the target displays the target in silhouette and makes it appear farther away than it is. If the target is silhouetted, it is easier to see and easier to engage.
- Air illumination devices are in constant motion as they descend to the ground. This movement creates changing shadows on any illuminated target causing a stationary target to appear as if it is moving.

10010. Engaging Targets while Wearing the Field Protective Mask

While engaging targets in a combat environment, a Marine is under considerable stress caused by fear, fatigue, and the noise of battle. His stress is further aggravated by the fear and uncertainty associated with a nuclear, biological, and chemical (NBC) threat. However, a Marine must be able to operate under any battlefield condition, including an NBC environment. If a Marine wears the field protective mask, its bulk and reduced visibility can affect his firing position which in turn affects the rifle's zero and his ability to engage the target. A Marine must make adjustments to his firing position and the application of marksmanship fundamentals to counter the additional gear worn in an NBC environment. Therefore, a Marine should practice wearing his field protective mask when he is not in a combat environment. This allows him to overcome any restrictions caused by the mask, develop confidence in his ability to execute well-aimed shots while wearing the mask, and develop a plan of action. This plan should address how the rifle is presented to the target, how long the mask is worn, and the likelihood of enemy contact. If a Marine expects to wear the mask for an extended period and enemy contact is likely, he should consider adjusting the rifle sights so that his first rounds are on target.

Marksmanship Fundamentals

Wearing the field protective mask requires a Marine to make modifications to his aiming and breath control techniques.

Aiming

Wearing the field protective mask affects the aiming process and the ability to locate targets. The bulk of the mask may require an adjustment to stock weld, eye relief, head position, and placement of the buttstock in the shoulder.

Breath Control

Wearing the field protective mask affects breath control because breathing may be more difficult. Temporary fogging of the lens also may be experienced. If fogging occurs, a Marine should take a deep breath and fire while holding a full breath of air (inhaling clears the fog).

Firing Position

A good firing position provides balance, control, and stability during firing. The field protective mask's added bulk and other restrictions may require a Marine to make changes to his firing position. The adjustments are unique to each Marine and based on his body size and shape and his ability to adapt to the mask. Adjustments should be minor. However, all firing positions will be affected in the following areas:

Stock Weld

Changing the placement of the cheek on the stock may affect the rifle's BZO. Therefore, a Marine should obtain a BZO for the rifle in full mission-oriented protective posture gear. Stock weld will not be as comfortable or feel as solid as it does without the field protective mask. The loss of sensitivity between the cheek and the stock, due to the mask, may cause the cheek to be pressed too firmly against the stock. Pressing the cheek too firmly against the stock can cause the seal of the field protective mask to break. If this occurs, quickly clear the mask and resume a firing position. If the lens of the field protective mask fogs up while in a firing position, this indicates that the mask's seal has been broken. Clear the mask and resume the firing position.

Eye Relief

The added bulk of the field protective mask may increase eye relief because the head is farther back along the stock. If the eye is too far from the rear sight aperture it may be difficult to acquire the target and to maintain sight picture; if the eye is too close, the rear sight can hit the mask, possibly breaking a lens or its seal.

Head Position

The mask's shape and bulk can make sight alignment difficult to achieve. The restrictive vision caused by the mask may force a Marine to roll or tilt his head over the stock to achieve sight alignment. The Marine should keep his head as erect as possible while maintaining sight alignment.

Placement of Buttstock in the Shoulder

Placement of the buttstock in the shoulder pocket may have to be altered due to the mask's added bulk. If the rifle is canted, a Marine may place the buttstock of the rifle just outside the pocket of the shoulder to achieve sight alignment. Holding the rifle straight is the preferred method of obtaining sight alignment. However, if sight alignment cannot be achieved in this position, a Marine may alter the hold of the rifle to bring the aiming eye in line with the sights. Canting the rifle drastically affects the rifle's zero. A Marine should cant the rifle only as much as is needed to obtain a good stock weld and proper sight alignment. If the rifle is canted, the point of impact may not coincide with the point of aim. For example, when wearing the mask, a right-handed Marine's point of impact is usually high and to the left of center mass (for a left-handed Marine, high and to the right of center mass). Therefore a Marine has to offset aim an equal and opposite distance low and to the right. See paragraph 10003 for a discussion on offset aiming and the known strike of the round technique.

Appendix A

Data Book

Note

The principles for recording data in the data book are the same for the Entry Level Rifle (ELR) Program and the Sustainment Level Rifle (SLR) Program. This appendix provides generic information for completing the data book. Specifics pertaining to each program are contained in the ELR and SLR lesson plans.

1. Data Book

Of all the tools that assist the Marine in firing accurately and consistently, the data book, if properly used, is the most valuable asset. It contains a complete record of every shot fired and the weather conditions and their effects on shooting. When used properly it will assist the Marine in establishing and maintaining a battlesight zero (BZO).

2. Recording Data Before Firing

Recording information in the data book prior to firing saves valuable time on the firing line that should be used to prepare for firing. Some information can be recorded before going to the firing line. In the BEFORE FIRING section of the data book, record the following (see fig. A-1):

True Zero

Front Elevation

Enter the front sight post setting by recording the number of clicks up (+) or down (-) under FRONT ELEV.

Rear Elevation

Circle the rear sight elevation knob setting under REAR ELEV for the yard line firing from—

- 200 yards: 8/3-2
- 300 yards: 8/3
- 500 yards: 5

Note

At 500 yards, the rear sight elevation knob setting may be plus or minus one or two clicks off of 5.

Wind

Under the WIND column under TRUE ZERO, the R represents clicks right on the rifle from the initial sight setting and the L represents clicks left on the rifle. Enter the rear sight windage knob setting by recording the number of clicks right (clockwise) or left (counterclockwise) under WIND.

Wind

Prior to firing, check the wind. If wind conditions are present, a sight adjustment will have to be made prior to firing to ensure shots impact the center of the target.

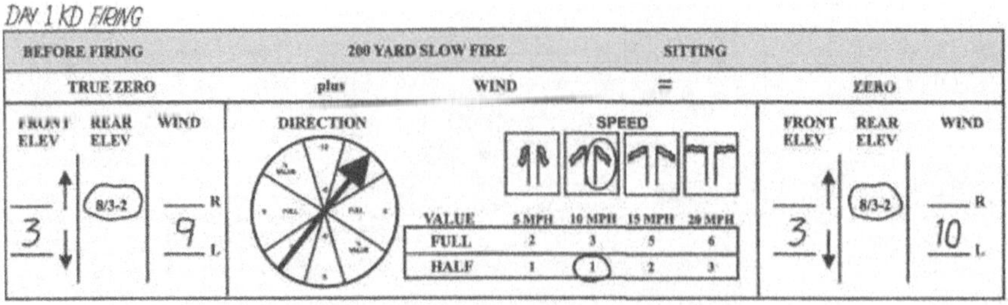

Figure A-1. Recording Data Before Firing.

Direction

Determine the direction of the wind and draw an arrow through the clock indicating the direction the wind is blowing. If the wind is not blowing, a data entry is not needed.

Note

Remember that your position is represented in the center of the clock and the target is at 12 o'clock.

Value

Look at the clock to determine if the wind is full, half, or no value wind. Under VALUE, circle FULL or HALF to indicate the wind value.

Speed

Observe the flag on the range and circle the appropriate flag indicating the wind's velocity (SPEED).

Determining Windage Adjustment

The chart beneath the flag indicates the number of clicks on the rear sight windage knob to offset the effects of the wind. Circle the number of clicks where the wind value and wind speed intersect.

Zero

Determine the zero you will place on your rifle to accommodate wind conditions to begin firing. This ZERO will be the TRUE ZERO plus the rear sight windage setting to compensate for the effects of wind.

Front Elev and Rear Elevation

Elevation adjustments are not affected by wind so the same settings are carried over from TRUE ZERO.

Wind

Wind will affect the strike of the round right or left on the target. Therefore, if wind is a factor, the rear sight windage knob must be adjusted to compensate for the effects of wind.

If the wind is blowing, add the number of clicks circled under the flags to the rear sight windage knob setting from TRUE ZERO.

Once the windage setting is determined, record it in the WIND column.

3. Recording Data During Firing

Recording Data During Slow Fire Stages

The method for calling and plotting slow fire shots in the data book is called "the shot behind method." It allows the Marine to spend less time recording data and more time firing on the target. This is because all the calling and plotting is done while the target is in the pits being marked. This information is recorded in the DURING FIRING portion of the data book page. The proper and most efficient method for recording data during KD slow fire stages is as follows (see fig. A-2):

Fire the First Shot

Fire the first shot. Then immediately check the wind flag to see if the speed or direction of the wind changed. If the wind direction changed, indicate the windage adjustment needed to compensate for it in the WIND row under CALL number 1. This windage adjustment will have to be applied to the rifle prior to firing the second shot.

Call the Shot Accurately

As soon as the shot is fired and the target is pulled into the pits, record the exact location where the tip of the front sight was on the target at the exact instant the shot was fired (assuming sight alignment was maintained). Plot this location on the target provided under number 1 in the block marked CALL.

Prepare to Fire the Second Shot

As soon as you have recorded the call for the first shot, prepare to fire the second shot.

Look at Where the First Shot Hit

As the target reappears out of the pits, look where the first shot hit the target. Remember this location so it can be plotted after firing the second shot.

Fire the Second Shot

Fire the second shot. Then check the wind flag to see if the wind changed speed or direction.

Call the Second Shot and Plot the First Shot

As soon as the second shot is fired and the target is pulled into the pits, record the call of the second shot.

Now plot the precise location of the first shot by writing the numeral 1 on the large target diagram provided in the block marked PLOT.

Prepare to Fire the Third Shot

Repeat steps until all shots have been fired. Indicate each slow fire shot with the appropriate number (e.g., 1, 2, 3, 4, 5).

Make a Sight Adjustment if Required

Sight adjustments should be made off of a shot group, not a single shot. Determine if a sight adjustment is necessary off of the first three shots fired. If the shots form a group (i.e, a group that fits inside the "A" target bull's-eye or the center scoring ring of the "D" target), but are not where they were called, make the necessary sight adjustment.

CAUTION

Generally, major sight adjustments from established sight settings are caused by poor application of the fundamentals, inconsistencies in firing positions, inconsistencies in sight picture at different ranges and different positions, and inconsistent tension on the sling. Every effort should be made to correct shooting errors prior to making a sight adjustment on the rifle.

Note

The plotting targets in the data book for the KD Course of Fire are in inches, requiring you to calculate the number of clicks to center your shot groups.

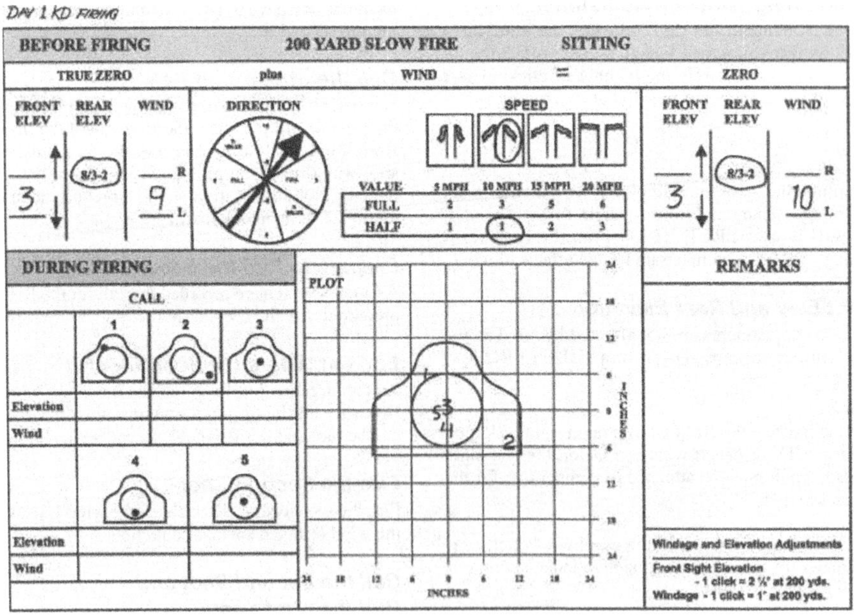

Figure A-2. Recording Slow Fire Data During Firing.

Elevation. Locate the closest horizontal grid line to the center of the plotted shot group. Follow the line across to the numbered vertical scale to determine the number of inches of elevation the shot group is off of target center. Calculate the number of clicks on your front sight post (for 200 and 300 yards) or rear sight elevation knob (for 500 yards) to bring your shot group center.

Windage. Locate the closest vertical grid line to the center of the plotted shot group. Follow the line down to the numbered horizontal scale to determine the number of inches of windage the shot group is off of target center. Calculate the number of clicks on your rear sight windage knob to bring shot group center.

Recording Data During Rapid Fire Stages

In the DURING FIRING section of the data book, record the following (see fig. A-3):

Mentally Call Shots While Firing

While firing the rapid fire string, make a mental note of any shots called out of the group.

Plot the Shot Group

After firing the rapid fire string, and when the target is marked, plot all visible hits with a dot precisely where they appear on the large target diagram in the block marked PLOT.

Make a Sight Adjustment if Required

Locate the center of the shot group. If the shots form a group, make the necessary sight adjustments off of the center of the group. If shots do not form a group and do not just contain a poor shot, do not make a sight adjustment. Determine the sight adjustment by locating the center of the shot group and using the grid lines on the target in the data book. These grid lines represent the number of inches to bring a shot group center. Looking at the shot group:

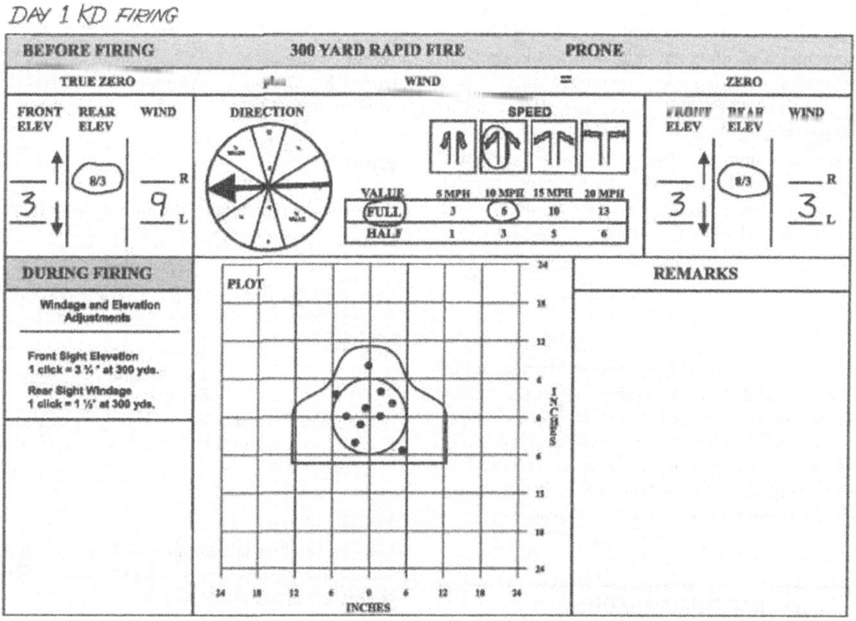

Figure A-3. Recording Rapid Fire Data During Firing.

CAUTION

Generally, major sight adjustments from established sight settings are caused by poor application of the fundamentals, inconsistencies in firing positions, inconsistencies in sight picture at different ranges and different positions, and inconsistent tension on the sling. Every effort should be made to correct shooting errors prior to making a sight adjustment on the rifle.

Note

The plotting targets in the data book for the KD Course of Fire are in inches, requiring you to calculate the number of clicks to center your shot groups.

Elevation. Locate the closest horizontal grid line to the center of the plotted shot group. Follow the line across to the numbered vertical scale to determine the number of inches of elevation the shot group is off of target center. Calculate the number of clicks on your front sight post to bring your shot group center.

Windage. Locate the closest vertical grid line to the center of the plotted shot group. Follow the line down to the numbered horizontal scale to determine the number of inches of windage the shot group is off of target center. Calculate the number of clicks on your rear sight windage knob to bring your shot group center.

Remarks

After firing a stage, record any data or information that can be helpful in improving shooting in the future. Anything done or observed should be recorded. These items will be helpful when analyzing daily shooting performance. Record this information in the REMARKS column. What the Marine fails to record may be the information he will need to improve.

4. Recording Data After Firing

In the AFTER FIRING section of the data book, record the following (see fig. A-4 on page A-6):

Zero

Upon completion of firing, determine the elevation and windage required to center the shot group (if necessary) and record this sight setting in the ZERO block of the AFTER FIRING section. If no adjustments are needed to center the shot group, record the sight settings currently on the rifle.

Front Elevation

Under the column FRONT ELEV, record the final elevation setting made on the front sight post.

Rear Elevation

Under the column REAR ELEV, record the rear sight elevation knob setting for the yard line firing from.

Note

If firing from 500 yards, record the final elevation setting made on the rear sight elevation knob.

Wind

Under the column WIND under ZERO, record final windage setting made on the rear sight windage knob.

Wind

Calculate the prevailing wind during the string of fire.

Direction

Determine the direction of the wind and draw an arrow through the clock indicating the direction the wind is blowing.

Note

Remember that your position is represented in the center of the clock and the target is at 12 o'clock.

Value

Look at the clock to determine if the wind is full, half, or no value wind. Under VALUE, circle FULL or HALF to indicate the wind value.

Speed

Observe the flag on the range and circle the appropriate flag indicating the wind's velocity (SPEED).

Determine Windage Adjustment

The chart beneath the flag indicates the number of clicks on the rear sight windage knob to offset the effects of the wind. Circle the number of clicks where the wind value and wind speed intersect.

True Zero

A true zero is the established zero without the windage adjustments to compensate for the effects of the string of fire's wind. A true zero is calculated because, the next time you fire, the wind conditions will probably be different. Therefore, the rear sight windage knob adjustments made to compensate for the string of fire's wind will not be the correct setting for wind conditions on other strings of fire or other days of firing.

Front Elevation and Rear Elevation

Because elevation adjustments are not affected by wind, the same settings are carried over from ZERO.

Wind

Calculate the windage adjustment to compensate for the string of fire's wind conditions the same way it was calculated in the BEFORE FIRING information of the data book. The only exception is now windage adjustments are being removed from the rifle rather than added to the rifle. Because the windage setting is being removed from the rifle, remove the number of clicks of windage right or left from the ZERO windage setting. Once the windage setting is determined, record it in the WIND column.

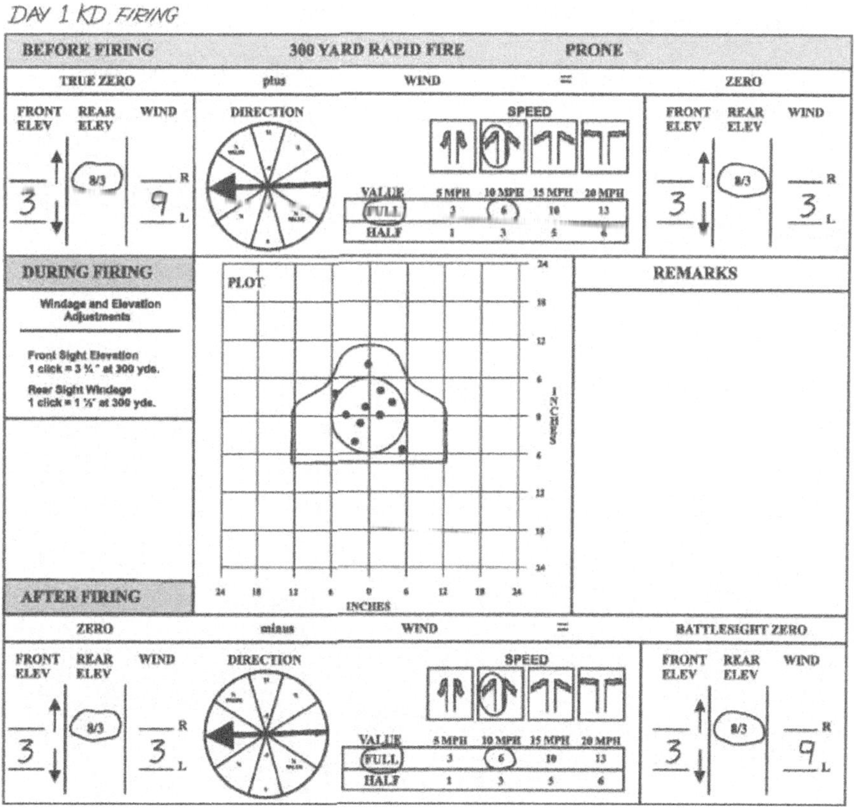

Figure A-4. Recording Data After Firing.

APPENDIX B

GLOSSARY

SECTION I. ACRONYMS

BZO . . . . battlesight zero

CLP . . . . cleaner, lubricant, and preservative

KD . . . . known distance

LAW . . . . lubricating oil, arctic weapons

NBC . . . . nuclear, biological, and chemical

ROE . . . . rules of engagement

SOP . . . . standing operating procedures

TM . . . . technical manual

SECTION II. DEFINITIONS

A

aiming point – The aiming point is the precise point where the tip of the front sight post is placed in relationship to the target.

alibi – Any condition caused by the weapon, ammunition, or range operation that causes the shooter not to have an equal opportunity to complete a string of fire as all other shooters on the range.

B

battlesight zero (BZO) – The elevation and windage settings required to engage point targets from 0-300 yards/meters under ideal weather conditions (i.e., no wind).

bone support – The body's skeletal structure supporting the rifle's weight.

breath control – Procedure used to fire the rifle at the moment of least movement in the body and the rifle.

C

canting – An angular deviation of the weapon to the left or right from a vertical position during firing.

center of mass – A point that is horizontally and vertically centered on the target.

center of mass hold – The placement of the tip of the front sight on the target center of mass prior to the shot breaking.

centerline of the bore – An imaginary straight line beginning at the chamber end of the barrel and proceeding out of the muzzle.

chamber check – Procedure used to determine a weapon's condition.

D

double feed – Attempted simultaneous feeding of two or more rounds from the magazine.

dry fire – Cocking, aiming, and squeezing the trigger of an unloaded rifle in order to practice the fundamentals of marksmanship.

detailed search – Method for conducting a systematic search of an area for specific target indicators.

E

eye relief – The distance between from the rear sight aperture to the aiming eye.

F

function check – Procedure used to ensure the selector lever operates properly.

flag method – Procedure used to determine wind velocity and direction on a Known Distance (KD) range.

field expedient battlesight zeroing – Process used to zero the rifle at 36 yards or 30 meters when a 300-yard/-meter range is not available.

G

gas operated – A self-loading firearm that utilizes the expanding force of the propellant's gases to operate the action of the weapon.

H

hasty search – Method for quickly searching an area for enemy activity.

hasty sight setting – A zero established for distances out to 800 yards/meters.

I

immediate threat target – A target engaged at a short range (50 meters or less) with little or no warning.

initial sight setting – Sight setting placed on a rifle that serves as the starting point from which all sight adjustments are made for the initial zeroing process.

L

limited technical inspection (LTI) – An inspection performed by an armorer on a weapon to determine its operational status (safety and function, not accuracy).

line of sight – An imaginary line extending from the shooter's eye through the rifle's sights and onto an aiming point on a target.

load – Command/Procedure used to take a weapon from Condition 4 to Condition 3 by inserting a magazine with rounds.

M

magazine – A container that holds ammunition in a position to be chambered.

magazine fed – A mechanical, automatic means of supplying a firearm with ammunition to be chambered.

make ready – Command/Procedure used to take a weapon from Condition 3 to Condition 1 by chambering a round.

muscular relaxation – The state of tension required to properly control the rifle. The shooter's muscles are in a relaxed state of control—tightened but not tensed.

N

natural point of aim – The location at which the rifle's sights settle if bone support and muscular relaxation are achieved.

O

observation method – Procedure used to determine wind velocity and direction in a tactical situation.

P

pie technique – A technique used to locate and engage targets from behind cover, while minimizing the Marines exposure to enemy fire.

R

range – **1.** A designated place where live fire weapons training is conducted. **2.** The horizontal distance to which a projectile can be propelled to a specified target. **3.** The horizontal distance between a weapon and target.

recoil management – The ability to manage or control the recoil of the rifle for a given shooting position.

rollout – A technique used to locate and engage targets from behind cover, while minimizing the Marines exposure to enemy fire.

S

semiautomatic – One full cycle of operation. The firing, extraction, and ejection of the spent cartridge, the cocking of the weapon and chambering of the succeeding round of ammunition after the trigger is pulled.

shooter error — Any action induced by the shooter that causes the weapon to fail to operate properly or miss the intended target.

sight alignment – The placement of the tip of the front sight post in the center of the rear sight aperture.

sight picture — The placement of the tip of the front sight post in the center of the target while maintaining sight alignment.

sling – When properly attached to the rifle, the sling provides maximum stability for the weapon and helps reduce the effects of the rifle's recoil. Also used for individual weapon transport.

stability of hold – The ability to acquire a stable position and to hold the rifle steady for any given rifle position.

stock weld – The firm, consistent contact of the cheek with the weapon's buttstock.

stoppage – Any condition that causes the rifle to fail to fire.

T

target indicators – Anything that reveals the enemy's position.

trajectory – The path of a projectile through the air and to a target.

triangulation process – Process used to determine the vertical and horizontal sight adjustments that must be made to center a shot group.

trigger control – The skillful manipulation of the trigger that causes the rifle to fire without disturbing sight alignment or sight picture.

U

unload – Command/Procedure used to take a weapon from any condition to Condition 4.

unload and show clear – Command/Procedure used to take a weapon from any condition to Condition 4 while requiring a second individual to check the weapon to verify that no ammunition is present.

user serviceability inspection – Procedure used to ensure a weapon is in an acceptable operating condition, conducted by the shooter.

V

velocity – Speed at which the projectile travels.

W

weapons carry – Procedure used to effectively handle the rifle while remaining alert to possible threat levels.

weapons condition – Describes a weapon's readiness for live fire.

weapons transport – Procedure used to carry the rifle for long periods of time and when one or both hands are needed for other work.

windage and elevation rules – Rules that define how far the strike of the round will move on the target for each click of front and rear sight elevation or rear sight windage for each 100 yards/meters of range to the target.

Z

zeroing – The process used to adjust the rifle sights that cause it to shoot to point of aim at a desired range.

Made in United States
North Haven, CT
04 November 2026

21774389R10183